A Problems Based Course in Advanced Calculus

Pure and *Applied*
The *Sally* SERIES
UNDERGRADUATE // TEXTS · 32

A Problems Based Course in Advanced Calculus

John M. Erdman

AMERICAN
MATHEMATICAL
SOCIETY
Providence, Rhode Island USA

EDITORIAL COMMITTEE

2010 *Mathematics Subject Classification*. Primary 00A07; Secondary 00-01.

For additional information and updates on this book, visit
www.ams.org/bookpages/amstext-32

Library of Congress Cataloging-in-Publication Data

Names: Erdman, John M., 1935- author.
Title: A problems based course in advanced calculus / John M. Erdman.
Other titles: Advanced calculus
Description: Providence, Rhode Island : American Mathematical Society, [2018] | Series: Pure
　　and applied undergraduate texts ; volume 32 | Includes bibliographical references and index.
Identifiers: LCCN 2017051409 | ISBN 9781470442460 (alk. paper)
Subjects: LCSH: Calculus–Textbooks. | Calculus–Study and teaching (Graduate) | AMS: General
　　– General and miscellaneous specific topics – Problem books. msc | General – Instructional
　　exposition (textbooks, tutorial papers, etc.). msc
Classification: LCC QA303.2 .E73 2018 | DDC 515–dc23
LC record available at https://lccn.loc.gov/2017051409

To my wife Argentina

Contents

Preface

In American universities two distinct types of courses are often called Advanced Calculus: one, largely for engineers, emphasizes advanced computational techniques in calculus; the other, a more theoretical course, usually taken by majors in mathematics and physical sciences (and often called Elementary Analysis or Intermediate Analysis), concentrates on conceptual development and proofs. This Problems Based Course is of the latter type. It is not a place to look for post-calculus material on Fourier series, Laplace transforms, and the like. It is intended for students of mathematics, and others, who have completed (or nearly completed) a standard introductory calculus sequence and who wish to understand where all those rules and formulas come from.

Many advanced calculus texts contain more topics than this one. When students are encouraged to develop much of the subject matter for themselves, it is not possible to "cover" material at the same breathtaking pace that can be achieved by a truly determined lecturer. But, while no attempt has been made to make the book encyclopedic, I do think it nevertheless provides an integrated overview of calculus and, for those who continue, a solid foundation for a first-year graduate course in real analysis.

As the title of the present document is intended to suggest, *A Problems Based Course in Advanced Calculus*, is as much an extended problem set as a textbook. The proofs of most of the major results are either *exercises* or *problems*. The distinction here is that solutions to exercises are available online while solutions to problems are not given. I hope that this arrangement will provide flexibility for instructors who wish to use it as a text. For those who prefer a (modified) Moore-style development, where students work out and present most of the material, there is a quite large collection of problems for them to hone their skills on. For instructors who prefer a lecture format, it should be easy to base a coherent series of lectures on the presentation of solutions to thoughtfully chosen problems.

I have tried to make this document (in a rather highly qualified sense discussed below) self-contained. In it we investigate how the edifice of calculus can be grounded in carefully developed substrata of sets, logic, and numbers. Will it be a complete or totally rigorous development of the subject? Absolutely not. I am not aware of any serious enthusiasm among mathematicians I know for requiring rigorous courses in mathematical logic and axiomatic set theory as prerequisites for a first introduction to analysis. In the use of the tools from set theory and formal logic, there are many topics that because of their complexity and depth are cheated or not even mentioned. (For example, though used often, the *axiom of choice* is mentioned only once.) Even everyday topics such as arithmetic (see Appendix G) are not developed in any great detail.

Before embarking on the main ideas of calculus proper, one *ideally* should have a good background in all sorts of things: quantifiers, logical connectives, set operations, writing proofs, the arithmetic and order properties of the real numbers, mathematical induction, least upper bounds, functions, composition of functions, images and inverse images of sets under functions, finite and infinite sets, countable and uncountable sets. On the one hand all these are technically prerequisite to a careful discussion of the foundations of calculus. On the other hand any attempt to do all this business systematically at the beginning of a course will defer the discussion of anything concerning calculus proper to the middle of the academic year and may very well both bore and discourage students. Furthermore, in many schools there may be students who have already studied much of this material (in a Proofs course, for example). In the spirit of compromise and flexibility I have relegated this material to the appendices. Treat it any way you like. I teach in a large university where students show up for Advanced Calculus with a wide variety of backgrounds, so it is my practice to go over the appendices first, covering many of them in a quite rapid and casual way, my goal being to provide just enough detail so that everyone will know where to find the relevant material when they need it later in the course. After a rapid traversal of the appendices, I start Chapter 1.

For this text to be useful, a student should have previously studied introductory calculus, more for mathematical maturity than anything else. Familiarity with properties of elementary functions and techniques of differentiation and integration may be assumed and made use of in a few *examples*—but is never relied upon in the logical development of the material.

One motive for my writing this text is to make available in fairly simple form material that I think of as "calculus done right". For example, differential calculus as it appears in many texts is a morass of tedious epsilon-delta arguments and partial derivatives, the net effect of which is to almost totally obscure the beauty and elegance which results from a careful and patient elaboration of the concept of tangency. On the other hand texts in which things are done right (for example Loomis and Sternberg [**LS90**]) tend to be rather forbidding. I have tried to write a text which will be helpful to a determined student with an average background. (I seriously doubt that it will be of much use to those who are chronically lazy or totally unengaged.)

In my mind one aspect of doing calculus "correctly" is arranging things so that there is nothing to unlearn later. For example, in this text topological properties of the real line are discussed early on. Later, topological things (continuity, compactness, connectedness, and so on) are discussed in the context of metric spaces (because they unify the material conceptually and greatly simplify subsequent arguments). But the important thing is that *definitions* in the single variable case and the metric space case are the same. Students do not have to unlearn material as they go to more general settings. Similarly, the differential calculus is eventually developed in its natural habitat of normed linear spaces. But here again, the student who has mastered the one-dimensional case, which occurs earlier in the text, will encounter definitions and theorems and proofs that are virtually identical to the ones with which he/she is already familiar. There is nothing to unlearn.

In the process of writing this document, I have rethought the proofs of many standard theorems. Although some, perhaps most, results in advanced calculus have reached a final, optimal form, there are many others that, despite dozens of different approaches over the years, have proofs that are genuinely confusing to most students. To mention just one example, there is the theorem concerning change of order of differentiation whose conclusion is

$$\frac{\partial^2 f}{\partial x \, \partial y} = \frac{\partial^2 f}{\partial y \, \partial x} \, .$$

This well-known result to this day still receives clumsy, impenetrable, and even incorrect proofs. I make an attempt, here and elsewhere in the text, to lead students to proofs that are conceptually clear, even when they may not be the shortest or most elegant. And throughout I try very hard to show that mathematics is about ideas and not about manipulation of symbols.

I extend my gratitude to the many students over the years who have endured various versions of this text and to colleagues and reviewers who have generously nudged me when they found me napping. I want especially to thank Dan Streeter, who provided me with much help in the technical aspects of getting this document produced. The text was prepared using $\mathcal{AMS}$-LaTeX. For the diagrams I used the macro package X͜y-pic by Kristoffer H. Rose and Ross Moore supplemented by additional macros in the *diagxy* package by Michael Barr.

Finally it remains only to say that I will be delighted to receive, and will consider, any comments, suggestions, or error reports. My e-mail address is `erdmanj@comcast.net`.

John M. Erdman
Portland State University
Portland, Oregon

For students:
How to use this book

Some years ago at a national meeting of mathematicians, many of the convention-eers went about wearing pins which proclaimed, "Mathematics is Not a Spectator Sport". It is hard to overemphasize the importance of this observation. The idea behind it has been said in many ways by many people; perhaps it was said best by Paul Halmos [**Hal67**]: *The only way to learn mathematics is to do mathematics.*

In most respects learning mathematics is more like learning to play tennis than learning history. It is principally an activity, and only secondarily a body of knowl-edge. Although no one would try to learn tennis just by watching others play the game and perhaps reading a book or two about it, thousands of mathematics stu-dents in American universities every year attempt to master mathematical subjects by reading textbooks and passively watching their instructors do mathematics on a blackboard. There are, of course, reasons why this is so, but it is unfortunate nevertheless. This book is designed to encourage *you* to do mathematics.

Perhaps the most important thing to know about the structure of this book is that, logically, the appendices come first. The appendices contain a summary of what is generally regarded as prerequisite information that, ideally, one should have before venturing into advanced calculus. In the real world, students' backgrounds differ enormously. If, at one extreme, you are quite familiar with nearly all of the appended material, you are ready for Chapter 1. Use the appendices for reference and notation. At the other extreme, a very poorly prepared student may profit from spending considerable time working on these sections. In between, where most students find themselves, you have choices. For example, if you are unfamiliar with the distinction between "countable" and "uncountable" sets, you can master the concepts right away, knowing that you will use them later, or you can wait until one of the words crops up in your reading, and learn the concepts when you need them.

When you sit down to work at mathematics it is important to have a sizeable block of time at your disposal during which you will not be interrupted. As you read, pay especially close attention to definitions. (After all, before you can think about a mathematical concept you must know what it means.) Read until you arrive at a result (results are labeled Theorem, Proposition, Example, Problem, Lemma, etc.). Every result requires justification. The proof of a result may appear in the body of the text, or it may be left to you as an *exercise* or a *problem*.

When you encounter a result, stop and try to prove it. Make a serious attempt. If a hint appears after the statement of the result, at first do not read it. Do not try to find the result elsewhere; and do not ask for help. Halmos [**Hal67**] points out, "To the passive reader, a routine computation and a miracle of ingenuity come with equal ease, and later, when he must depend on himself, he will find that they went as easily as they came." Of course, it is true that sometimes, even after considerable effort, you will not have discovered a proof. What then?

If a hint is given, and if you have tried seriously but unsuccessfully to derive the result, then (and only then) should you read the hint. Now try again. Seriously.

What if the hint fails to help, or if there is no hint? If you are stuck on a result whose proof is labeled Exercise, then consult the online solution. Turning to the solution should be regarded as a last resort. Even then do not read the whole proof; read just the first line or two, enough to get you started. Now try to complete the proof on your own. If you can do a few more steps, fine. If you get stuck again in midstream, read some more of the proof. Use as little of the online proof as possible.

What if you are stuck on a result whose proof is a Problem but there is no online solution? After a really serious attempt to solve the problem, go on. You cannot bring your mathematical education to a halt because of one refractory problem. Work on the next result. After a day or two go back and try again. Problems often solve themselves; frequently, an intractably murky result, after having been allowed to rest for a few days, will suddenly, and inexplicably become entirely clear. In the worst case, if repeated attempts fail to produce a solution, you may have to discuss the problem with someone else—instructor, friend, mother, ...

A question that students frequently ask is, "When I'm stuck and I have no idea at all what to do next, how can I continue to work on a problem?" I know of only one really good answer. It is advice due to Pólya, "If you can't solve a problem, then there is an easier problem you can solve: find it."

Consider examples. After all, mathematical theorems are usually generalizations of things that happen in interesting special cases. Try to prove the result in some concrete cases. If you succeed, try to generalize your argument. Are you stuck on a theorem about general metric spaces? Try to prove it for Euclidean n-space. No luck? How about the plane? Can you get the result for the real line? The unit interval?

Add hypotheses. If you cannot prove the result as stated, can you prove it under more restrictive assumptions? If you are having no success with a theorem concerning general matrices, can you prove it for symmetric ones? How about diagonal ones? What about the 2×2 case?

Finally, one way or another, you succeed in producing a proof of the stated result. Is that the end of the story? By no means. Now you should look carefully at the hypotheses. Can any of them be weakened? Eliminated altogether? If not, construct counterexamples to show that each of the hypotheses is necessary. Is the conclusion true in a more general setting? If so, prove it. If not, give a counterexample. Is the converse true? Again, prove or disprove. Can you find applications of the result? Does it add anything to what you already know about other mathematical objects and facts?

Once you have worked your way through several results (a section or a chapter, say) it is a good idea to consider the organization of the material. Is the order in which the definitions, theorems, etc., occur a good one? Or can you think of a more perspicuous ordering? Rephrase definitions (being careful, of course, not to change meanings!), recast theorems, reorganize material, add examples. Do anything you can to make the results into a clear and coherent body of material. In effect you should end up writing your own advanced calculus text.

The fruit of this labor is understanding. After serious work on the foregoing items, you will begin to feel that you understand the body of material in question. Beginning calculus presents one with techniques, formulas, and algorithms, none of which, usually, are long remembered. Advanced calculus can bring understanding that remains. The quest for understanding, by the way, is pretty much what mathematicians do with their lives.

Chapter 1

Intervals

The three great realms of calculus are differential calculus, integral calculus, and the theory of infinite series. Central to each of these is the notion of *limit*: derivatives, integrals, and infinite series can be defined as limits of appropriate objects. Before entering these realms, however, one must have some background. There are the basic facts about the algebra, the order properties, and the topology of the real line $\mathbb{R}$. One must know about functions and the various ways in which they can be combined (addition, composition, and so on). And, finally, one needs some basic facts about continuous functions, in particular the *intermediate value theorem* and the *extreme value theorem*.

Much of this elementary material appears in the appendices. The remainder (the topology of $\mathbb{R}$, continuity, and the *intermediate* and *extreme value theorems*) occupies the first six chapters. After these have been completed, we proceed with limits and the differential calculus of functions of a single variable.

As you will recall from beginning calculus, we are accustomed to calling certain intervals "open" (for example, $(0,1)$ and $(-\infty, 3)$ are open intervals) and other intervals "closed" (for example, $[0,1]$ and $[1,\infty)$ are closed intervals). In this chapter and the next we investigate the meaning of the terms open and closed. These concepts turn out to be rather more important than one might at first expect. It will become clear, after our discussion of such matters as continuity, connectedness, and compactness in subsequent chapters, just how important they really are.

1.1. Distance and neighborhoods

1.1.1. Definition. If x and a are real numbers, the DISTANCE BETWEEN x and a, which we denote by $d(x,a)$, is defined to be $|x - a|$.

1.1.2. Example. There are exactly two real numbers whose distance from the number 3 is 7.

Proof. We are looking for all numbers $x \in \mathbb{R}$ such that $d(x, 3) = 7$. In other words we want solutions to the equation $|x - 3| = 7$. There are two such solutions. If $x - 3 \geq 0$, then $|x - 3| = x - 3$; so $x = 10$ satisfies the equation. On the other hand, if $x - 3 < 0$, then $|x - 3| = -(x - 3) = 3 - x$, in which case $x = -4$ is a solution. $\qquad\square$

1.1.3. Exercise. Find the set of all points on the real line that are within five units of the number -2.

1.1.4. Problem. Find the set of all real numbers whose distance from 4 is greater than 15.

1.1.5. Definition. Let a be a point in $\mathbb{R}$, and let $\epsilon > 0$. The open interval $(a - \epsilon, a + \epsilon)$ centered at a is called the ϵ-NEIGHBORHOOD of a and is denoted by $J_\epsilon(a)$. Notice that this neighborhood consists of all numbers x whose distance from a is less than ϵ; that is, such that $|x - a| < \epsilon$.

1.1.6. Example. The $\frac{1}{2}$-neighborhood of 3 is the open interval $\left(\frac{5}{2}, \frac{7}{2}\right)$.

Proof. We have $d(x, 3) < \frac{1}{2}$ only if $|x - 3| < \frac{1}{2}$. Solve this inequality to obtain $J_{\frac{1}{2}}(3) = \left(\frac{5}{2}, \frac{7}{2}\right)$. $\qquad\square$

1.1.7. Example. The open interval $(1, 4)$ is an ϵ-neighborhood of an appropriate point.

Proof. The midpoint of $(1, 4)$ is the point $\frac{5}{2}$. The distance from this point to either end of the interval is $\frac{3}{2}$. Thus $(1, 4) = J_{\frac{3}{2}}\left(\frac{5}{2}\right)$. $\qquad\square$

1.1.8. Problem. Find, if possible, a number ϵ such that the ϵ-neighborhood of $\frac{1}{3}$ contains both $\frac{1}{4}$ and $\frac{1}{2}$ but does not contain $\frac{17}{30}$. If such a neighborhood does not exist, explain why.

1.1.9. Problem. Find, if possible, a number ϵ such that the ϵ-neighborhood of $\frac{1}{3}$ contains $\frac{11}{12}$ but does not contain either $\frac{1}{2}$ or $\frac{5}{8}$. If such a neighborhood does not exist, explain why.

1.1.10. Problem. Let $U = \left(\frac{1}{4}, \frac{2}{3}\right)$ and $V = \left(\frac{1}{2}, \frac{6}{5}\right)$. Write U and V as ϵ-neighborhoods of appropriate points. (That is, find numbers a and ϵ such that $U = J_\epsilon(a)$, and find numbers b and δ such that $V = J_\delta(b)$.) Also write the sets $U \cup V$ and $U \cap V$ as ϵ-neighborhoods of appropriate points.

1.1.11. Problem. Generalize your solution to the preceding problem to show that the union and the intersection of any two ϵ-neighborhoods that overlap is itself an ϵ-neighborhood of some point. *Hint.* Since ϵ-neighborhoods are open intervals of finite length, we can write the given neighborhoods as (a, b) and (c, d). There are really just two distinct cases. One neighborhood may contain the other; say, $a \leq c < d \leq b$. Or each may have points that are not in the other; say $a < c < b < d$. Deal with the two cases separately.

1.1.12. Proposition. *If $a \in \mathbb{R}$ and $0 < \delta \leq \epsilon$, then $J_\delta(a) \subseteq J_\epsilon(a)$.*

Proof. Exercise.

1.2. Interior of a set

1.2.1. Definition. Let $A \subseteq \mathbb{R}$. A point a is an INTERIOR POINT of A if some ϵ-neighborhood of a lies entirely in A. That is, a is an interior point of A if and only if there exists $\epsilon > 0$ such that $J_\epsilon(a) \subseteq A$. The set of all interior points of A is denoted by A° and is called the INTERIOR of A.

1.2.2. Example. Every point of the interval $(0, 1)$ is an interior point of that interval. Thus $(0, 1)^\circ = (0, 1)$.

Proof. Let a be an arbitrary point in $(0, 1)$. Choose ϵ to be the smaller of the two (positive) numbers a and $1 - a$. Then $J_\epsilon(a) = (a - \epsilon, a + \epsilon) \subseteq (0, 1)$ (because $\epsilon \leq a$ implies $a - \epsilon \geq 0$, and $\epsilon \leq 1 - a$ implies $a + \epsilon \leq 1$). □

1.2.3. Example. If $a < b$, then every point of the interval (a, b) is an interior point of the interval. Thus $(a, b)^\circ = (a, b)$.

Proof. Problem.

1.2.4. Example. The point 0 is not an interior point of the interval $[0, 1)$.

Proof. Argue by contradiction. Suppose 0 belongs to the interior of $[0, 1)$. Then for some $\epsilon > 0$, the interval $(-\epsilon, \epsilon) = J_\epsilon(0)$ is contained in $[0, 1)$. But this is impossible since the number $-\frac{1}{2}\epsilon$ belongs to $(-\epsilon, \epsilon)$ but not to $[0, 1)$. □

1.2.5. Example. Let $A = [a, b]$ where $a < b$. Then $A^\circ = (a, b)$.

Proof. Problem.

1.2.6. Example. Let $A = \{x \in \mathbb{R} \colon x^2 - x - 6 \geq 0\}$. Then $A^\circ \neq A$.

Proof. Exercise.

1.2.7. Problem. Let $A = \{x \in \mathbb{R} \colon x^3 - 2x^2 - 11x + 12 \leq 0\}$. Find A°.

1.2.8. Example. The interior of the set $\mathbb{Q}$ of rational numbers is empty.

Proof. No open interval contains only rational numbers. □

1.2.9. Proposition. *If A and B are sets of real numbers with $A \subseteq B$, then $A^\circ \subseteq B^\circ$.*

Proof. Let $a \in A^\circ$. Then there is an $\epsilon > 0$ such that $J_\epsilon(a) \subseteq A \subseteq B$. This shows that $a \in B^\circ$. □

1.2.10. Proposition. *If A is a set of real numbers, then $A^{\circ\circ} = A^\circ$.*

Proof. Problem.

1.2.11. Proposition. *If A and B are sets of real numbers, then*
$$(A \cap B)^\circ = A^\circ \cap B^\circ.$$

Proof. Exercise. *Hint.* Show separately that $(A \cap B)^\circ \subseteq A^\circ \cap B^\circ$ and that $A^\circ \cap B^\circ \subseteq (A \cap B)^\circ$.

1.2.12. Proposition. *If A and B are sets of real numbers, then*

$$(A \cup B)^\circ \supseteq A^\circ \cup B^\circ .$$

Proof. Exercise.

1.2.13. Example. Equality may fail in the preceding proposition.

Proof. Problem. *Hint.* See if you can find sets A and B in $\mathbb{R}$ both of which have empty interior but whose union is all of $\mathbb{R}$.

Chapter 2

Topology of the real line

It is clear from the definition of interior that the interior of a set is always contained in the set. Those sets for which the reverse inclusion also holds are called *open sets*.

2.1. Open subsets of $\mathbb{R}$

2.1.1. Definition. A subset U of $\mathbb{R}$ is OPEN if $U^\circ = U$. That is, a set is open if and only if every point of the set is an interior point of the set. If U is an open subset of $\mathbb{R}$, we write $U \overset{\circ}{\subseteq} \mathbb{R}$.

Notice, in particular, that the empty set is open. This is a consequence of the way implication is defined in Appendix D.2: the condition that each point of $\emptyset$ be an interior point is *vacuously satisfied* because there *are no points* in $\emptyset$. (One argues that *if* an element x belongs to the empty set, *then* it is an interior point of the set. The hypothesis is false; so the implication is true.) Also notice that $\mathbb{R}$ itself is an open subset of $\mathbb{R}$.

2.1.2. Example. Bounded open intervals are open sets. That is, if $a < b$, then the open interval (a, b) is an open set.

Proof. Example 1.2.3. □

2.1.3. Example. The interval $(0, \infty)$ is an open set.

Proof. Problem.

One way of seeing that a set is open is to verify that each of its points is an interior point of the set. That is what the definition says. Often it is easier to observe that the set can be written as a union of bounded open intervals. That this happens exactly when a set is open is the point of the next proposition.

2.1.4. Proposition. *A nonempty subset of $\mathbb{R}$ is open if and only if it is a union of bounded open intervals.*

Proof. Let $U \subseteq \mathbb{R}$. First, let us suppose U is a nonempty open subset of $\mathbb{R}$. Each point of U is then an interior point of U. So for each $x \in U$, we may choose a bounded open interval $J(x)$ centered at x that is entirely contained in U. Since $x \in J(x)$ for each $x \in U$, we see that

$$(1) \qquad U = \bigcup_{x \in U} \{x\} \subseteq \bigcup_{x \in U} J(x).$$

On the other hand, since $J(x) \subseteq U$ for each $x \in U$, we have (see Proposition F.1.8)

$$(2) \qquad \bigcup_{x \in U} J(x) \subseteq U.$$

Together (1) and (2) show that U is a union of bounded open intervals.

For the converse suppose $U = \bigcup \mathfrak{J}$, where $\mathfrak{J}$ is a family of open bounded intervals. Let x be an arbitrary point of U. We need only show that x is an interior point of U. To this end choose an interval $J \in \mathfrak{J}$ that contains x. Since J is a bounded open interval, we may write $J = (a, b)$ where $a < b$. Choose ϵ to be the smaller of the numbers $x - a$ and $b - x$. Then it is easy to see that $\epsilon > 0$ and that $x \in J_\epsilon(x) = (x - \epsilon, x + \epsilon) \subseteq (a, b)$. Thus x is an interior point of U. $\qquad\square$

2.1.5. Example. Every interval of the form $(-\infty, a)$ is an open set. So is every interval of the form (a, ∞). (Notice that this and Example 2.1.2 give us the very comforting result that the things we are accustomed to calling open intervals are indeed open sets.)

Proof. Problem.

The study of calculus has two main ingredients: algebra and topology. Algebra deals with operations and their properties and with the resulting structure of groups, fields, vector spaces, algebras, and the like. Topology, on the other hand, is concerned with closeness, ϵ-neighborhoods, and open sets and with the associated structure of metric spaces and various kinds of topological spaces. Almost everything in calculus results from the interplay between algebra and topology.

2.1.6. Definition. The word "topology" has a technical meaning. A family $\mathfrak{T}$ of subsets of a set X is a TOPOLOGY on X if

(a) $\emptyset$ and X belong to $\mathfrak{T}$;

(b) if $\mathfrak{S} \subseteq \mathfrak{T}$ (that is, if $\mathfrak{S}$ is a subfamily of $\mathfrak{T}$), then $\bigcup \mathfrak{S} \in \mathfrak{T}$; and

(c) if $\mathfrak{S}$ is a *finite* subfamily of $\mathfrak{T}$, then $\bigcap \mathfrak{S} \in \mathfrak{T}$.

We can paraphrase this definition by saying that a family of subsets of X that contains both $\emptyset$ and X is a topology on X if it is closed under arbitrary unions and finite intersections. If this definition doesn't make any sense to you at first reading, don't fret. This kind of abstract definition, although easy enough to remember, is irritatingly difficult to understand. Staring at it doesn't help. It appears that a bewildering array of entirely different things might turn out to be topologies. And this is indeed the case. An understanding and appreciation of the definition come only gradually. You will notice as you advance through this material that many important concepts, such as continuity, compactness, and connectedness, are defined by (or characterized by) open sets. Thus, theorems that involve these ideas will rely on properties of open sets for their proofs. This is true not only in the present realm of the real line but in the much wider world of metric spaces, which we will shortly encounter in all their fascinating variety. You will notice that two properties of open sets are used over and over: that unions of open sets are open, and that finite intersections of open sets are open. Nothing else about open sets turns out to be of much importance. Gradually one comes to see that these two facts completely dominate the discussion of continuity, compactness, and so on. Ultimately, it becomes clear that nearly everything in the proofs goes through in situations where *only* these properties are available—that is, in topological spaces.

Our goal at the moment is quite modest: we show that the family of all open subsets of ℝ is indeed a topology on ℝ.

2.1.7. Proposition. *Let* 𝔖 *be a family of open sets in* ℝ. *Then*

(a) *the union of* 𝔖 *is an open subset of* ℝ; *and*

(b) *if* 𝔖 *is finite, the intersection of* 𝔖 *is an open subset of* ℝ.

Proof. Exercise.

2.1.8. Example. The set $U = \{x \in \mathbb{R} \colon x < -2\} \cup \{x > 0 \colon x^2 - x - 6 < 0\}$ is an open subset of ℝ.

Proof. Problem.

2.1.9. Example. The set $\mathbb{R} \setminus \mathbb{N}$ is an open subset of ℝ.

Proof. Problem.

2.1.10. Example. The family 𝔗 of open subsets of ℝ is not closed under arbitrary intersections. (That is, there exists a family 𝔖 of open subsets of ℝ such that $\bigcap \mathfrak{S}$ is *not* open.)

Proof. Problem.

2.2. Closed subsets of ℝ

Next we will investigate the closed subsets of ℝ. These will turn out to be the complements of open sets. But initially we will approach them from a different perspective.

2.2.1. Definition. A point b in ℝ is an ACCUMULATION POINT of a set $A \subseteq \mathbb{R}$ if every ϵ-neighborhood of b contains at least one point of A distinct from b. (We do *not* require that b belong to A, although, of course, it may.) The set of all accumulation points of A is called the DERIVED SET of A and is denoted by A'. The CLOSURE of A, denoted by $\overline{A}$, is $A \cup A'$.

2.2.2. Example. Let $A = \{1/n \colon n \in \mathbb{N}\}$. Then 0 is an accumulation point of A. Furthermore, $\overline{A} = \{0\} \cup A$.

Proof. Problem.

2.2.3. Example. Let A be $(0,1) \cup \{2\} \subseteq \mathbb{R}$. Then $A' = [0,1]$ and $\overline{A} = [0,1] \cup \{2\}$.

Proof. Problem.

2.2.4. Example. Every real number is an accumulation point of the set $\mathbb{Q}$ of rational numbers (since every open interval in $\mathbb{R}$ contains infinitely many rationals); so $\overline{\mathbb{Q}}$ is all of $\mathbb{R}$.

2.2.5. Exercise. Let $A = \mathbb{Q} \cap (0,\infty)$. Find A°, A', and $\overline{A}$.

2.2.6. Problem. Let $A = (0,1] \cup \big([2,3] \cap \mathbb{Q}\big)$. Find

(a) A°;

(b) $\overline{A}$;

(c) $\overline{A^\circ}$;

(d) $\left(\overline{A}\right)^\circ$;

(e) $\overline{A^c}$;

(f) $\left(\overline{A^c}\right)^\circ$;

(g) $\left(A^c\right)^\circ$; and

(h) $\overline{\left(A^c\right)^\circ}$.

2.2.7. Example. Let A be a nonempty subset of $\mathbb{R}$. If A is bounded above, then $\sup A$ belongs to the closure of A. Similarly, if A is bounded below, then $\inf A$ belongs to $\overline{A}$.

Proof. Problem.

2.2.8. Problem. Starting with a set A, what is the greatest number of *different* sets you can get by applying successively the operations of closure, interior, and complement? Apply them as many times as you wish and in any order. For example, starting with the empty set doesn't produce much. We get only $\emptyset$ and $\mathbb{R}$. If we start with the closed interval $[0,1]$, we get four sets: $[0,1]$, $(0,1)$, $(-\infty,0] \cup [1,\infty)$, and $(-\infty,0) \cup (1,\infty)$. By making a more cunning choice of A, how many different sets can you get?

2.2.9. Proposition. *Let $A \subseteq \mathbb{R}$. Then*

(a) $\left(A^\circ\right)^c = \overline{A^c}$*; and*

(b) $\left(A^c\right)^\circ = \left(\overline{A}\right)^c$.

Proof. Exercise. *Hint.* Part (b) is a very easy consequence of (a).

2.2.10. Definition. A subset A of $\mathbb{R}$ is CLOSED if $\overline{A} = A$.

2.2.11. Example. Every closed interval $\big($that is, intervals of the form $[a,b]$ or $(-\infty,a]$ or $[a,\infty)$ or $(-\infty,\infty)\big)$ is closed.

Proof. Problem.

CAUTION. It is a common mistake to treat subsets of $\mathbb{R}$ as if they were doors or windows, and to conclude, for example, that a set is closed because it is not open, or that it cannot be closed because it is open. These "conclusions" are wrong! A subset of $\mathbb{R}$ may be open and not closed, or closed and not open, or both open and closed, or neither. For example, in $\mathbb{R}$,

(a) $(0, 1)$ is open but not closed;

(b) $[0, 1]$ is closed but not open;

(c) $\mathbb{R}$ is both open and closed; and

(d) $[0, 1)$ is neither open nor closed.

This is not to say, however, that there is no relationship between these properties. In the next proposition we discover that the correct relation has to do with complements.

2.2.12. Proposition. *A subset of* $\mathbb{R}$ *is open if and only if its complement is closed.*

Proof. Problem. *Hint.* Use Proposition 2.2.9.

2.2.13. Proposition. *The intersection of an arbitrary family of closed subsets of* $\mathbb{R}$ *is closed.*

Proof. Let $\mathfrak{A}$ be a family of closed subsets of $\mathbb{R}$. By *De Morgan's law* (see Proposition F.3.5) $\bigcap \mathfrak{A}$ is the complement of $\bigcup \{A^c \colon A \in \mathfrak{A}\}$. Since each set A^c is open (by Proposition 2.2.12), the union of $\{A^c \colon A \in \mathfrak{A}\}$ is open (by Proposition 2.1.7(a)); and its complement $\bigcap \mathfrak{A}$ is closed (Proposition 2.2.12 again). □

2.2.14. Proposition. *The union of a finite family of closed subsets of* $\mathbb{R}$ *is closed.*

Proof. Problem.

2.2.15. Problem. Give an example to show that the union of an arbitrary family of closed subsets of $\mathbb{R}$ need not be closed.

2.2.16. Definition. Let a be a real number. Any open subset of $\mathbb{R}$ that contains a is a NEIGHBORHOOD of a. Notice that an ϵ-neighborhood of a is a very special type of neighborhood: it is an interval, and it is symmetric about a. For most purposes the extra internal structure possessed by ϵ-neighborhoods is irrelevant to the matter at hand. To see that we can operate as easily with general neighborhoods as with ϵ-neighborhoods, do the next problem.

2.2.17. Problem. Let A be a subset of $\mathbb{R}$. Prove the following.

(a) A point a is an interior point of A if and only if some neighborhood of a lies entirely in A.

(b) A point b is an accumulation point of A if and only if every neighborhood of b contains at least one point of A distinct from b.

Chapter 3

Continuous functions from $\mathbb{R}$ to $\mathbb{R}$

The continuous functions are perhaps the most important single class of functions studied in calculus. Very roughly, a function $f\colon \mathbb{R} \to \mathbb{R}$ is continuous at a point a in $\mathbb{R}$ if $f(x)$ can be made arbitrarily close to $f(a)$ by insisting that x be sufficiently close to a. In this chapter we define continuity for real valued functions of a real variable and derive some useful necessary and sufficient conditions for such a function to be continuous. Also we will show that composites of continuous functions are themselves continuous. We will postpone until Chapter 4 the proofs that other combinations (sums, products, and so on) of continuous functions are continuous. The first applications of continuity will come in Chapters 5 and 6 on connectedness and compactness.

3.1. Continuity—as a local property

The definition of continuity uses the notion of the inverse image of a set under a function. The notation used here is neither original nor entirely standard. The notations for images and inverse images of sets under a function are defined in Definitions L.1.1 and L.1.4. More relevant information occurs in Appendices L and M.

3.1.1. Definition. A function $f\colon \mathbb{R} \to \mathbb{R}$ is CONTINUOUS AT a point a in $\mathbb{R}$ if $f^{\leftarrow}(V)$ contains a neighborhood of a whenever V is a neighborhood of $f(a)$. Here is another way of saying exactly the same thing: f is continuous at a if every neighborhood of $f(a)$ contains the image under f of a neighborhood of a. (If it is not entirely clear that these two assertions are equivalent, use Propositions M.1.22 (a) and M.1.23 (a) to prove that $U \subseteq f^{\leftarrow}(V)$ if and only if $f^{\rightarrow}(U) \subseteq V$.)

As we saw in Chapter 2, it seldom matters whether we work with general neighborhoods (as in the preceding definition) or with the more restricted ϵ-neighborhoods (as in the next proposition).

3.1.2. Proposition. *A function* $f \colon \mathbb{R} \to \mathbb{R}$ *is continuous at* $a \in \mathbb{R}$ *if and only if for every* $\epsilon > 0$ *there exists* $\delta > 0$ *such that*

$$(3) \qquad\qquad J_\delta(a) \subseteq f^{\leftarrow}\big(J_\epsilon(f(a))\big),$$

Before starting on a proof it is always a good idea to be certain that the meaning of the proposition is entirely clear. In the present case the "if and only if" tells us that we are being given a condition *characterizing* continuity at the point a; that is, a condition that is both necessary and sufficient in order for the function to be continuous at a. The condition states that no matter what positive number ϵ we are given, we can find a corresponding positive number δ such that property (3) holds. This property is the heart of the matter and should be thought of conceptually rather than in terms of symbols. Learning mathematics is easiest when we regard the content of mathematics to be ideas; it is hardest when mathematics is thought of as a game played with symbols. Thus property (3) says that if x is any number in the open interval $(a-\delta, a+\delta)$, then the corresponding value $f(x)$ lies in the open interval between $f(a) - \epsilon$ and $f(a) + \epsilon$. Once we are clear about what it is that we wish to establish, it is time to turn to the proof.

Proof. Suppose f is continuous at a. If $\epsilon > 0$, then $J_\epsilon(f(a))$ is a neighborhood of $f(a)$ and therefore $f^{\leftarrow}\big(J_\epsilon(f(a))\big)$ contains a neighborhood U of a. Since a is an interior point of U, there exists $\delta > 0$ such that $J_\delta(a) \subseteq U$. Then

$$J_\delta(a) \subseteq U \subseteq f^{\leftarrow}\big(J_\epsilon(f(a))\big).$$

Conversely, suppose that for every $\epsilon > 0$ there exists $\delta > 0$ such that $J_\delta(a) \subseteq f^{\leftarrow}\big(J_\epsilon(f(a))\big)$. Let V be a neighborhood of $f(a)$. Then $J_\epsilon(f(a)) \subseteq V$ for some $\epsilon > 0$. By hypothesis, there exists $\delta > 0$ such that $J_\delta(a) \subseteq f^{\leftarrow}\big(J_\epsilon(f(a))\big)$. Then $J_\delta(a)$ is a neighborhood of a and $J_\delta(a) \subseteq f^{\leftarrow}(V)$; so f is continuous at a. $\qquad\square$

We may use the remarks preceding the proof of Proposition 3.1.2 to give a second characterization of continuity at a point. Even though the condition given is algebraic in form, it is best to think of it geometrically. Think of it in terms of distance. In the next corollary read "$|x - a| < \delta$" as "x is within δ units of a" (not as "the absolute value of x minus a is less than δ"). If you follow this advice, statements about continuity (when we get to metric spaces) will sound familiar; otherwise, everything will appear new and strange.

3.1.3. Corollary. *A function* $f \colon \mathbb{R} \to \mathbb{R}$ *is continuous at* a *if and only if for every* $\epsilon > 0$ *there exists* $\delta > 0$ *such that*

$$(4) \qquad\qquad |x - a| < \delta \implies |f(x) - f(a)| < \epsilon.$$

Proof. This is just a restatement of Proposition 3.1.2 because condition (3) holds if and only if

$$x \in J_\delta(a) \implies f(x) \in J_\epsilon(f(a)).$$

But x belongs to the interval $J_\delta(a)$ if and only if $|x - a| < \delta$, and $f(x)$ belongs to $J_\epsilon(f(a))$ if and only if $|f(x) - f(a)| < \epsilon$. Thus (3) and (4) say exactly the same thing. $\qquad\square$

Technically, Definition 3.1.1 is the definition of continuity at a point, while Proposition 3.1.2 and Corollary 3.1.3 are characterizations of this property. Nevertheless, it is not altogether wrong-headed to think of them as alternative (but equivalent) definitions of continuity. It really doesn't matter which one we choose to be *the* definition. Each of them has its uses. For example, consider the result: if f is continuous at a and g is continuous at $f(a)$, then $g \circ f$ is continuous at a (see Proposition 3.2.17). The simplest proof of this uses Definition 3.1.1. On the other hand, when we wish to verify that some particular function is continuous at a point (see, for example, Example 3.2.2), then, usually, Corollary 3.1.3 is best. There are other characterizations of continuity (see Propositions 3.2.12–3.2.16). Before embarking on a particular problem involving this concept, it is wise to take a few moments to reflect on which choice among the assorted characterizations is likely to produce the simplest and most direct proof. This is a favor both to your reader and to yourself.

3.2. Continuity—as a global property

3.2.1. Definition. A function $f \colon \mathbb{R} \to \mathbb{R}$ is CONTINUOUS if it is continuous at every point in its domain.

The purpose of the next few examples is to give you practice in showing that particular functions are (or are not) continuous. If you did a lot of this in beginning calculus, you may wish to skip to the paragraph preceding Proposition 3.2.12.

3.2.2. Example. The function $f \colon \mathbb{R} \to \mathbb{R} \colon x \mapsto -2x + 3$ is continuous.

Proof. We use Corollary 3.1.3. Let $a \in \mathbb{R}$. Given $\epsilon > 0$ choose $\delta = \frac{1}{2}\epsilon$. If $|x - a| < \delta$, then

$$|f(x) - f(a)| = |(-2x + 3) - (-2a + 3)|$$
$$= 2|x - a| < 2\delta = \epsilon. \qquad \square$$

3.2.3. Example. The function $f \colon \mathbb{R} \to \mathbb{R}$ defined by

$$f(x) = \begin{cases} 0 & \text{for } x \leq 0, \\ 1 & \text{for } x > 0 \end{cases}$$

is not continuous at $a = 0$.

Proof. Use Proposition 3.1.2; the denial of the condition given there is that there exists a number ϵ such that for all $\delta > 0$ property (3) fails. (See Example D.4.5.) Let $\epsilon = \frac{1}{2}$. Then $f^{\leftarrow}\big(J_{1/2}(f(0))\big) = f^{\leftarrow}(J_{1/2}(0)) = f^{\leftarrow}(-\frac{1}{2}, \frac{1}{2}) = (-\infty, 0]$. Clearly, this contains no δ-neighborhood of 0. Thus (3) is violated, and f is not continuous at 0. $\qquad \square$

3.2.4. Example. The function $f \colon \mathbb{R} \to \mathbb{R} \colon x \mapsto 5x - 8$ is continuous.

Proof. Exercise.

3.2.5. Example. The function $f \colon x \mapsto x^3$ is continuous at the point $a = -1$.

Proof. Exercise.

3.2.6. Example. The function $f \colon x \mapsto 2x^2 - 5$ is continuous.

Proof. Exercise.

3.2.7. Example. Let $f(x) = 7 - 5x$ for all $x \in \mathbb{R}$. Then f is continuous at the point $a = -2$.

Proof. Problem.

3.2.8. Example. Let $f(x) = \sqrt{|x + 2|}$ for all $x \in \mathbb{R}$. Then f is continuous at the point $a = 0$.

Proof. Problem.

3.2.9. Example. If $f(x) = 3x - 5$ for all $x \in \mathbb{R}$, then f is continuous.

Proof. Problem.

3.2.10. Example. The function f defined by

$$f(x) = \begin{cases} -x^2 & \text{for } x < 0, \\ x + \frac{1}{10} & \text{for } x \geq 0 \end{cases}$$

is not continuous.

Proof. Problem.

3.2.11. Example. For $x > 0$, sketch the functions $x \mapsto \sin \frac{1}{x}$ and $x \mapsto x \sin \frac{1}{x}$. Then verify the following.

(a) The function f defined by

$$f(x) = \begin{cases} 0 & \text{for } x \leq 0, \\ \sin \frac{1}{x}, & \text{for } x > 0 \end{cases}$$

is not continuous.

(b) The function f defined by

$$f(x) = \begin{cases} 0 & \text{for } x \leq 0, \\ x \sin \frac{1}{x}, & \text{for } x > 0 \end{cases}$$

is continuous at 0.

Proof. Problem.

The continuity of a function f at a point is a LOCAL property; that is, it is entirely determined by the behavior of f in arbitrarily small neighborhoods of the point. The continuity of f, on the other hand, is a GLOBAL property; it can be determined only if we know how f behaves everywhere on its domain. In Definition 3.1.1, Proposition 3.1.2, and Corollary 3.1.3 we gave three equivalent conditions for local continuity. In Propositions 3.2.12–3.2.16 we give equivalent conditions for the corresponding global concept. The next proposition gives the most useful of these conditions; it is the one that becomes the *definition* of continuity in arbitrary topological spaces. It says a necessary and sufficient condition for f to be continuous is that the inverse image under f of open sets be open. This shows that continuity is a purely TOPOLOGICAL PROPERTY; that is, it is entirely determined by the topologies (families of all open sets) of the domain and the codomain of the function.

3.2.12. Proposition. *A function* $f \colon \mathbb{R} \to \mathbb{R}$ *is continuous if and only if* $f^{\leftarrow}(U)$ *is open whenever* U *is open in* $\mathbb{R}$.

Proof. Exercise.

3.2.13. Proposition. *A function* $f \colon \mathbb{R} \to \mathbb{R}$ *is continuous if and only if* $f^{\leftarrow}(C)$ *is a closed set whenever* C *is a closed subset of* $\mathbb{R}$.

Proof. Problem.

3.2.14. Proposition. *A function* $f \colon \mathbb{R} \to \mathbb{R}$ *is continuous if and only if*

$$f^{\leftarrow}(B^{\circ}) \subseteq (f^{\leftarrow}(B))^{\circ}$$

for all $B \subseteq \mathbb{R}$.

Proof. Problem.

3.2.15. Proposition. *A function* $f \colon \mathbb{R} \to \mathbb{R}$ *is continuous if and only if*

$$f^{\rightarrow}(\overline{A}) \subseteq \overline{f^{\rightarrow}(A)}$$

for all $A \subseteq \mathbb{R}$.

Proof. Problem. *Hint.* This isn't so easy. Use Proposition 3.2.13 and the fact (see Propositions M.1.22 and M.1.23) that for any sets A and B

$$f^{\rightarrow}(f^{\leftarrow}(B)) \subseteq B \quad \text{and} \quad A \subseteq f^{\leftarrow}(f^{\rightarrow}(A)).$$

Show that if f is continuous, then $\overline{A} \subseteq f^{\leftarrow}\big(\overline{f^{\rightarrow}(A)}\big)$. Then apply $f^{\rightarrow}$. For the converse, apply the hypothesis to the set $f^{\leftarrow}(C)$ where C is a closed subset of $\mathbb{R}$. Then apply $f^{\leftarrow}$.

3.2.16. Proposition. *A function* $f \colon \mathbb{R} \to \mathbb{R}$ *is continuous if and only if*

$$\overline{f^{\leftarrow}(B)} \subseteq f^{\leftarrow}(\overline{B})$$

for all $B \subseteq \mathbb{R}$.

Proof. Problem.

3.2.17. Proposition. *Let* $f, g \colon \mathbb{R} \to \mathbb{R}$. *If* f *is continuous at* a *and* g *is continuous at* $f(a)$, *then the composite function* $g \circ f$ *is continuous at* a.

Proof. Let W be a neighborhood of $g(f(a))$. We wish to show that the inverse image of W under $g \circ f$ contains a neighborhood of a. Since g is continuous at $f(a)$, the set $g^{\leftarrow}(W)$ contains a neighborhood V of $f(a)$. And since f is continuous at a, the set $f^{\leftarrow}(V)$ contains a neighborhood U of a. Then

$$\begin{aligned}
(g \circ f)^{\leftarrow}(W) &= f^{\leftarrow}(g^{\leftarrow}(W)) \\
&\supseteq f^{\leftarrow}(V) \\
&\supseteq U,
\end{aligned}$$

which is what we wanted.   □

This important result has an equally important but entirely obvious consequence.

3.2.18. Corollary. *The composite of two continuous functions is continuous.*

3.2.19. Problem. Give a direct proof of Corollary 3.2.18. (That is, give a proof that does not rely on Proposition 3.2.17.)

3.3. Functions defined on subsets of $\mathbb{R}$

The remainder of this chapter is devoted to a small but important technical problem. Thus far the definitions and propositions concerning continuity have dealt with functions whose domain is $\mathbb{R}$. What do we do about functions whose domain is a proper subset of $\mathbb{R}$? After all, many old friends—the functions $x \mapsto \sqrt{x}$, $x \mapsto \frac{1}{x}$, and $x \mapsto \tan x$, for example—have domains that are not all of $\mathbb{R}$. The difficulty is that if we were to attempt to apply Proposition 3.1.2 to the square root function $f \colon x \mapsto \sqrt{x}$ (which would of course be improper since the hypothesis $f \colon \mathbb{R} \to \mathbb{R}$ is not satisfied), we would come to the unwelcome conclusion that f is not continuous at 0: if $\epsilon > 0$, then the set $f^{\leftarrow}\big(J_\epsilon(f(0))\big) = f^{\leftarrow}\big(J_\epsilon(0)\big) = f^{\leftarrow}(-\epsilon, \epsilon) = [0, \epsilon^2)$ contains no neighborhood of 0 in $\mathbb{R}$.

Now this can't be right. What we must do is provide an appropriate definition for the continuity of functions whose domains are proper subsets of $\mathbb{R}$. And we wish to do it in such a way that we make as few changes as possible in the resulting propositions.

The source of our difficulty is the demand (in Definition 3.1.1) that $f^{\leftarrow}(V)$ contain a neighborhood of the point a—and neighborhoods have been defined only in $\mathbb{R}$. But why $\mathbb{R}$? That is not the domain of our function; the set $A = [0, \infty)$ is. We *should* be talking about neighborhoods in A. So the question we now face is, How should we define neighborhoods in (and open subsets of) proper subsets of $\mathbb{R}$? The best answer is astonishingly simple. An open subset of A is the intersection of an open subset of $\mathbb{R}$ with A.

3.3.1. Definition. Let $A \subseteq \mathbb{R}$. A set U contained in A is OPEN IN A if there exists an open subset V of $\mathbb{R}$ such that $U = V \cap A$. Briefly, the open subsets of A are restrictions to A of open subsets of $\mathbb{R}$. If U is an open subset of A, we write $U \overset{\circ}{\subseteq} A$. A NEIGHBORHOOD OF a IN A is an open subset of A that contains a.

3.3.2. Example. The set $[0, 1)$ is an open subset of $[0, \infty)$.

Proof. Let $V = (-1, 1)$. Then $V \overset{\circ}{\subseteq} \mathbb{R}$ and $[0, 1) = V \cap [0, \infty)$; so $[0, 1) \overset{\circ}{\subseteq} [0, \infty)$. $\square$

Since, as we have just seen, $[0, 1)$ is open in $[0, \infty)$ but is not open in $\mathbb{R}$, there is a possibility for confusion. Openness is not an intrinsic property of a set. When we say that a set is open, the answer to the question, "Open in what?", must be either clear from context or else specified. Since the topology (that is, collection of open subsets) that a set A inherits from $\mathbb{R}$ is often called the RELATIVE TOPOLOGY on A, emphasis may be achieved by saying that a subset B of A is *relatively open* in A. Thus, for example, we may say that $[0, 1)$ is relatively open in $[0, \infty)$; or we may say that $[0, 1)$ is a relative neighborhood of 0 (or any other point in the interval). The question here is emphasis and clarity, not logic.

3.3.3. Example. The set $\{1\}$ is an open subset of $\mathbb{N}$.

Proof. Problem.

3.3.4. Example. The set of all rational numbers x such that $x^2 \leq 2$ is an open subset of $\mathbb{Q}$.

Proof. Problem.

3.3.5. Example. The set of all rational numbers x such that $x^2 \leq 4$ is *not* an open subset of $\mathbb{Q}$.

Proof. Problem.

3.3.6. Definition. Let $A \subseteq \mathbb{R}$, $a \in A$, and $\epsilon > 0$. The ϵ-NEIGHBORHOOD OF a IN A is $(a - \epsilon, a + \epsilon) \cap A$.

3.3.7. Example. In $\mathbb{N}$ the $\frac{1}{2}$-neighborhood of 1 is $\{1\}$.

Proof. Since $(1 - \frac{1}{2}, 1 + \frac{1}{2}) \cap \mathbb{N} = (\frac{1}{2}, \frac{3}{2}) \cap \mathbb{N} = \{1\}$, we conclude that the $\frac{1}{2}$-neighborhood of 1 is $\{1\}$. □

Now is the time to show that the family of relatively open subsets of A (that is, the relative topology on A) is in fact a topology on A. In particular, we must show that this family is closed under unions and finite intersections.

3.3.8. Proposition. *Let* $A \subseteq \mathbb{R}$. *Then*

(a) *$\emptyset$ and A are relatively open in A;*

(b) *if $\mathfrak{U}$ is a family of relatively open sets in A, then $\bigcup \mathfrak{U}$ is relatively open in A; and*

(c) *if $\mathfrak{U}$ is a finite family of relatively open sets in A, then $\bigcap \mathfrak{U}$ is relatively open in A.*

Proof. Problem.

3.3.9. Example. Every subset of $\mathbb{N}$ is relatively open in $\mathbb{N}$.

Proof. Problem.

Now we are in a position to consider the continuity of functions defined on proper subsets of $\mathbb{R}$.

3.3.10. Definition. Let $a \in A \subseteq \mathbb{R}$ and $f : A \to \mathbb{R}$. The function f is CONTINUOUS AT a if $f^{\leftarrow}(V)$ contains a neighborhood of a in A whenever V is a neighborhood of $f(a)$. The function f is CONTINUOUS if it is continuous at each point in its domain.

It is important to notice that we discuss the continuity of a function only at points where it is defined. We will not, for example, make the claim found in so many beginning calculus texts that the function $x \mapsto 1/x$ is discontinuous at zero. Nor will we try to decide whether the sine function is continuous at the Bronx Zoo.

The next proposition tells us that the crucial characterization of continuity in terms of inverse images of open sets (see Proposition 3.2.12) still holds under the definition we have just given. Furthermore, codomains don't matter; that is, it doesn't matter whether we start with open subsets of $\mathbb{R}$ or with sets open in the range of the function.

3.3.11. Proposition. *Let A be a subset of $\mathbb{R}$. A function $f\colon A \to \mathbb{R}$ is continuous if and only if $f^{\leftarrow}(V)$ is open in A whenever V is open in* ran f.

Proof. Problem. *Hint.* Notice that if $W \overset{\circ}{\subseteq} \mathbb{R}$ and $V = W \cap \operatorname{ran} f$, then $f^{\leftarrow}(V) = f^{\leftarrow}(W)$.

3.3.12. Problem. Discuss the changes (if any) that must be made in Proposition 3.1.2, Corollary 3.1.3, Proposition 3.2.17, and Corollary 3.2.18, in order to accommodate functions whose domain is not all of $\mathbb{R}$.

3.3.13. Problem. Let $A \subseteq \mathbb{R}$. Explain what "(relatively) closed subset of A" should mean. Suppose further that $B \subseteq A$. Decide how to define "the closure of B in A" and "the interior of B with respect to A". Explain what changes these definitions will require in Propositions 3.2.13–3.2.16 so that the results hold for functions whose domains are proper subsets of $\mathbb{R}$.

3.3.14. Problem. Let $A = (0,1) \cup (1,2)$. Define $f\colon A \to \mathbb{R}$ by

$$f(x) = \begin{cases} 0 & \text{for } 0 < x < 1, \\ 1 & \text{for } 1 < x < 2. \end{cases}$$

Is f a continuous function?

3.3.15. Example. The function f, defined by

$$f(x) = \frac{1}{x} \qquad \text{for } x \neq 0,$$

is continuous.

Proof. Problem.

3.3.16. Example. The function f, defined by

$$f(x) = \sqrt{x+2} \qquad \text{for } x \geq -2,$$

is continuous at $x = -2$.

Proof. Problem.

3.3.17. Example. The square root function is continuous.

Proof. Problem.

3.3.18. Problem. Define $f\colon (0,1) \to \mathbb{R}$ by setting $f(x) = 0$ if x is irrational and $f(x) = 1/n$ if $x = m/n$, where m and n are natural numbers with no common factors. Where is f continuous?

3.3.19. Example. Inclusion mappings between subsets of $\mathbb{R}$ are continuous. That is, if $A \subseteq B \subseteq \mathbb{R}$, then the inclusion mapping $\iota\colon A \to B\colon a \mapsto a$ is continuous; see Definition L.3.1.

Proof. Let $U \overset{\circ}{\subseteq} B$. By the definition of the relative topology on B (see Definition 3.3.1), there exists an open subset V of $\mathbb{R}$ such that $U = V \cap B$. Then $\iota^{\leftarrow}(U) = \iota^{\leftarrow}(V \cap B) = V \cap B \cap A = V \cap A \overset{\circ}{\subseteq} A$. Since the inverse image under ι of each open set is open, ι is continuous. $\qquad\square$

It is amusing to note how easy it is with the preceding example in hand to show that restrictions of continuous functions are continuous.

3.3.20. Proposition. *Let* $A \subseteq B \subseteq \mathbb{R}$. *If* $f \colon B \to \mathbb{R}$ *is continuous, then* $f|_A$, *the restriction of* f *to* A, *is continuous.*

Proof. Recall (see Appendix L.5) that $f|_A = f \circ \iota$, where ι is the inclusion mapping of A into B. Since f is continuous (by hypothesis) and ι is continuous (by Example 3.3.19), their composite $f|_A$ is also continuous (by the generalization of Corollary 3.2.18 in Problem 3.3.12). $\qquad\square$

We conclude this chapter with the observation that if a continuous function is positive at a point, it is positive nearby.

3.3.21. Proposition. *Let* $A \subseteq \mathbb{R}$, *and let* $f \colon A \to \mathbb{R}$ *be continuous at the point* a. *If* $f(a) > 0$, *then there exists a neighborhood* J *of* a *in* A *such that* $f(x) > \frac{1}{2}f(a)$ *for all* $x \in J$.

Proof. Problem.

Chapter 4

Sequences of real numbers

Sequences are an extremely useful tool in dealing with topological properties of sets in $\mathbb{R}$ and, as we will see later, in general metric spaces. A major goal of this chapter is to illustrate this usefulness by showing how sequences may be used to characterize open sets, closed sets, and continuous functions.

4.1. Convergence of sequences

A SEQUENCE is a function whose domain is the set $\mathbb{N}$ of natural numbers. In this chapter the sequences we consider will all be sequences of real numbers, that is, functions from $\mathbb{N}$ into $\mathbb{R}$. If a is a sequence, it is conventional to write its value at a natural number n as a_n rather than as $a(n)$. The sequence itself may be written in a number of ways:

$$a = (a_n)_{n=1}^{\infty} = (a_n) = (a_1, a_2, a_3, \dots).$$

Care should be exercised in using the last of these notations. It would not be clear, for example, whether $(\frac{1}{3}, \frac{1}{5}, \frac{1}{7}, \dots)$ is intended to be the sequence of reciprocals of odd primes (in which case the next term would be $\frac{1}{11}$) or the sequence of reciprocals of odd integers greater than 1 (in which case the next term would be $\frac{1}{9}$). The element a_n in the range of a sequence is the nth TERM of the sequence.

It is important to distinguish in one's thinking between a sequence and its range. Think of a sequence $(x_1, x_2, x_3, \dots)$ as an *ordered* set: there is a first element x_1, and a second element x_2, and so on. The range $\{x_1, x_2, x_3, \dots\}$ is just a set. There is no "first" element. For example, the sequences $(1, 2, 3, 4, 5, 6, \dots)$ and $(2, 1, 4, 3, 6, 5, \dots)$ are different, whereas the sets $\{1, 2, 3, 4, 5, 6, \dots\}$ and $\{2, 1, 4, 3, 6, 5, \dots\}$ are exactly the same (both are $\mathbb{N}$). A sequence may have repeated entries; its range, a set, may not. For example, the sequence $(1, -1, 1, -1, 1, -1, \dots)$ is an infinite list, whereas its range, the set $\{1, -1\}$, has only two elements.

Remark. Occasionally, in what follows, it will be convenient to alter the preceding definition a bit to allow the domain of a sequence to be the set $\mathbb{N} \cup \{0\}$ of all positive integers. It is worth noticing as we go along that this in no way affects the correctness of the results we prove in this chapter.

4.1.1. Definition. A sequence (x_n) of real numbers is EVENTUALLY IN a set A if there exists a natural number n_0 such that $x_n \in A$ whenever $n \geq n_0$.

4.1.2. Example. For each $n \in \mathbb{N}$, let $x_n = 3 + \frac{7-n}{2}$. Then the sequence $(x_n)_{n=1}^{\infty}$ is eventually strictly negative; that is, the sequence is eventually in the interval $(-\infty, 0)$.

Proof. If $n \geq 14$, then $\frac{7-n}{2} \leq -\frac{7}{2}$ and $x_n = 3 + \frac{7-n}{2} \leq -\frac{1}{2} < 0$. So $x_n \in (-\infty, 0)$ whenever $n \geq 14$. $\qquad\square$

4.1.3. Example. For each $n \in \mathbb{N}$, let $x_n = \frac{2n-3}{n}$. Then the sequence $(x_n)_{n=1}^{\infty}$ is eventually in the interval $(\frac{3}{2}, 2)$.

Proof. Problem.

4.1.4. Definition. A sequence $(x_n)_{n=1}^{\infty}$ of real numbers CONVERGES to $a \in \mathbb{R}$ if it is eventually in every ϵ-neighborhood of a. When the sequence converges to a, we write

$$(5) \qquad\qquad\qquad x_n \to a \text{ as } n \to \infty.$$

These symbols may be read as "x_n approaches a as n gets large". If (x_n) converges to a, the number a is the LIMIT of the sequence (x_n). It would not be proper to refer to *the* limit of a sequence if it were possible for a sequence to converge to two different points. We show now that this cannot happen; limits of sequences are unique.

4.1.5. Proposition. *Let (x_n) be a sequence in $\mathbb{R}$. If $x_n \to a$ and $x_n \to b$ as $n \to \infty$, then $a = b$.*

Proof. Argue by contradiction. Suppose $a \neq b$, and let $\epsilon = |a - b|$. Then $\epsilon > 0$. Since (x_n) is eventually in $J_{\epsilon/2}(a)$, there exists $n_0 \in \mathbb{N}$ such that $x_n \in J_{\epsilon/2}(a)$ for $n \geq n_0$. That is, $|x_n - a| < \frac{\epsilon}{2}$ for $n \geq n_0$. Similarly, since (x_n) is eventually in $J_{\epsilon/2}(b)$, there is an $m_0 \in \mathbb{N}$ such that $|x_n - b| < \frac{\epsilon}{2}$ for $n \geq m_0$. Now if n is any integer larger than both n_0 and m_0, then

$$\epsilon = |a - b| = |a - x_n + x_n - b| \leq |a - x_n| + |x_n - b| < \frac{\epsilon}{2} + \frac{\epsilon}{2} = \epsilon.$$

But $\epsilon < \epsilon$ is impossible. Therefore, our initial supposition was wrong, and $a = b$. $\quad\square$

Since limits are unique, we may use an alternative notation to (5): if (x_n) converges to a, we may write

$$\lim_{n \to \infty} x_n = a.$$

(Notice how inappropriate this notation would be if limits were not unique.)

4.1.6. Definition. If a sequence (x_n) does not converge (that is, if there exists no $a \in \mathbb{R}$ such that (x_n) converges to a), then the sequence DIVERGES. Sometimes a divergent sequence (x_n) has the property that it is eventually in every interval of the form (p, ∞) where $p \in \mathbb{N}$. In this case we write

$$x_n \to \infty \text{ as } n \to \infty \qquad \text{or} \qquad \lim_{n \to \infty} x_n = \infty.$$

If a divergent sequence (x_n) is eventually in every interval of the form $(-\infty, -p)$ for $p \in \mathbb{N}$, we write

$$x_n \to -\infty \text{ as } n \to \infty \qquad \text{or} \qquad \lim_{n \to \infty} x_n = -\infty.$$

CAUTION. It is *not* true that every divergent sequence satisfies either $x_n \to \infty$ or $x_n \to -\infty$. See Example 4.1.9 below.

It is easy to rephrase the definition of convergence of a sequence in slightly different language. The next problem gives two such variants. Sometimes one of these alternative "definitions" is more convenient than Definition 4.1.4.

4.1.7. Problem. Let (x_n) be a sequence of real numbers and $a \in \mathbb{R}$.

(a) Show that $x_n \to a$ if and only if for every $\epsilon > 0$ there exists $n_0 \in \mathbb{N}$ such that $|x_n - a| < \epsilon$ whenever $n \geq n_0$.

(b) Show that $x_n \to a$ if and only if (x_n) is eventually in every neighborhood of a.

4.1.8. Example. The sequence $(\frac{1}{n})_{n=1}^{\infty}$ converges to 0.

Proof. Let $\epsilon > 0$. Use the *Archimedean property* Proposition J.4.1 of the real numbers to choose $N \in \mathbb{N}$ large enough that $N > \frac{1}{\epsilon}$. Then

$$\left| \frac{1}{n} - 0 \right| = \frac{1}{n} \leq \frac{1}{N} < \epsilon$$

whenever $n \geq N$. $\square$

4.1.9. Example. The sequence $\left((-1)^n\right)_{n=0}^{\infty}$ diverges.

Proof. Argue by contradiction. If we assume that $(-1)^n \to a$ for some $a \in \mathbb{R}$, then there exists $N \in \mathbb{N}$ such that $n \geq N$ implies $|(-1)^n - a| < 1$. Thus for every $n \geq N$,

$$
\begin{aligned}
2 &= |(-1)^n - (-1)^{n+1}| \\
&= |(-1)^n - a + a - (-1)^{n+1}| \\
&\leq |(-1)^n - a| + |a - (-1)^{n+1}| \\
&< 1 + 1 = 2,
\end{aligned}
$$

which is not possible. $\square$

4.1.10. Example. The sequence $\left(\dfrac{n}{n+1}\right)_{n=1}^{\infty}$ converges to 1.

Proof. Problem.

4.1.11. Proposition. *Let (x_n) be a sequence in $\mathbb{R}$. Then $x_n \to 0$ if and only if $|x_n| \to 0$.*

Proof. Exercise.

4.2. Algebraic combinations of sequences

As you will recall from beginning calculus, one standard way of computing the limit of a complicated expression is to find the limits of the constituent parts of the expression and then combine them algebraically. Suppose, for example, we are given sequences (x_n) and (y_n) and are asked to find the limit of the sequence (x_n+y_n). What do we do? First we try to find the limits of the individual sequences (x_n) and (y_n). Then we add. This process is justified by a proposition that says, roughly, that *the limit of a sum is the sum of the limits.* Limits with respect to other algebraic operations behave similarly.

The aim of the following problem is to develop the theory detailing how limits of sequences interact with algebraic operations.

4.2.1. Problem. Suppose that (x_n) and (y_n) are sequences of real numbers, that $x_n \to a$ and $y_n \to b$, and that $c \in \mathbb{R}$. For the case where a and b are real numbers, derive

(a) $x_n + y_n \to a + b$;

(b) $x_n - y_n \to a - b$;

(c) $x_n y_n \to ab$;

(d) $cx_n \to ca$; and

(e) $\dfrac{x_n}{y_n} \to \dfrac{a}{b}$ if $b \neq 0$.

Then consider what happens in case $a = \pm\infty$ or $b = \pm\infty$ (or both). What can you say (if anything) about the limits of the left-hand expressions of (a)–(e)? In those cases in which nothing can be said, give examples to demonstrate as many outcomes as possible. For example, if $a = \infty$ and $b = -\infty$, then nothing can be concluded about the limit of $x_n + y_n$ as n gets large. All of the following are possible.

(i) $x_n + y_n \to -\infty$. (Let $x_n = n$ and $y_n = -2n$.)

(ii) $x_n + y_n \to \alpha$, where α is any real number. (Let $x_n = \alpha + n$ and $y_n = -n$.)

(iii) $x_n + y_n \to \infty$. (Let $x_n = 2n$ and $y_n = -n$.)

(iv) None of the above. (Let $(x_n) = (1, 2, 5, 6, 9, 10, 13, 14, \dots)$ and $y_n = (0, -3, -4, -7, -8, -11, -12, -15, \dots)$.)

4.2.2. Example. If $x_n \to a$ in $\mathbb{R}$ and $p \in \mathbb{N}$, then $x_n{}^p \to a^p$.

Proof. Use induction on p. The conclusion obviously holds when $p = 1$. Suppose $x_n{}^{p-1} \to a^{p-1}$. Apply part (c) of Problem 4.2.1:

$$x_n{}^p = x_n{}^{p-1} \cdot x_n \to a^{p-1} \cdot a = a^p. \qquad \square$$

4.2.3. Example. If $x_n = \dfrac{2 - 5n + 7n^2 - 6n^3}{4 - 3n + 5n^2 + 4n^3}$ for each $n \in \mathbb{N}$, then $x_n \to -\frac{3}{2}$ as $n \to \infty$.

Proof. Problem.

4.2.4. Problem. Find $\lim_{n\to\infty}(\sqrt{n^2+5n}-n)$.

Another very useful tool in computing limits is the *sandwich theorem*, which says that a sequence sandwiched between two other sequences with a common limit has that same limit.

4.2.5. Proposition (Sandwich theorem). *Let a be a real number or one of the symbols $+\infty$ or $-\infty$. If $x_n \to a$ and $z_n \to a$ and if $x_n \leq y_n \leq z_n$ for all n, then $y_n \to a$.*

Proof. Problem.

4.2.6. Example. If $x_n = \dfrac{\sin\left(3+\pi^{n^2}\right)}{n^{3/2}}$ for each $n \in \mathbb{N}$, then $x_n \to 0$ as $n \to \infty$.

Proof. Problem. *Hint.* Use Proposition 4.1.11, Example 4.1.8, and Proposition 4.2.5.

4.2.7. Example. If (x_n) is a sequence in $(0,\infty)$ and $x_n \to a$, then $\sqrt{x_n} \to \sqrt{a}$.

Proof. Problem. *Hint.* There are two possibilities: treat the cases $a = 0$ and $a > 0$ separately. For the first use Problem 4.1.7(a). For the second use Problem 4.2.1(b) and Proposition 4.1.11; write $\sqrt{x_n} - \sqrt{a}$ as $|x_n - a|/(\sqrt{x_n} + \sqrt{a})$. Then find an inequality that allows you to use the sandwich theorem (Proposition 4.2.5).

4.2.8. Example. The sequence $\left(n^{1/n}\right)$ converges to 1.

Proof. Problem. *Hint.* For each n let $a_n = n^{1/n} - 1$. Apply the *binomial theorem* Problem I.1.17 to $(1+a_n)^n$ to obtain the inequality $n > \frac{1}{2}n(n-1)\,a_n{}^2$ and hence to conclude that $0 < a_n < \sqrt{\frac{2}{n-1}}$ for every n. Use the sandwich theorem, Proposition 4.2.5.

4.3. Sufficient condition for convergence

An amusing situation sometimes arises in which we know what the limit of a sequence must be *if* it exists, but we have no idea whether the sequence actually converges. Here is an example of this odd behavior. The sequence will be recursively defined. Sequences are said to be RECURSIVELY defined if only the first term or first few terms are given explicitly and the remaining terms are defined by means of the preceding term(s).

Consider the sequence (x_n) defined so that $x_1 = 1$, and for $n \in \mathbb{N}$,

(6) $$x_{n+1} = \frac{3(1+x_n)}{3+x_n}.$$

It is obvious that *IF the sequence converges* to a limit, say ℓ, then ℓ must satisfy

$$\ell = \frac{3(1+\ell)}{3+\ell}.$$

(This is obtained by taking limits as $n \to \infty$ on both sides of equation (6).) Cross-multiplying and solving for ℓ leads us to the conclusion that $\ell = \pm\sqrt{3}$. Since the first term x_1 is positive, equation (6) makes it clear that all the terms will be positive; so ℓ cannot be $-\sqrt{3}$. Thus, it is entirely clear that if the limit of the sequence exists, it must be $\sqrt{3}$. What is *not* clear is whether the limit exists at all. This indicates how useful it is to know some conditions sufficient to guarantee convergence of a sequence. The next proposition gives one important example of such a condition: it says that bounded monotone sequences converge.

First, we present the relevant definitions.

4.3.1. Definition. A sequence (x_n) of real numbers is BOUNDED if its range is a bounded subset of $\mathbb{R}$. Another way of saying the same thing is that a sequence (x_n) in $\mathbb{R}$ is bounded if there exists a number M such that $|x_n| \leq M$ for all $n \in \mathbb{N}$.

4.3.2. Definition. A sequence (a_n) of real numbers is INCREASING if $a_{n+1} \geq a_n$ for every $n \in \mathbb{N}$; it is STRICTLY INCREASING if $a_{n+1} > a_n$ for every n. A sequence is DECREASING if $a_{n+1} \leq a_n$ for every n, and it is STRICTLY DECREASING if $a_{n+1} < a_n$ for every n. A sequence is MONOTONE if it is either increasing or decreasing.

4.3.3. Proposition. *Every bounded monotone sequence of real numbers converges. In fact, if a sequence is bounded and increasing, then it converges to the least upper bound of its range. Similarly, if it is bounded and decreasing, then it converges to the greatest lower bound of its range.*

Proof. Let (x_n) be a bounded increasing sequence. Let R be the range of the sequence, and let ℓ be the least upper bound of R; that is, $\ell = \sup\{x_n : n \in \mathbb{N}\}$. We have seen in Example 2.2.7 that the least upper bound of a nonempty subset of $\mathbb{R}$ belongs to the closure of that set. In particular, $\ell \in \overline{R}$. Thus, given any $\epsilon > 0$, we can find a number x_{n_0} of R that lies in the interval $J_\epsilon(\ell)$. In fact, $x_{n_0} \in (\ell - \epsilon, \ell]$. Since the sequence is increasing and bounded above by ℓ, we have $x_n \in (\ell - \epsilon, \ell] \subseteq J_\epsilon(\ell)$ for every $n \geq n_0$. What we have just proved is that the sequence (x_n) is eventually in every ϵ-neighborhood of ℓ. That is, $x_n \to \ell$.

If (x_n) is bounded and decreasing, then the sequence $(-x_n)$ is bounded and increasing. If R is the range of (x_n) and g is the greatest lower bound of R, then $-R$ is the range of $(-x_n)$ and $-g$ is the least upper bound of $-R$. By what we have just proved, $-x_n \to -g$. So $x_n \to g$ as desired. $\square$

4.3.4. Example. Let (x_n) be the sequence recursively defined above: $x_1 = 1$ and (6) holds for all $n \geq 1$. Then $x_n \to \sqrt{3}$ as $n \to \infty$.

Proof. We have argued previously that the limit of the sequence is $\sqrt{3}$ *if* it exists. So what we must show is that the sequence does converge. We have thus far one tool—Proposition 4.3.3. Thus, we *hope* that we will be able to prove that the sequence is bounded and monotone. If the sequence starts at 1, is monotone, and approaches $\sqrt{3}$, then it must be increasing. How do we prove that the sequence is bounded? Bounded by what? We observed earlier that $x_1 \geq 0$ and equation (6) then guarantees that all the remaining x_n's will be positive. So 0 is a lower bound for the sequence. If the sequence is increasing and approaches $\sqrt{3}$, then $\sqrt{3}$ will be an upper bound for the terms of the sequence. Thus it appears that the way to make use of Proposition 4.3.3 is to prove two things: (1) $0 \leq x_n \leq \sqrt{3}$ for all n; and (2) (x_n) is increasing. If we succeed in establishing both of these claims, we are done.

Claim 1. $0 \le x_n \le \sqrt{3}$ for all $n \in \mathbb{N}$.

Proof. We have already observed that $x_n \ge 0$ for all n. We use mathematical induction to show that $x_n \le \sqrt{3}$ for all n. Since $x_1 = 1$, it is clear that $x_1 \le \sqrt{3}$. For the inductive hypothesis suppose that for some particular k we have

$$(7) \qquad\qquad x_k \le \sqrt{3}\,.$$

We wish to show that $x_{k+1} \le \sqrt{3}$. To start, multiply both sides of (7) by $3 - \sqrt{3}$. (If you are wondering, "How did you know to do that?", consult the next problem.) This gives us

$$3x_k - \sqrt{3}x_k \le 3\sqrt{3} - 3.$$

Rearrange so that all terms are positive

$$3 + 3x_k \le 3\sqrt{3} + \sqrt{3}x_k = \sqrt{3}(3 + x_k).$$

But then clearly

$$x_{k+1} = \frac{3 + 3x_k}{3 + x_k} \le \sqrt{3},$$

which was to be proved.

Claim 2. The sequence (x_n) is increasing.

Proof. We will show that $x_{n+1} - x_n \ge 0$ for all n. For each n

$$x_{n+1} - x_n = \frac{3 + 3x_n}{3 + x_n} - x_n = \frac{3 - x_n{}^2}{3 + x_n} \ge 0.$$

(We know that $3 - x_n{}^2 \ge 0$ from Claim 1.) $\qquad\qquad\qquad\qquad\qquad \square$

4.3.5. Problem. A student, Fred R. Dimm, tried on an exam to prove the claims made in Example 4.3.4. For the inductive proof of Claim 1 that $x_n \le \sqrt{3}$, he offered the following "proof":

$$x_{k+1} \le \sqrt{3},$$
$$\frac{3 + 3x_k}{3 + x_k} \le \sqrt{3},$$
$$3 + 3x_k \le \sqrt{3}(3 + x_k),$$
$$3x_k - \sqrt{3}x_k \le 3\sqrt{3} - 3,$$
$$(3 - \sqrt{3})x_k \le (3 - \sqrt{3})\sqrt{3},$$
$$x_k \le \sqrt{3}, \qquad \text{which is true by hypothesis.}$$

(a) Aside from his regrettable lack of explanation, Fred seems to be making a serious logical error. Explain to poor Fred why his offering is not a proof. *Hint.* What would you say about a proof of the assertion $1 = 2$, that goes as follows?

$$1 = 2,$$
$$0 \cdot 1 = 0 \cdot 2,$$
$$0 = 0, \qquad \text{which is true.}$$

(b) Now explain why Fred's computation in (a) is really quite useful scratch work, even if it is not a proof. *Hint.* In the correct proof of Claim 1, how *might* its author have been inspired to "multiply both sides of (7) by $3 - \sqrt{3}$"?

4.3.6. Example. The condition (bounded and monotone) given in Proposition 4.3.3, while sufficient to guarantee the convergence of a sequence, is not necessary.

Proof. Problem. (Give an explicit example.)

Expressed as a conditional, Proposition 4.3.3 says that if a sequence is bounded and monotone, then it converges. Example 4.3.6 shows that the converse of this conditional is not correct. A partial converse does hold however: if a sequence converges, it must be bounded.

4.3.7. Proposition. *Every convergent sequence in $\mathbb{R}$ is bounded.*

Proof. Exercise.

We will encounter many situations when it is important to know the limit as n gets large of r^n, where r is a number in the interval $(-1, 1)$, and the limit of $r^{1/n}$, where r is a number greater than 0. The next two propositions show that the respective limits are always 0 and 1.

4.3.8. Proposition. *If $|r| < 1$, then $r^n \to 0$ as $n \to \infty$. If $|r| > 1$, then (r^n) diverges.*

Proof. Suppose that $|r| < 1$. If $r = 0$, the proof is trivial, so we consider only $r \neq 0$. Since $0 < |r| < 1$ whenever $-1 < r < 0$, Proposition 4.1.11 allows us to restrict our attention to numbers lying in the interval $(0, 1)$. Thus we suppose that $0 < r < 1$ and prove that $r^n \to 0$. Let R be the range of the sequence (r^n). That is, let $R = \{r^n \colon n \in \mathbb{N}\}$. Let $g = \inf R$. Notice that $g \geq 0$ since 0 is a lower bound for R. We use an argument by contradiction to prove that $g = 0$. Assume to the contrary that $g > 0$. Since $0 < r < 1$, it must follow that $gr^{-1} > g$. Since g is the greatest lower bound of R, the number gr^{-1} is not a lower bound for R. Thus there exists a member, say r^k, of R such that $r^k < gr^{-1}$. But then $r^{k+1} < g$, which contradicts the choice of g as a lower bound of R. We conclude that $g = 0$.

The sequence (r^n) is bounded and decreasing. Thus by Proposition 4.3.3 it converges to the greatest lower bound of its range; that is, $r^n \to 0$ as $n \to \infty$.

Now suppose $r > 1$. Again we argue by contradiction. Suppose that (r^n) converges. Then its range R is a bounded subset of $\mathbb{R}$.

Let $\ell = \sup R$. Since $r > 1$, it is clear that $\ell r^{-1} < \ell$. Since ℓ is the least upper bound of R, there exists a number r^k of R such that $r^k > \ell r^{-1}$. Then $r^{k+1} > \ell$, contrary to the choice of ℓ as an upper bound for R.

Finally, suppose that $r < -1$. If (r^n) converges, then its range is bounded. In particular $\{r^{2n} \colon n \in \mathbb{N}\}$ is bounded. As in the preceding paragraph, this is impossible.   □

4.3.9. Proposition. *If $r > 0$, then $r^{1/n} \to 1$ as $n \to \infty$.*

Proof. Problem. *Hint.* Show that $\frac{1}{n} < r < n$ for some natural number n. Then use Example 4.2.8 and Proposition 4.2.5. (You will also make use of a standard arithmetic fact—one that arises in Problem J.4.5—that if $0 < a < b$, then $a^{1/n} < b^{1/n}$ for every natural number n.)

4.3.10. Problem. Find $\lim_{n\to\infty} \dfrac{5^n + 3n + 1}{7^n - n - 2}$.

4.4. Subsequences

As Example 4.1.9 shows, boundedness of a sequence of real numbers does not suffice to guarantee convergence. It is interesting to note, however, that although the sequence $\left((-1)^n\right)_{n=1}^{\infty}$ does not converge, it does have subsequences that converge. The odd numbered terms form a constant sequence $(-1, -1, -1, \ldots)$, which of course converges. The even terms converge to $+1$. It is an interesting, if not entirely obvious, fact that every bounded sequence has a convergent subsequence. This is a consequence of our next proposition.

Before proving Proposition 4.4.3 we discuss the notion of *subsequence*. The basic idea here is simple enough. Let $a = (a_1, a_2, a_3, \ldots)$ be a sequence of real numbers, and suppose that we construct a new sequence b by taking some but not necessarily all of the terms of a and listing them in the same order in which they appear in a. Then we say that this new sequence b is a *subsequence* of a. We might, for example, choose every fifth member of a thereby obtaining the subsequence $b = (b_1, b_2, b_3, \ldots) = (a_5, a_{10}, a_{15}, \ldots)$. The following definition formalizes this idea.

4.4.1. Definition. Let (a_n) be a sequence of real numbers. If $(n_k)_{k=1}^{\infty}$ is a strictly increasing sequence in $\mathbb{N}$, then the sequence

$$(a_{n_k}) = (a_{n_k})_{k=1}^{\infty} = (a_{n_1}, a_{n_2}, a_{n_3}, \ldots)$$

is a SUBSEQUENCE of the sequence (a_n).

4.4.2. Example. If $a_k = 2^{-k}$ and $b_k = 4^{-k}$ for all $k \in \mathbb{N}$, then (b_k) is a subsequence of (a_k). Intuitively, this is clear, since the second sequence $(\frac{1}{4}, \frac{1}{16}, \frac{1}{64}, \ldots)$ just picks out the even-numbered terms of the first sequence $(\frac{1}{2}, \frac{1}{4}, \frac{1}{8}, \frac{1}{16}, \ldots)$. This "picking out" is implemented by the strictly increasing function $n(k) = 2k$ (for $k \in \mathbb{N}$). Thus $b = a \circ n$ since

$$a_{n_k} = a(n(k)) = a(2k) = 2^{-2k} = 4^{-k} = b_k$$

for all k in $\mathbb{N}$.

4.4.3. Proposition. *Every sequence of real numbers has a monotone subsequence.*

Proof. Exercise. *Hint.* A definition may be helpful. Say that a term a_m of a sequence in $\mathbb{R}$ is a PEAK TERM if it is greater than or equal to every succeeding term (that is, if $a_m \geq a_{m+k}$ for all $k \in \mathbb{N}$). There are only two possibilities: either there is a subsequence of the sequence (a_n) consisting of peak terms, or else there is a last peak term in the sequence. (Include in this second case the situation in which there are no peak terms.) Show in each case how to select a monotone subsequence of (a_n).

4.4.4. Corollary. *Every bounded sequence of real numbers has a convergent subsequence.*

Proof. This is an immediate consequence of Propositions 4.3.3 and 4.4.3. $\square$

Our immediate purpose in studying sequences is to facilitate our investigation of the topology of the set of real numbers. We will first prove a key result, Proposition 4.4.9, usually known as *Bolzano's theorem*, which tells us that bounded infinite subsets of $\mathbb{R}$ have accumulation points. We then proceed to make available for future work sequential characterizations of open sets (Proposition 4.4.10), closed sets (Corollary 4.4.12), and continuity of functions (Proposition 4.4.13).

4.4.5. Definition. A sequence $(A_1, A_2, A_3, \dots)$ of sets is NESTED if $A_{k+1} \subseteq A_k$ for every k.

4.4.6. Proposition. *The intersection of a nested sequence of nonempty closed bounded intervals whose lengths approach 0 contains exactly one point.*

Proof. Problem. *Hint.* Suppose that $J_n = [a_n, b_n] \neq \emptyset$ for each $n \in \mathbb{N}$, that $J_{n+1} \subseteq J_n$ for each n, and that $b_n - a_n \to 0$ as $n \to \infty$. Show that $\bigcap_{n=1}^{\infty} J_n = \{c\}$ for some $c \in \mathbb{R}$.

4.4.7. Problem. Show by example that the intersection of a nested sequence of nonempty closed intervals may be empty.

4.4.8. Problem. Show by example that Proposition 4.4.6 no longer holds if "closed" is replaced by "open".

4.4.9. Proposition (Bolzano's theorem). *Every bounded infinite subset of $\mathbb{R}$ has at least one accumulation point.*

Proof. Let A be a bounded infinite subset of $\mathbb{R}$. Since it is bounded, it is contained in some interval $J_0 = [a_0, b_0]$. Let c_0 be the midpoint of J_0. Then at least one of the intervals $[a_0, c_0]$ or $[c_0, b_0]$ contains infinitely many members of A (see Problem O.1.17). Choose one that does and call it J_1. Now divide J_1 in half and choose J_2 to be one of the resulting closed subintervals whose intersection with A is infinite. Proceed inductively to obtain a nested sequence of closed intervals

$$J_0 \supseteq J_1 \supseteq J_2 \supseteq J_3 \supseteq \cdots ,$$

each one of which contains infinitely many points of A. By Proposition 4.4.6 the intersection of all the intervals J_k consists of exactly one point c. Every open interval about c contains some interval J_k and hence infinitely many points of A. Thus c is an accumulation point of A. $\square$

4.4.10. Proposition. *A subset U of $\mathbb{R}$ is open if and only if every sequence that converges to an element of U is eventually in U.*

Proof. Suppose U is open in $\mathbb{R}$. Let (x_n) be a sequence that converges to a point a in U. Since U is a neighborhood of a, (x_n) is eventually in U by Problem 4.1.7(b).

Conversely, suppose that U is not open. Then there is at least one point a of U that is not an interior point of U. Then for each $n \in \mathbb{N}$ there is a point x_n that belongs to $J_{1/n}(a)$ but *not* to U. Then the sequence (x_n) converges to a but no term of the sequence belongs to U. ☐

4.4.11. Proposition. *A point b belongs to the closure of a set A in $\mathbb{R}$ if and only if there exists a sequence (a_n) in A that converges to b.*

Proof. Exercise.

4.4.12. Corollary. *A subset A of $\mathbb{R}$ is closed if and only if b belongs to A whenever there is a sequence (a_n) in A that converges to b.*

4.4.13. Proposition. *Let $A \subseteq \mathbb{R}$. A function $f\colon A \to \mathbb{R}$ is continuous at a if and only if $f(x_n) \to f(a)$ whenever (x_n) is a sequence in A such that $x_n \to a$.*

Proof. Problem.

4.4.14. Problem. Discuss in detail the continuity of algebraic combinations of continuous real valued functions defined on subsets of $\mathbb{R}$. Show, for example, that if functions f and g are continuous at a point a in $\mathbb{R}$, then such combinations as $f + g$ and fg are also continuous at a. What can you say about the continuity of polynomials? *Hint.* Use Problem 4.2.1 and Proposition 4.4.13.

We conclude this chapter with some problems. The last six of these concern the convergence of recursively defined sequences. Most of these are pretty much like Example 4.3.4 and require more in the way of patience than ingenuity.

4.4.15. Problem. If A is a nonempty subset of $\mathbb{R}$ that is bounded above, then there exists an increasing sequence of elements of A that converges to $\sup A$. Similarly, if A is nonempty and bounded below, then there is a decreasing sequence in A converging to $\inf A$.

4.4.16. Problem (Geometric series). Let $|r| < 1$ and $a \in \mathbb{R}$. For each $n \geq 0$, let $s_n = \sum_{k=0}^{n} ar^k$.

(a) Show that the sequence $(s_n)_{n=0}^{\infty}$ converges. *Hint.* Consider $s_n - rs_n$.

(b) The limit found in part (a) is usually denoted by $\sum_{k=0}^{\infty} ar^k$. (This is the SUM of a geometric series.) Use (a) to find $\sum_{k=0}^{\infty} 2^{-k}$.

(c) Show how (a) may be used to write the decimal $0.15242\overline{424}\cdots$ as the quotient of two natural numbers.

4.4.17. Exercise. Suppose a sequence (x_n) of real numbers satisfies

$$4x_{n+1} = x_n{}^3$$

for all $n \in \mathbb{N}$. For what values of x_1 does the sequence (x_n) converge? For each such x_1 what is $\lim_{n\to\infty} x_n$? *Hint.* First establish that *if* the sequence (x_n) converges, its limit must be -2, 0, or 2. This suggests looking at several special cases: $x_1 < -2$, $x_1 = -2$, $-2 < x_1 < 0$, $x_1 = 0$, $0 < x_1 < 2$, $x_1 = 2$, and $x_1 > 2$. In case $x_1 < -2$, for example, show that $x_n < -2$ for all n. Use this to show that the sequence (x_n) is decreasing and that it has no limit. The other cases can be treated in a similar fashion.

4.4.18. Problem. Suppose a sequence (x_n) in $\mathbb{R}$ satisfies

$$5x_{n+1} = 3x_n + 4$$

for all $n \in \mathbb{N}$. Show that (x_n) converges. *Hint.* First answer an easy question: If $x_n \to \ell$, what is ℓ? Then look at three cases: $x_1 < \ell$, $x_1 = \ell$, and $x_1 > \ell$. Show, for example, that if $x_1 < \ell$, then (x_n) is bounded and increasing.

4.4.19. Problem. Suppose a sequence (x_n) in $\mathbb{R}$ satisfies

$$x_{n+1} = \sqrt{2 + x_n}$$

for all $n \in \mathbb{N}$. Show that (x_n) converges. To what does it converge?

4.4.20. Problem. Suppose that a sequence (x_n) in $\mathbb{R}$ satisfies

$$x_{n+1} = 1 - \sqrt{1 - x_n}$$

for all $n \in \mathbb{N}$. Show that (x_n) converges. To what does it converge? Does $\left(\frac{x_{n+1}}{x_n}\right)$ converge?

4.4.21. Problem. Suppose that a sequence (x_n) of real numbers satisfies

$$3x_{n+1} = x_n{}^3 - 2$$

for all $n \in \mathbb{N}$. For what choices of x_1 does the sequence converge? To what? *Hint.* Compute $x_{n+1} - x_n$.

4.4.22. Problem. Let $a > 1$. Suppose that a sequence (x_n) in $\mathbb{R}$ satisfies

$$(a + x_n)x_{n+1} = a(1 + x_n)$$

for all n. Show that if $x_1 > 0$, then (x_n) converges. In this case find $\lim x_n$.

Chapter 5

Connectedness and the intermediate value theorem

5.1. Connected subsets of $\mathbb{R}$

There appears to be no way of finding an exact solution to such an equation as

$$(8) \qquad\qquad \sin x = x - 1.$$

No algebraic trick or trigonometric identity seems to help. To be entirely honest, we must ask what reason we have for thinking that (8) *has* a solution. True, even a rough sketch of the graphs of the functions $x \mapsto \sin x$ and $x \mapsto x-1$ seems to indicate that the two curves cross not far from $x = 2$. But is it not possible in the vast perversity of the-way-things-are that the two curves might somehow skip over one another without actually having a point in common? To say that it seems unlikely is one thing, to say that we *know* it cannot happen is another. The solution to our dilemma lies in a result familiar from nearly every beginning calculus course: the *intermediate value theorem.* Despite the complicated symbol-ridden formulations in most calculus texts, what this theorem really says is that continuous functions from $\mathbb{R}$ to $\mathbb{R}$ take intervals to intervals. In order to prove this fact, we will first need to introduce an important topological property of all intervals, *connectedness.* Once we have proved the intermediate value theorem not only will we have the intellectual satisfaction of saying that we *know* (8) has a solution, but we can utilize this same theorem to find approximations to the solution of (8) whose accuracy is limited only by the computational means at our disposal.

So, first we define "connected". Or rather, we define "disconnected", and then agree that sets are connected if they are not disconnected.

5.1.1. Definition. A subset A of $\mathbb{R}$ is DISCONNECTED if there exist disjoint nonempty sets U and V, both open in A, whose union is A. In this case we say that the sets U and V *disconnect* A. A subset A of $\mathbb{R}$ is CONNECTED if it is not disconnected.

5.1.2. Proposition. *A set $A \subseteq \mathbb{R}$ is connected if and only if the only subsets of A that are both open in A and closed in A are the null set and A itself.*

Proof. Exercise.

5.1.3. Example. The set $\{1, 4, 8\}$ is disconnected.

Proof. Let $A = \{1, 4, 8\} \subseteq \mathbb{R}$. Notice that the sets $\{1\}$ and $\{4, 8\}$ are open subsets of A. Reason: $(-\infty, 2) \cap A = \{1\}$ and $(2, \infty) \cap A = \{4, 8\}$; so $\{1\}$ is the intersection of an open subset of $\mathbb{R}$ with A, and so is $\{4, 8\}$ (see Definition 3.3.1). Thus $\{1\}$ and $\{4, 8\}$ are disjoint nonempty open subsets of A whose union is A. That is, the sets $\{1\}$ and $\{4, 8\}$ disconnect A. $\qquad\square$

5.1.4. Example. The set $\mathbb{Q}$ of rational numbers is disconnected.

Proof. The sets $\{x \in \mathbb{Q} \colon x < \sqrt{2}\}$ and $\{x \in \mathbb{Q} \colon x > \sqrt{2}\}$ disconnect $\mathbb{Q}$. $\qquad\square$

5.1.5. Example. The set $\{\frac{1}{n} \colon n \in \mathbb{N}\}$ is disconnected.

Proof. Problem.

If A is a subset of $\mathbb{R}$, it is somewhat awkward that "connected" is defined in terms of (relatively) open subsets of A. In general, it is a nuisance to deal with relatively open subsets. It would be much more convenient to deal only with the familiar topology of $\mathbb{R}$. Happily, this can be arranged.

It is an elementary observation that A is disconnected if we can find disjoint sets U and V, both open in $\mathbb{R}$, whose intersections with A are nonempty and whose union contains A. (For then the sets $U \cap A$ and $V \cap A$ disconnect A.) Somewhat less obvious is the fact that A is disconnected if and only if it can be written as the union of two nonempty subsets C and D of $\mathbb{R}$ such that

$$(9) \qquad\qquad C \cap \overline{D} = \overline{C} \cap D = \emptyset.$$

(The indicated closures are in $\mathbb{R}$.)

5.1.6. Definition. Two nonempty subsets C and D of $\mathbb{R}$ that satisfy equation (9) are said to be MUTUALLY SEPARATED in $\mathbb{R}$.

5.1.7. Proposition. *A subset A of $\mathbb{R}$ is disconnected if and only if it is the union of two nonempty sets mutually separated in $\mathbb{R}$.*

Proof. If A is disconnected, it can be written as the union of two disjoint nonempty sets U and V that are open in A. (These sets need not, of course, be open in $\mathbb{R}$.) We show that U and V are mutually separated. It suffices to prove that $U \cap \overline{V}$ is empty, that is, $U \subseteq \mathbb{R} \setminus \overline{V}$. To this end suppose that $u \in U$. Since U is open in A, there exists $\delta > 0$ such that

$$A \cap J_\delta(u) = \{x \in A \colon |x - u| < \delta\} \subseteq U \subseteq \mathbb{R} \setminus V.$$

The interval $J_\delta(u) = (u - \delta, u + \delta)$ is the union of two sets: $A \cap J_\delta(u)$ and $A^c \cap J_\delta(u)$. We have just shown that the first of these belongs to $\mathbb{R} \setminus V$. Certainly the second piece contains no points of A and therefore no points of V. Thus $J_\delta(u) \subseteq \mathbb{R} \setminus V$. This shows that u does not belong to the closure (in $\mathbb{R}$) of the set V; so $u \in \mathbb{R} \setminus \overline{V}$. Since u was an arbitrary point of U, we conclude that $U \subseteq \mathbb{R} \setminus \overline{V}$.

Conversely, suppose that $A = U \cup V$ where U and V are nonempty sets mutually separated in $\mathbb{R}$. To show that the sets U and V disconnect A, we need only show that they are open in A, since they are obviously disjoint.

Let us prove that U is open in A. Let $u \in U$, and notice that since $U \cap \overline{V}$ is empty, u cannot belong to $\overline{V}$.

Thus there exists $\delta > 0$ such that $J_\delta(u)$ is disjoint from V. Then certainly $A \cap J_\delta(u)$ is disjoint from V. Thus $A \cap J_\delta(u)$ is contained in U. Conclusion: U is open in A. $\qquad\square$

The importance of Proposition 5.1.7 is that it allows us to avoid the definition of connected, which involves *relatively* open subsets, and replace it with an equivalent condition that refers only to closures in $\mathbb{R}$. There is no important idea here, only a matter of convenience. We will use this characterization in the proof of our next proposition, which identifies the connected subsets of the real line. (To appreciate the convenience that Proposition 5.1.7 brings our way, try to prove the next result using only the definition of connected and not Proposition 5.1.7.) As mentioned earlier, we need to know that in the real line the intervals are the connected sets. This probably agrees with your intuition in the matter. It is plausible, and it is true; but it is not obvious, and its proof requires a little thought.

5.1.8. Definition. A subset J of $\mathbb{R}$ is an INTERVAL provided that it satisfies the following condition: if $c, d \in J$ and $c < z < d$, then $z \in J$. (Thus, in particular, the empty set and sets containing a single point are intervals.)

5.1.9. Proposition. *A subset J of $\mathbb{R}$ is connected if and only if it is an interval.*

Proof. To show that intervals are connected, argue by contradiction. Assume that there exists an interval J in $\mathbb{R}$ that is not connected. By Proposition 5.1.7 there exist nonempty sets C and D in $\mathbb{R}$ such that $J = C \cup D$ and

$$C \cap \overline{D} = \overline{C} \cap D = \emptyset$$

(closures are taken in $\mathbb{R}$). Choose $c \in C$ and $d \in D$. Without loss of generality suppose $c < d$. Let $A = (-\infty, d) \cap C$ and $z = \sup A$. Certainly $c \le z \le d$. Now $z \in \overline{C}$. (We know from Example 2.2.7 that z belongs to $\overline{A}$ and therefore to $\overline{C}$.) Furthermore $z \in \overline{D}$. (If $z = d$, then $z \in D \subseteq \overline{D}$. If $z < d$, then the interval (z, d) is contained in J and, since z is an upper bound for A, this interval contains no point of C. Thus $(z, d) \subseteq D$ and $z \in \overline{D}$.) Finally, since z belongs to J, it is a member of either C or D. But $z \in C$ implies $z \in C \cap \overline{D} = \emptyset$, which is impossible; and $z \in D$ implies $z \in \overline{C} \cap D = \emptyset$, which is also impossible.

The converse is easy. If J is not an interval, then there exist numbers $c < z < d$ such that $c, d \in J$ and $z \notin J$. It is easy to see that the sets $(-\infty, z) \cap J$ and $(z, \infty) \cap J$ disconnect J. $\qquad\square$

5.2. Continuous images of connected sets

Some of the facts we prove in this chapter are quite specific to the real line $\mathbb{R}$. When we move to more complicated metric spaces, no reasonable analogue of these facts is true. For example, even in the plane $\mathbb{R}^2$ nothing remotely like Proposition 5.1.9

holds. While it is not unreasonable to guess that the connected subsets of the plane are those in which we can move continuously between any two points of the set without leaving the set, this conjecture turns out to be wrong. The latter property, *arcwise connectedness*, is sufficient for connectedness to hold—but it is not necessary. In Chapter 17 we will give an example of a connected set that is not arcwise connected.

Despite the fact that some of our results are specific to $\mathbb{R}$, others will turn out to be true in very general settings. The next theorem, for example, which says that continuity preserves connectedness, is true in $\mathbb{R}$, in $\mathbb{R}^n$, in metric spaces, and even in general topological spaces. More important, the same proof works in all these cases! Thus when you get to Chapter 17, where connectedness in metric spaces is discussed, you will already know the proof that the continuous image of a connected set is itself connected.

5.2.1. Theorem. *The continuous image of a connected set is connected.*

Proof. Exercise. *Hint.* Prove the contrapositive. Let $f\colon A \to \mathbb{R}$ be a continuous function where $A \subseteq \mathbb{R}$. Show that if ran f is disconnected, then so is A.

The important intermediate value theorem is an obvious corollary of the preceding theorem.

5.2.2. Theorem (Intermediate value theorem: conceptual version). *The continuous image in $\mathbb{R}$ of an interval is an interval.*

Proof. Obvious from Proposition 5.1.9 and Theorem 5.2.1. $\qquad\square$

A slight variant of this theorem, familiar from beginning calculus, helps explain the *name* of the result. It says that if a continuous real valued function defined on an interval takes on two values, then it takes on every intermediate value, that is, every value between them. It is useful in establishing the existence of solutions to certain equations and also in approximating these solutions.

5.2.3. Theorem (Intermediate value theorem: calculus text version). *Let $f\colon J \to \mathbb{R}$ be a continuous function defined on an interval $J \subseteq \mathbb{R}$. If a, $b \in$ ran f and $a < z < b$, then $z \in$ ran f.*

Proof. Exercise.

Here is a typical application of the intermediate value theorem.

5.2.4. Example. The equation

$$(10) \qquad x^{27} + 5x^{13} + x = x^3 + x^5 + \frac{2}{\sqrt{1+3x^2}}$$

has at least one real solution.

Proof. Exercise. *Hint.* Consider the function f whose value at x is the left side minus the right side of (10). What can you say without much thought about $f(2)$ and $f(-2)$?

As another application of the intermediate value theorem, we prove a fixed point theorem. (**Definition:** Let $f\colon S \to S$ be a mapping from a set S into itself. A point c in S is a FIXED POINT of the function f if $f(c) = c$.) The next result is a (very) special case of the celebrated Brouwer fixed point theorem, which says that every continuous map from the closed unit ball of $\mathbb{R}^n$ into itself has a fixed point. The proof of this more general result is rather complicated and will not be given here.

5.2.5. Proposition. *Let $a < b$ in $\mathbb{R}$. Every continuous map of the interval $[a, b]$ into itself has a fixed point.*

Proof. Exercise.

5.2.6. Example. The equation

$$x^{180} + \frac{84}{1 + x^2 + \cos^2 x} = 119$$

has at least two solutions in $\mathbb{R}$.

Proof. Problem.

5.2.7. Problem. Show that the equation

$$\frac{1}{\sqrt{4x^2 + x + 4}} - 1 = x - x^5$$

has at least one real solution. Locate such a solution between consecutive integers.

5.2.8. Problem. We return to the problem we discussed at the beginning of this chapter. Use the intermediate value theorem to find a solution to the equation

$$\sin x = 1 - x$$

accurate to within 10^{-5}. *Hint.* You may assume, for the purposes of this problem, that the function $x \mapsto \sin x$ is continuous. You will not want to do the computations by hand; write a program for a computer or programmable calculator. Notice to begin with that there is a solution in the interval $[0, 1]$. Divide the interval in half and decide which half, $[0, \frac{1}{2}]$ or $[\frac{1}{2}, 1]$, contains the solution. Then take the appropriate half and divide *it* in half. Proceed in this way until you have achieved the desired accuracy. Alternatively, you may find it convenient to divide each interval into tenths rather than halves.

5.3. Homeomorphisms

5.3.1. Definition. Two subsets A and B of $\mathbb{R}$ are HOMEOMORPHIC if there exists a continuous bijection f from A onto B such that f^{-1} is also continuous. In this case the function f is a HOMEOMORPHISM.

Notice that if two subsets A and B are homeomorphic, then there is a one-to-one correspondence between the open subsets of A and those of B. In terms of topology, the two sets are identical. Thus if we know that the open intervals $(0, 1)$ and $(3, 7)$ are homeomorphic (see the next problem), then we treat these two intervals for all topological purposes as indistinguishable. (Of course, a concept such as distance is another matter; it is *not* a topological property. When we consider the distance between points of our two intervals, we can certainly distinguish between them: one is four times the length of the other.) A homeomorphism is sometimes called a TOPOLOGICAL ISOMORPHISM.

5.3.2. Problem. Discuss the homeomorphism classes of intervals in $\mathbb{R}$. That is, tell, as generally as you can, which intervals in $\mathbb{R}$ are homeomorphic to which others—and, of course, explain why. It might be easiest to start with some examples. Show that

(a) the open intervals $(0, 1)$ and $(3, 7)$ are homeomorphic; and

(b) the three intervals $(0, 1)$, $(0, \infty)$, and $\mathbb{R}$ are homeomorphic.

Then do some counterexamples. Show that

(c) no two of the intervals $(0, 1)$, $(0, 1]$, and $[0, 1]$ are homeomorphic.

Hint. For (a) consider the function $x \mapsto 4x + 3$. For part of (c) suppose that $f : (0, 1] \to (0, 1)$ is a homeomorphism. What can you say about the restriction of f to $(0, 1)$?

When you feel comfortable with the examples, then try to prove more general statements. For example, show that any two bounded open intervals are homeomorphic.

Finally, try to find the *most general* possible homeomorphism classes. (A HOMEOMORPHISM CLASS is a family of intervals any two of which are homeomorphic.)

5.3.3. Problem. Describe the class of all continuous mappings from $\mathbb{R}$ into $\mathbb{Q}$.

Chapter 6

Compactness and
the extreme value theorem

One of the most important results in beginning calculus is the *extreme value theorem*: a continuous function on a closed and bounded subset of the real line achieves both a maximum and a minimum value. In the present chapter we prove this result. Central to understanding the extreme value theorem is a curious observation: while neither boundedness nor the property of being closed is preserved by continuity (see Problems 6.1.1 and 6.1.2), the property of being closed *and* bounded is preserved. Once we have proved this result, it is easy to see that a continuous function defined on a closed and bounded set attains a maximum and minimum on the set.

Nevertheless, there are some complications along the way. To begin with, the proof that continuity preserves the property of being closed and bounded turns out to be awkward and unnatural. Furthermore, although this result can be generalized to $\mathbb{R}^n$, it does not hold in more general metric spaces. This suggests—even if it is not conclusive evidence—that we are looking at the wrong concept. One of the mathematical triumphs of the early twentieth century was the recognition that indeed the very concept of closed-and-bounded is a manifestation of the veil of *māyā*, a seductively simple vision that obscures the "real" topological workings behind the scenes. Enlightenment, at this level, consists in piercing this veil of illusion and seeing behind it the "correct" concept—compactness. There is now overwhelming evidence that compactness *is* the appropriate concept. First and most rewarding is the observation that the proofs of the preservation of compactness by continuity and of the extreme value theorem now become extremely natural. Also, the same proofs work not only for $\mathbb{R}^n$ but for general metric spaces and even arbitrary topological spaces. Furthermore, the property of compactness is an intrinsic one—if $A \subseteq B \subseteq C$, then A is compact in B if and only if it is compact in C. (The property of being closed is not intrinsic: the interval $(0, 1]$ is closed in $(0, \infty)$ but not in $\mathbb{R}$.) Finally, there is the triumph of products. In the early 1900s there were other contenders for the honored place ultimately held by compactness—sequential compactness and countable compactness. Around 1930 the great Russian mathematician Tychonov was able to show that arbitrary products of compact spaces are compact—a powerful and useful property not shared by the competitors.

There is, however, a price to be paid for the wonders of compactness—a frustratingly unintuitive definition. It is doubtless this lack of intuitive appeal that explains why it took workers in the field so long to come up with the optimal notion. Do not be discouraged if you feel you don't understand the definition. I'm not sure anyone really "understands" it. What is important is to be able to *use* it. And that is quite possible, with a little practice. In this chapter we first define compactness (Definition 6.1.3) and give two important examples of compact sets: finite sets (see Example 6.2.1) and the interval $[0, 1]$ (see Example 6.2.4). Then we give a number of ways of creating new compact sets from old ones (see Propositions 6.2.2 and 6.2.8, Problem 6.2.9(b), and Theorem 6.3.2). In Theorem 6.3.2 we show that the continuous image of a compact set is compact, and in Theorem 6.3.3 we prove the extreme value theorem. Finally (in Example 6.3.6) we prove the Heine–Borel theorem for $\mathbb{R}$: the compact subsets of $\mathbb{R}$ are those that are both closed and bounded.

6.1. Compactness

6.1.1. Problem. Give an example to show that if f is a continuous real valued function of a real variable and A is a closed subset of $\mathbb{R}$ that is contained in the domain of f, then it is not necessarily the case that $f^{\to}(A)$ is a closed subset of $\mathbb{R}$.

6.1.2. Problem. Give an example to show that if f is a continuous real valued function of a real variable and A is a bounded subset of the domain of f, then it is not necessarily the case that $f^{\to}(A)$ is bounded.

6.1.3. Definition. A family $\mathfrak{U}$ of sets is said to COVER a set A if $\bigcup \mathfrak{U} \supseteq A$. The phrases "$\mathfrak{U}$ covers A", "$\mathfrak{U}$ is a cover for A", "$\mathfrak{U}$ is a covering of A", and "A is covered by $\mathfrak{U}$" are used interchangeably. If A is a subset of $\mathbb{R}$ and $\mathfrak{U}$ is a cover for A that consists entirely of open subsets of $\mathbb{R}$, then $\mathfrak{U}$ is an OPEN COVER for A. If $\mathfrak{U}$ is a family of sets that covers A and $\mathfrak{V}$ is a subfamily of $\mathfrak{U}$ that also covers A, then $\mathfrak{V}$ is a SUBCOVER of $\mathfrak{U}$ for A. A subset A of $\mathbb{R}$ is COMPACT if *every* open cover of A has a finite subcover.

6.1.4. Example. For each integer n, let U_n be the open interval $(n, n + \frac{4}{3})$. Then the family $\mathfrak{U} = \bigcup_{-\infty}^{\infty} U_n$ covers the real line $\mathbb{R}$. Notice that there is no proper subset of $\mathfrak{U}$ which covers $\mathbb{R}$; that is, $\mathfrak{U}$ has no proper subcover.

6.1.5. Example. For each natural number n, let U_n be the open interval $(-n, n)$. Then the family $\mathfrak{U} = \bigcup_{1}^{\infty} U_n$ is a cover for the real line $\mathbb{R}$. Notice that the family $\mathfrak{V} = \bigcup_{1}^{\infty} U_{2n}$ is a proper subcover for $\mathbb{R}$, but there is no finite subcover of $\mathfrak{U}$ for $\mathbb{R}$.

6.1.6. Example. Let $A = [0, 1]$, $U_1 = (-3, \frac{2}{3})$, $U_2 = (-1, \frac{1}{2})$, $U_3 = (0, \frac{1}{2})$, $U_4 = (\frac{1}{3}, \frac{2}{3})$, $U_5 = (\frac{1}{2}, 1)$, $U_6 = (\frac{9}{10}, 2)$, and $U_7 = (\frac{2}{3}, \frac{3}{2})$. Then the family $\mathfrak{U} = \{U_k : 1 \leq k \leq 7\}$ is an open cover for A (because each U_k is open and $\bigcup_{k=1}^{7} U_k = (-3, 2) \supseteq A$). The subfamily $\mathfrak{V} = \{U_1, U_5, U_6\}$ of $\mathfrak{U}$ is a subcover for A (because

$U_1 \cup U_5 \cup U_6 = (-3, 2) \supseteq A)$. The subfamily $\mathfrak{W} = \{U_1, U_2, U_3, U_7\}$ of $\mathfrak{U}$ is *not* a subcover for A (because $U_1 \cup U_2 \cup U_3 \cup U_7 = (-3, \frac{2}{3}) \cup (\frac{2}{3}, \frac{3}{2})$ does not contain A).

6.1.7. Problem. Let J be the open unit interval $(0, 1)$. For each a let $U_a = \left(a, a + \frac{1}{4}\right)$, and let $\mathfrak{U} = \{U_a : 0 \le a \le \frac{3}{4}\}$. Then certainly $\mathfrak{U}$ covers J.

(a) Find a finite subfamily of $\mathfrak{U}$ that covers J.

(b) Explain why a solution to (a) does not suffice to show that J is compact.

(c) Show that J is not compact.

6.2. Examples of compact subsets of $\mathbb{R}$

It is easy to give examples of sets that are *not* compact. As we have seen in Examples 6.1.4 and 6.1.5, $\mathbb{R}$ itself is not compact. What is usually a lot trickier, because the definition is hard to apply directly, is proving that some particular compact set really is compact. The simplest examples of compact spaces are the finite ones; see Example 6.2.1. Finding nontrivial examples is another matter.

In this section we guarantee ourselves a generous supply of compact subsets of $\mathbb{R}$ by specifying some rather powerful methods for creating new compact sets from old ones. In particular, we will show that a set is compact if it is

(a) a closed subset of a compact set (Proposition 6.2.2);

(b) a finite union of compact sets (Problem 6.2.9(b)); or

(c) the continuous image of a compact set (Theorem 6.3.2).

Nevertheless, it is clear that we need something to start with. In Example 6.2.4 we prove directly from the definition that the closed unit interval $[0, 1]$ is compact. It is fascinating to see how this single example together with conditions (a)–(c) above can be used to generate a great variety of compact sets in general metric spaces. This will be done in Chapter 15.

6.2.1. Example. Every finite subset of $\mathbb{R}$ is compact.

Proof. Let $A = \{x_1, x_2, \ldots, x_n\}$ be a finite subset of $\mathbb{R}$. Let $\mathfrak{U}$ be a family of open sets that covers A. For each $k = 1, \ldots, n$ there is at least one set U_k in $\mathfrak{U}$ that contains x_k. Then the family $\{U_1, \ldots, U_n\}$ is a finite subcover of $\mathfrak{U}$ for A. $\square$

6.2.2. Proposition. *Every closed subset of a compact set is compact.*

Proof. Problem. *Hint.* Let A be a closed subset of a compact set K, and let $\mathfrak{U}$ be a cover of A by open sets. Consider $\mathfrak{U} \cup \{A^c\}$.

6.2.3. Proposition. *Every compact subset of* $\mathbb{R}$ *is closed and bounded.*

Proof. Exercise. *Hint.* To show that a compact set A is closed, show that its complement is open. To this end let y be an arbitrary point in A^c. For each x in A take disjoint open intervals about x and y.

As we will see later, the preceding result holds in arbitrary metric spaces. In fact, it is true in very general topological spaces. Its converse, the celebrated Heine–Borel theorem (see Example 6.3.6), although also true in $\mathbb{R}^n$ (see Theorem 16.4.1), does *not* hold in all metric spaces (see Problems 6.3.9, 16.4.7, and 16.4.8). Thus it is important at a conceptual level not to regard the property closed-and-bounded as being identical with compactness.

Now, finally, we give a nontrivial example of a compact space. The proof requires verifying some details, which at first glance may make it seem complicated. The basic idea behind the proof, however, is quite straightforward. It repays close study since it involves an important technique of proof that we will encounter again. The first time you read the proof, try to see its structure, to understand its basic logic. Postpone for a second reading the details that show that the conditions labeled (1) and (2) hold. Try to understand instead why verification of these two conditions is really all we need in order to prove that $[0, 1]$ is compact.

6.2.4. Example. The closed unit interval $[0, 1]$ is compact.

Proof. Let $\mathfrak{U}$ be a family of open subsets of $\mathbb{R}$ that covers $[0, 1]$, and let A be the set of all x in $[0, 1]$ such that the closed interval $[0, x]$ can be covered by finitely many members of $\mathfrak{U}$. It is clear that A is nonempty (since it contains 0), and that if a number y belongs to A, then so does any number in $[0, 1]$ less than y. We prove two more facts about A:

(1) If $x \in A$ and $x < 1$, then there exists a number $y > x$ such that $y \in A$.

(2) If y is a number in $[0, 1]$ such that $x \in A$ for all $x < y$, then $y \in A$.

To prove (1) suppose that $x \in A$ and $x < 1$. Since $x \in A$, there exists sets $U_1, \ldots, U_n$ in $\mathfrak{U}$ that cover the interval $[0, x]$. The number x belongs to at least one of these sets, say U_1. Since U_1 is an open subset of $\mathbb{R}$, there is an interval (a, b) such that $x \in (a, b) \subseteq U_1$. Since $x < 1$ and $x < b$, there exists a number $y \in (0, 1)$ such that $x < y < b$. From $[0, x] \subseteq \bigcup_{k=1}^{n} U_k$ and $[x, y] \subseteq (a, b) \subseteq U_1$, it follows that $U_1, \ldots, U_n$ cover the interval $[0, y]$. Thus $y > x$ and y belongs to A.

The proof of (2) is similar. Suppose that $y \in [0, 1]$ and that $[0, y) \subseteq A$. The case $y = 0$ is trivial, so we suppose that $y > 0$. Then y belongs to at least one member of $\mathfrak{U}$, say V. Choose an open interval (a, b) in $[0, 1]$ such that $y \in (a, b) \subseteq V$. Since $a \in A$, there is a finite collection of sets $U_1, \ldots, U_n$ in $\mathfrak{U}$ that covers $[0, a]$. Then clearly $\{U_1, \ldots, U_n, V\}$ is a cover for $[0, y]$. This shows that y belongs to A.

Finally, let $u = \sup A$. We are done if we can show that $u = 1$. Suppose to the contrary that $u < 1$. Then $[0, u) \subseteq A$. We conclude from (2) that $u \in A$ and then from (1) that there is a point greater than u that belongs to A. This contradicts the choice of u as the supremum of A. □

6.2.5. Problem. Let $A = \{0\} \cup \{1/n : n \in \mathbb{N}\}$, and let $\mathfrak{U}$ be the family $\{U_n : n \geq 0\}$, where $U_0 = (-1, 0.1)$ and $U_n = \left(\frac{5}{6n}, \frac{7}{6n}\right)$ for $n \geq 1$.

(a) Find a finite subfamily of $\mathfrak{U}$ that covers A.

(b) Explain why a solution to (a) does not suffice to show that A is compact.

(c) Use the definition of compactness to show that A is compact.

(d) Use Proposition 6.2.2 to show that A is compact.

6.2.6. Example. Show that the set $A = \{1/n : n \in \mathbb{N}\}$ is not a compact subset of $\mathbb{R}$.

Proof. Problem.

6.2.7. Problem. Give two proofs that the interval $[0, 1)$ is not compact—one making use of Proposition 6.2.3 and one not.

6.2.8. Proposition. *The intersection of a nonempty collection of compact subsets of $\mathbb{R}$ is itself compact.*

Proof. Problem.

6.2.9. Problem. Let $\mathfrak{K}$ be the family of compact subsets of $\mathbb{R}$.

(a) Show that $\bigcup \mathfrak{K}$ need not be compact.

(b) Show that if $\mathfrak{K}$ contains only finitely many sets, then $\bigcup \mathfrak{K}$ is compact.

6.3. The extreme value theorem

6.3.1. Definition. A real valued function f on a set A is said to have a MAXIMUM at a point a in A if $f(a) \geq f(x)$ for every x in A; the number $f(a)$ is the MAXIMUM VALUE of f. The function has a MINIMUM at a if $f(a) \leq f(x)$ for every x in A, and in this case $f(a)$ is the MINIMUM VALUE of f. A number is an EXTREME VALUE of f if it is either a maximum or a minimum value. It is clear that a function may fail to have maximum or minimum values. For example, on the open interval $(0, 1)$ the function $f : x \mapsto x$ assumes neither a maximum nor a minimum.

The concepts we have just defined are frequently called GLOBAL (or ABSOLUTE) MAXIMUM and GLOBAL (or ABSOLUTE) MINIMUM. This is to distinguish them from two different ideas *local* (or *relative*) *maximum* and *local* (or *relative*) *minimum*, which we will encounter later. In this text, "maximum" and "minimum" without qualifiers will be the global concepts defined above.

Our goal now is to show that every continuous function on a compact set attains both a maximum and a minimum. This turns out to be an easy consequence of the fact, which we prove next, that the continuous image of a compact set is compact.

6.3.2. Theorem. *Let A be a subset of $\mathbb{R}$. If A is compact and $f \colon A \to \mathbb{R}$ is continuous, then $f^{\to}(A)$ is compact.*

Proof. Exercise.

6.3.3. Theorem (Extreme value theorem). *If A is a compact subset of $\mathbb{R}$ and $f \colon A \to \mathbb{R}$ is continuous, then f assumes both a maximum and a minimum value on A.*

Proof. Exercise.

6.3.4. Example. The closed interval $[-3, 7]$ is a compact subset of $\mathbb{R}$.

Proof. Let $A = [0, 1]$ and $f(x) = 10x - 3$. Since A is compact and f is continuous, Theorem 6.3.2 tells us that the set $[-3, 7] = f^{\to}(A)$ is compact. $\qquad\square$

6.3.5. Example. If $a < b$, then the closed interval $[a, b]$ is a compact subset of $\mathbb{R}$.

Proof. Problem.

6.3.6. Example (Heine–Borel theorem for $\mathbb{R}$). Every closed and bounded subset of $\mathbb{R}$ is compact.

Proof. Problem. *Hint.* Use Example 6.3.5 and Proposition 6.2.2.

6.3.7. Example. Define $f\colon [-1,1] \to \mathbb{R}$ by

$$f(x) = \begin{cases} x \sin\left(\dfrac{1}{x}\right) & \text{if } x \neq 0, \\ 0 & \text{if } x = 0. \end{cases}$$

The set of all x such that $f(x) = 0$ is a compact subset of $[-1,1]$.

Proof. Problem.

6.3.8. Problem. Let $f\colon A \to B$ be a continuous bijection between subsets of $\mathbb{R}$.

(a) Show by example that f need not be a homeomorphism.

(b) Show that if A is compact, then f must be a homeomorphism.

6.3.9. Problem. Find in $\mathbb{Q}$ a set that is both relatively closed and bounded but is not compact.

6.3.10. Problem. Show that the interval $[0,\infty)$ is not compact using each of the following:

(a) the definition of compactness;

(b) Proposition 6.2.3;

(c) the extreme value theorem.

6.3.11. Problem. Let f and g be two continuous functions mapping the interval $[0,1]$ into itself. Show that if $f \circ g = g \circ f$, then f and g agree at some point of $[0,1]$. *Hint.* Argue by contradiction. Show that we may suppose, without loss of generality, that $f(x) - g(x) > 0$ for all x in $[0,1]$. Now try to show that there is a number $a > 0$ such that $f^n(x) \geq g^n(x) + na$ for every natural number n and every x in $[0,1]$. (Here $f^n = f \circ f \circ \cdots \circ f$ (n copies of f), and g^n is defined similarly.)

Chapter 7

Limits of real valued functions

In Chapter 4 we studied limits of sequences of real numbers. In this very short chapter we investigate limits of real valued functions of a real variable. Our principal result (Proposition 7.2.3) is a characterization of the continuity of a function f at a point in terms of the limit of f at that point.

Despite the importance of this characterization, there is one crucial difference between a function being continuous at a point and having a limit there. If f is continuous at a, then a must belong to the domain of f. In order for f to have a limit at a, it is not required that a be in the domain of f.

7.1. Definition

To facilitate the definition of "limit" we introduce the notion of a *deleted neighborhood* of a point.

7.1.1. Definition. If $J = (b, c)$ is a neighborhood of a point a (that is, if $b < a < c$), then J^*, the DELETED NEIGHBORHOOD associated with J, is just J with the point a deleted. That is, $J^* = (b, a) \cup (a, c)$. In particular, if $J_\delta(a)$ is the δ-neighborhood of a, then $J_\delta^*(a)$ denotes $(a - \delta, a) \cup (a, a + \delta)$.

7.1.2. Definition. Let A be a subset of $\mathbb{R}$, let $f \colon A \to \mathbb{R}$, let a be an accumulation point of A, and let l be a real number. We say that l is the LIMIT OF f AS x APPROACHES a (or the LIMIT OF f AT a) if, for every $\epsilon > 0$, there exists $\delta > 0$ such that $f(x) \in J_\epsilon(l)$ whenever $x \in A \cap J_\delta^*(a)$.

Using slightly different notation, we may write this condition as

$$(\forall \epsilon > 0)(\exists \delta > 0)(\forall x \in A)\, 0 < |x - a| < \delta \implies |f(x) - l| < \epsilon.$$

If this condition is satisfied, we write

$$f(x) \to l \text{ as } x \to a$$

or

$$\lim_{x \to a} f(x) = l.$$

(Notice that this last notation is a bit optimistic. It would not make sense if f could have two distinct limits as x approaches a. We will show in Proposition 7.1.3 that this cannot happen.)

The first thing we notice about the preceding definition is that the point a at which we take the limit need not belong to the domain A of the function f. Very often in practice it does not. Recall the definition in beginning calculus of the derivative of a function $f\colon \mathbb{R} \to \mathbb{R}$ at a point a. It is the limit as $h \to 0$ of the Newton quotient $\frac{f(a+h)-f(a)}{h}$. This quotient is not defined at the point $h = 0$. Nevertheless, we may still take the limit as h approaches 0.

Here is another example of the same phenomenon. The function on $(0, \infty)$ defined by $x \mapsto (1+x)^{1/x}$ is not defined at $x = 0$. But its limit at 0 exists: recall from beginning calculus that $\lim_{x\to 0}(1+x)^{1/x} = e$.

We have one last comment about the definition: Even if a function f is defined at a point a, the value of f at a is irrelevant to the question of the existence of a limit there. According to the definition, we consider only points x satisfying $0 < |x - a| < \delta$. The condition $0 < |x - a|$ says just one thing: $x \neq a$.

7.1.3. Proposition. *Let $f\colon A \to \mathbb{R}$ where $A \subseteq \mathbb{R}$, and let a be an accumulation point of A. If $f(x) \to b$ as $x \to a$ and if $f(x) \to c$ as $x \to a$, then $b = c$.*

Proof. Exercise.

7.1.4. Problem. Why do we require that the point at which we take limits be an accumulation point of the domain of the function? *Hint.* Proposition 7.1.3.

7.2. Continuity and limits

There is a close connection between the existence of a limit of a function at a point a and the continuity of the function at a. In Proposition 7.2.3 we state the precise relationship. But first we give two examples to show that in the absence of additional hypotheses neither of these implies the other.

7.2.1. Example. The inclusion function $f\colon \mathbb{N} \to \mathbb{R}\colon n \mapsto n$ is continuous (because *every* subset of $\mathbb{N}$ is open in $\mathbb{N}$, and thus every function defined on $\mathbb{N}$ is continuous). But the limit of f exists at no point (because $\mathbb{N}$ has no accumulation points).

7.2.2. Example. Let
$$f(x) = \begin{cases} 0 & \text{for } x \neq 0, \\ 1 & \text{for } x = 0. \end{cases}$$
Then $\lim_{x\to 0} f(x)$ exists (and equals 0), but f is not continuous at $x = 0$.

We have shown in the two preceding examples that a function f may be continuous at a point a without having a limit there and that it may have a limit at a without being continuous there. If we require the point a to belong to the domain of f and to be an accumulation point of the domain of f (these conditions are independent!), then a necessary and sufficient condition for f to be continuous at a is that the limit of f as x approaches a (exist and) be equal to $f(a)$.

7.2.3. Proposition. *Let* $f\colon A \to \mathbb{R}$ *where* $A \subseteq \mathbb{R}$, *and let* $a \in A \cap A'$. *Then* f *is continuous at* a *if and only if*

$$\lim_{x \to a} f(x) = f(a).$$

Proof. Exercise.

7.2.4. Proposition. *If* $f\colon A \to \mathbb{R}$, *where* $A \subseteq \mathbb{R}$, *and* $a \in A'$, *then*

$$\lim_{h \to 0} f(a + h) = \lim_{x \to a} f(x)$$

in the sense that if either limit exists, then so does the other and the two limits are equal.

Proof. Exercise.

7.2.5. Problem. Let $f(x) = 4 - x$ if $x < 0$ and $f(x) = (2 + x)^2$ if $x > 0$. Using the definition of "limit", show that $\lim_{x \to 0} f(x)$ exists.

7.2.6. Proposition. *Let* $f\colon A \to \mathbb{R}$ *where* $A \subseteq \mathbb{R}$, *and let* $a \in A'$. *Then*

$$\lim_{x \to a} f(x) = 0 \qquad \text{if and only if} \quad \lim_{x \to a} |f(x)| = 0.$$

Proof. Problem.

7.2.7. Proposition. *Let* f, g, $h\colon A \to \mathbb{R}$, *where* $A \subseteq \mathbb{R}$, *and let* $a \in A'$. *If* $f \leq g \leq h$ *and*

$$\lim_{x \to a} f(x) = \lim_{x \to a} h(x) = l,$$

then $\lim_{x \to a} g(x) = l$.

Proof. Problem. *Hint.* A slight modification of your proof of Proposition 4.2.5 should do the trick.

7.2.8. Problem (Limits of algebraic combinations of functions). Carefully formulate and prove the standard results from beginning calculus on the limits of sums, constant multiples, products, and quotients of functions.

7.2.9. Problem. Let A, $B \subseteq \mathbb{R}$, $a \in A$, $f\colon A \to B$, and $g\colon B \to \mathbb{R}$.

(a) If $l = \lim_{x \to a} f(x)$ exists and g is continuous at l, then

$$\lim_{x \to a} (g \circ f)(x) = g(l).$$

(b) Show by example that the following assertion need not be true: If $l = \lim_{x \to a} f(x)$ exists and $\lim_{y \to l} g(y)$ exists, then $\lim_{x \to a} (g \circ f)(x)$ exists.

Differentiation
of real valued functions

Differential calculus is a highly geometric subject—a fact that is not always made entirely clear in elementary texts, where the study of derivatives as numbers often usurps the place of the fundamental notion of linear approximation. The contemporary French mathematician Jean Dieudonné comments on the problem in Chapter 8 of his magisterial multivolume treatise on the *Foundations of Modern Analysis* [**Die62**]:

> ... the fundamental idea of calculus [is] the "local" approximation of functions by *linear* functions. In the classical teaching of Calculus, this idea is immediately obscured by the accidental fact that, on a one-dimensional vector space, there is a one-to-one correspondence between linear forms and numbers, and therefore the derivative at a point is defined as a *number* instead of a *linear form*. This slavish subservience to the shibboleth of numerical interpretation at any cost becomes much worse when dealing with functions of several variables ...

The goal of this chapter is to display as vividly as possible the geometric underpinnings of the differential calculus. The emphasis is on "tangency" and "linear approximation", not on number.

8.1. The families $\mathfrak{D}$ and o

8.1.1. Notation. Let $a \in \mathbb{R}$. We denote by $\mathcal{F}_a$ the family of all real valued functions defined on a neighborhood of a. That is, f belongs to $\mathcal{F}_a$ if there exists an open set U such that $a \in U \subseteq \operatorname{dom} f$.

Notice that for each $a \in \mathbb{R}$, the set $\mathcal{F}_a$ is closed under addition and multiplication. (We define the sum of two functions f and g in $\mathcal{F}_a$ to be the function $f + g$

whose value at x is $f(x) + g(x)$ whenever x belongs to dom $f \cap$ dom g. A similar convention holds for multiplication.)

Among the functions defined on a neighborhood of zero are two subfamilies of crucial importance; they are $\mathfrak{O}$ (the family of "*big-oh*" functions) and $\mathfrak{o}$ (the family of "*little-oh*" functions).

8.1.2. Definition. A function f in $\mathcal{F}_0$ belongs to $\mathfrak{O}$ if there exist numbers $c > 0$ and $\delta > 0$ such that
$$|f(x)| \le c\,|x|$$
whenever $|x| < \delta$.

A function f in $\mathcal{F}_0$ belongs to $\mathfrak{o}$ if for every $c > 0$ there exists $\delta > 0$ such that
$$|f(x)| \le c\,|x|$$
whenever $|x| < \delta$. Notice that f belongs to $\mathfrak{o}$ if and only if $f(0) = 0$ and
$$\lim_{h \to 0} \frac{|f(h)|}{|h|} = 0\,.$$

8.1.3. Example. Let $f(x) = \sqrt{|x|}$. Then f belongs to neither $\mathfrak{O}$ nor $\mathfrak{o}$. (A function belongs to $\mathfrak{O}$ only if in some neighborhood of the origin its graph lies between two lines of the form $y = cx$ and $y = -cx$.)

8.1.4. Example. Let $g(x) = |x|$. Then g belongs to $\mathfrak{O}$ but not to $\mathfrak{o}$.

8.1.5. Example. Let $h(x) = x^2$. Then h is a member of both $\mathfrak{O}$ and $\mathfrak{o}$.

Much of the elementary theory of differential calculus rests on a few simple properties of the families $\mathfrak{O}$ and $\mathfrak{o}$. These are given in Propositions 8.1.8–8.1.14.

8.1.6. Definition. A function $L \colon \mathbb{R} \to \mathbb{R}$ is LINEAR if
$$L(x + y) = L(x) + L(y)$$
and
$$L(cx) = cL(x)$$
for all x, y, $c \in \mathbb{R}$. The family of all linear functions from $\mathbb{R}$ into $\mathbb{R}$ will be denoted by $\mathfrak{L}$.

The collection of linear functions from $\mathbb{R}$ into $\mathbb{R}$ is not very impressive, as the next problem shows. When we get to spaces of higher dimension, the situation will become more interesting.

8.1.7. Example. A function $f \colon \mathbb{R} \to \mathbb{R}$ is linear if and only if its graph is a (nonvertical) line through the origin.

Proof. Problem.

CAUTION. Since linear functions must pass through the origin, straight lines are not in general graphs of linear functions.

8.1.8. Proposition. *Every member of $\mathfrak{o}$ belongs to $\mathfrak{O}$; so does every member of $\mathfrak{L}$. Every member of $\mathfrak{O}$ is continuous at 0.*

Proof. Obvious from the definitions.   □

8.1.9. Proposition. *Other than the constant function zero, no linear function belongs to $\mathfrak{o}$.*

Proof. Exercise.

8.1.10. Proposition. *The family $\mathfrak{O}$ is closed under addition and multiplication by constants.*

Proof. Exercise.

8.1.11. Proposition. *The family $\mathfrak{o}$ is closed under addition and multiplication by constants.*

Proof. Problem.

The next two propositions say that the composite of a function in $\mathfrak{O}$ with one in $\mathfrak{o}$ (in either order) ends up in $\mathfrak{o}$.

8.1.12. Proposition. *If $g \in \mathfrak{O}$ and $f \in \mathfrak{o}$, then $f \circ g \in \mathfrak{o}$.*

Proof. Problem.

8.1.13. Proposition. *If $g \in \mathfrak{o}$ and $f \in \mathfrak{O}$, then $f \circ g \in \mathfrak{o}$.*

Proof. Exercise.

8.1.14. Proposition. *If $\phi, f \in \mathfrak{O}$, then $\phi f \in \mathfrak{o}$.*

Proof. Exercise.

Remark. The preceding propositions can be summarized rather concisely. (Notation: $\mathcal{C}_0$ is the set of all functions in $\mathcal{F}_0$ that are continuous at 0.)

$$(1) \qquad \mathfrak{L} \cup \mathfrak{o} \subseteq \mathfrak{O} \subseteq \mathcal{C}_0 \,.$$
$$(2) \qquad \mathfrak{L} \cap \mathfrak{o} = 0 \,.$$
$$(3) \qquad \mathfrak{O} + \mathfrak{O} \subseteq \mathfrak{O}; \qquad \alpha \mathfrak{O} \subseteq \mathfrak{O} \,.$$
$$(4) \qquad \mathfrak{o} + \mathfrak{o} \subseteq \mathfrak{o}; \qquad \alpha \mathfrak{o} \subseteq \mathfrak{o} \,.$$
$$(5) \qquad \mathfrak{o} \circ \mathfrak{O} \subseteq \mathfrak{o} \,.$$
$$(6) \qquad \mathfrak{O} \circ \mathfrak{o} \subseteq \mathfrak{o} \,.$$
$$(7) \qquad \mathfrak{O} \cdot \mathfrak{O} \subseteq \mathfrak{o} \,.$$

8.1.15. Problem. Show that $\mathfrak{O} \circ \mathfrak{O} \subseteq \mathfrak{O}$. That is, if $g \in \mathfrak{O}$ and $f \in \mathfrak{O}$, then $f \circ g \in \mathfrak{O}$. (As usual, the domain of $f \circ g$ is taken to be $\{x \colon g(x) \in \mathrm{dom}\, f\}$.)

8.2. Tangency

The fundamental idea of differential calculus is the local approximation of a "smooth" function by a translate of a linear one. Certainly, the expression "local approximation" could be taken to mean many different things. One sense of this expression that has stood the test of usefulness over time is *tangency*. Two functions are said to be tangent at zero if their difference lies in the family $\mathfrak{o}$. We can of course define tangency of functions at an arbitrary point (see Problem 8.2.12 below); but for our purposes, "tangency at 0" will suffice. All the facts we need to know concerning this relation turn out to be trivial consequences of the results we have just proved.

8.2.1. Definition. Two functions f and g in $\mathcal{F}_0$ are TANGENT AT ZERO, in which case we write $f \simeq g$, if $f - g \in \mathfrak{o}$.

8.2.2. Example. Let $f(x) = x$ and $g(x) = \sin x$. Then $f \simeq g$, since $f(0) = g(0) = 0$ and $\lim_{x \to 0} \frac{x - \sin x}{x} = \lim_{x \to 0} \left(1 - \frac{\sin x}{x}\right) = 0$.

8.2.3. Example. If $f(x) = x^2 - 4x - 1$ and $g(x) = \left(3x^2 + 4x - 1\right)^{-1}$, then $f \simeq g$.

Proof. Exercise.

8.2.4. Proposition. *The relation "tangency at zero" is an equivalence relation on $\mathcal{F}_0$.*

Proof. Exercise.

The next result shows that at most one linear function can be tangent at zero to a given function.

8.2.5. Proposition. *Let $S, T \in \mathfrak{L}$ and $f \in \mathcal{F}_0$. If $S \simeq f$ and $T \simeq f$, then $S = T$.*

Proof. Exercise.

8.2.6. Proposition. *If $f \simeq g$ and $j \simeq k$, then $f + j \simeq g + k$, and, furthermore, $\alpha f \simeq \alpha g$ for all $\alpha \in \mathbb{R}$.*

Proof. Problem.

Suppose that f and g are tangent at zero. Under what circumstances are $h \circ f$ and $h \circ g$ tangent at zero? And when are $f \circ j$ and $g \circ j$ tangent at zero? We prove next that sufficient conditions are h is linear and j belongs to $\mathfrak{O}$.

8.2.7. Proposition. *Let $f, g \in \mathcal{F}_0$ and $T \in \mathfrak{L}$. If $f \simeq g$, then $T \circ f \simeq T \circ g$.*

Proof. Problem.

8.2.8. Proposition. *Let $h \in \mathfrak{O}$ and $f, g \in \mathcal{F}_0$. If $f \simeq g$, then $f \circ h \simeq g \circ h$.*

Proof. Problem.

8.2.9. Example. Let $f(x) = 3x^2 - 2x + 3$ and $g(x) = \sqrt{-20x + 25} - 2$ for $x \leq 1$. Then $f \simeq g$.

Proof. Problem.

8.2.10. Problem. Let $f(x) = x^3 - 6x^2 + 7x$. Find a linear function $T \colon \mathbb{R} \to \mathbb{R}$ that is tangent to f at 0.

8.2.11. Problem. Let $f(x) = |x|$. Show that there is no linear function $T \colon \mathbb{R} \to \mathbb{R}$ that is tangent to f at 0.

8.2.12. Problem. Let $T_a \colon x \mapsto x + a$. The mapping T_a is called TRANSLATION BY a. Note that it is *not* linear (unless, of course, $a = 0$). We say that functions f and g in $\mathcal{F}_a$ are TANGENT AT a if the functions $f \circ T_a$ and $g \circ T_a$ are tangent at 0.

(a) Let $f(x) = 3x^2 + 10x + 13$ and $g(x) = \sqrt{-20x - 15}$. Show that f and g are tangent at -2.

(b) Develop a theory for the relationship "tangency at a" that generalizes our work on "tangency at 0".

8.2.13. Problem. Each of the following is an abbreviated version of a proposition. Formulate precisely and prove.

(a) $\mathcal{C}_0 + \mathfrak{D} \subseteq \mathcal{C}_0$.

(b) $\mathcal{C}_0 + \mathfrak{o} \subseteq \mathcal{C}_0$.

(c) $\mathfrak{D} + \mathfrak{o} \subseteq \mathfrak{D}$.

8.2.14. Problem. Suppose that $f \simeq g$. Then the following hold.

(a) If g is continuous at 0, so is f.

(b) If g belongs to $\mathfrak{D}$, so does f.

(c) If g belongs to $\mathfrak{o}$, so does f.

8.3. Linear approximation

One often hears that differentiation of a smooth function f at a point a in its domain is the process of finding the best "linear approximation" to f at a. This informal assertion is not quite correct. For example, as we know from beginning calculus, the tangent line at $x = 1$ to the curve $y = 4 + x^2$ is the line $y = 2x + 3$, which is not a linear function since it does not pass through the origin. To rectify this rather minor shortcoming, we first translate the graph of the function f so that the point $(a, f(a))$ goes to the origin and *then* find the best linear approximation at the origin. The operation of translation is carried out by a somewhat notorious acquaintance from beginning calculus, Δy. The source of its notoriety is two-fold: first, in many texts it is inadequately defined; and second, the notation Δy fails to alert the reader to the fact that under consideration is a function of *two* variables. We will be careful on both counts.

8.3.1. Definition. Let $f \in \mathcal{F}_a$. Define the function Δf_a by

$$\Delta f_a(h) := f(a + h) - f(a)$$

for all h such that $a + h$ is in the domain of f. Notice that since f is defined in a neighborhood of a, the function Δf_a is defined in a neighborhood of 0; that is, Δf_a belongs to $\mathcal{F}_0$. Notice also that $\Delta f_a(0) = 0$.

8.3.2. Problem. Let $f(x) = \cos x$ for $0 \leq x \leq 2\pi$.

(a) Sketch the graph of the function f.

(b) Sketch the graph of the function Δf_π.

8.3.3. Proposition. *If* $f \in \mathcal{F}_a$ *and* $\alpha \in \mathbb{R}$, *then*

$$\Delta(\alpha f)_a = \alpha \, \Delta f_a \,.$$

Proof. To show that two functions are equal, show that they agree at each point in their domain. Here

$$\begin{aligned}
\Delta(\alpha f)_a(h) &= (\alpha f)(a+h) - (\alpha f)(a) \\
&= \alpha f(a+h) - \alpha f(a) \\
&= \alpha(f(a+h) - f(a)) \\
&= \alpha \, \Delta f_a(h)
\end{aligned}$$

for every h in the domain of Δf_a.   □

8.3.4. Proposition. *If* $f, g \in \mathcal{F}_a$, *then*

$$\Delta(f+g)_a = \Delta f_a + \Delta g_a \,.$$

Proof. Exercise.

The last two propositions prefigure the fact that differentiation is a linear operator; the next result will lead to *Leibniz's rule* for differentiating products.

8.3.5. Proposition. *If* $\phi, f \in \mathcal{F}_a$, *then*

$$\Delta(\phi f)_a = \phi(a) \cdot \Delta f_a + \Delta\phi_a \cdot f(a) + \Delta\phi_a \cdot \Delta f_a \,.$$

Proof. Problem.

Finally, we present a result that prepares the way for the *chain rule*.

8.3.6. Proposition. *If* $f \in \mathcal{F}_a$, $g \in \mathcal{F}_{f(a)}$, *and* $g \circ f \in \mathcal{F}_a$, *then*

$$\Delta(g \circ f)_a = \Delta g_{f(a)} \circ \Delta f_a \,.$$

Proof. Exercise.

8.3.7. Proposition. *Let* $A \subseteq \mathbb{R}$. *A function* $f \colon A \to \mathbb{R}$ *is continuous at the point* a *in* A *if and only if* Δf_a *is continuous at* 0.

Proof. Problem.

8.3.8. Proposition. *If* $f \colon U \to U_1$ *is a bijection between subsets of* $\mathbb{R}$, *then for each* a *in* U *the function* $\Delta f_a \colon U - a \to U_1 - f(a)$ *is invertible and*

$$\left(\Delta f_a\right)^{-1} = \Delta\left(f^{-1}\right)_{f(a)} \,.$$

Proof. Problem.

8.4. Differentiability

We now have developed enough machinery to talk sensibly about *differentiating* real valued functions.

8.4.1. Definition. Let $f \in \mathcal{F}_a$. We say that f is DIFFERENTIABLE AT a if there exists a linear function that is tangent at 0 to Δf_a. If such a function exists, it is called THE DIFFERENTIAL OF f at a and is denoted by df_a. (Don't be put off by the slightly complicated notation; df_a is just a member of $\mathfrak{L}$ satisfying $df_a \simeq \Delta f_a$.) We denote by $\mathcal{D}_a$ the family of all functions in $\mathcal{F}_a$ that are differentiable at a.

The next proposition justifies the use of the definite article that modifies "differential" in the preceding paragraph.

8.4.2. Proposition. *Let $f \in \mathcal{F}_a$. If f is differentiable at a, then its differential is unique. (That is, there is at most one linear map tangent at 0 to Δf_a.)*

Proof. Proposition 8.2.5. □

8.4.3. Example. It is instructive to examine the relationship between the differential of f at a, which we defined in Definition 8.4.1, and the derivative of f at a as defined in beginning calculus. For $f \in \mathcal{F}_a$ to be differentiable at a it is necessary that there be a linear function $T \colon \mathbb{R} \to \mathbb{R}$ that is tangent at 0 to Δf_a. According to Example 8.1.7, there must exist a constant c such that $Tx = cx$ for all x in $\mathbb{R}$. For T to be tangent to Δf_a, it must be the case that

$$\Delta f_a - T \in \mathfrak{o} \, ;$$

that is,

$$\lim_{h \to 0} \frac{\Delta f_a(h) - ch}{h} = 0 \, .$$

Equivalently,

$$\lim_{h \to 0} \frac{f(a+h) - f(a)}{h} = \lim_{h \to 0} \frac{\Delta f_a(h)}{h} = c \, .$$

In other words, the function T, which is tangent to Δf_a at 0, must be a line through the origin whose slope is

$$\lim_{h \to 0} \frac{f(a+h) - f(a)}{h} \, .$$

This is, of course, the familiar "derivative of f at a" from beginning calculus. Thus for any real valued function f that is differentiable at a in $\mathbb{R}$,

$$df_a(h) = f'(a) \cdot h$$

for all $h \in \mathbb{R}$.

8.4.4. Problem. Explain carefully the quotation from Dieudonné given at the beginning of the chapter.

8.4.5. Example. Let $f(x) = 3x^2 - 7x + 5$ and $a = 2$. Then f is differentiable at a. (Sketch the graph of the differential df_a.)

Proof. Problem.

8.4.6. Example. Let $f(x) = \sin x$ and $a = \pi/3$. Then f is differentiable at a. (Sketch the graph of the differential df_a.)

Proof. Problem.

8.4.7. Proposition. *Let $T \in \mathfrak{L}$ and $a \in \mathbb{R}$. Then $dT_a = T$.*

Proof. Problem.

8.4.8. Proposition. *If $f \in \mathcal{D}_a$, then $\Delta f_a \in \mathfrak{O}$.*

Proof. Exercise.

8.4.9. Corollary. *Every function that is differentiable at a point is continuous there.*

Proof. Exercise.

8.4.10. Proposition. *If f is differentiable at a and $\alpha \in \mathbb{R}$, then αf is differentiable at a and*
$$d(\alpha f)_a = \alpha\, df_a \,.$$

Proof. Exercise.

8.4.11. Proposition. *If f and g are differentiable at a, then $f + g$ is differentiable at a and*
$$d(f + g)_a = df_a + dg_a \,.$$

Proof. Problem.

8.4.12. Proposition (Leibniz's rule). *If $\phi, f \in \mathcal{D}_a$, then $\phi f \in \mathcal{D}_a$ and*
$$d(\phi f)_a = d\phi_a \cdot f(a) + \phi(a)\, df_a \,.$$

Proof. Exercise.

8.4.13. Theorem (The chain rule). *If $f \in \mathcal{D}_a$ and $g \in \mathcal{D}_{f(a)}$, then $g \circ f \in \mathcal{D}_a$ and*
$$d(g \circ f)_a = dg_{f(a)} \circ df_a \,.$$

Proof. Exercise.

8.4.14. Problem (A problem set on functions from $\mathbb{R}$ into $\mathbb{R}$). We are now in a position to derive the standard results, usually contained in the first semester of a beginning calculus course, concerning the differentiation of real valued functions of a single real variable. Having at our disposal the machinery developed earlier in this chapter, we may derive these results quite easily; and so the proof of each is a problem.

8.4.15. Definition. If $f \in \mathcal{D}_a$, the DERIVATIVE OF f AT a, denoted by $f'(a)$ or $Df(a)$, is defined to be $\lim_{h \to 0} \frac{f(a+h)-f(a)}{h}$. By Proposition 7.2.4 this is the same as $\lim_{x \to a} \frac{f(x)-f(a)}{x-a}$.

8.4.16. Proposition. *If $f \in \mathcal{D}_a$, then $Df(a) = df_a(1)$.*

Proof. Problem.

8.4.17. Proposition. *If f, $g \in \mathcal{D}_a$, then*
$$D(fg)(a) = Df(a) \cdot g(a) + f(a) \cdot Dg(a).$$

Proof. Problem. *Hint.* Use *Leibniz's rule* (Proposition 8.4.12) and Proposition 8.4.16.

8.4.18. Example. Let $r(t) = \frac{1}{t}$ for all $t \neq 0$. Then r is differentiable and $Dr(t) = -\frac{1}{t^2}$ for all $t \neq 0$.

Proof. Problem.

8.4.19. Proposition. *If $f \in \mathcal{D}_a$ and $g \in \mathcal{D}_{f(a)}$, then $g \circ f \in \mathcal{D}_a$ and*
$$D(g \circ f)(a) = (Dg)(f(a)) \cdot Df(a).$$

Proof. Problem.

8.4.20. Proposition. *If f, $g \in \mathcal{D}_a$ and $g(a) \neq 0$, then*
$$D\left(\frac{f}{g}\right)(a) = \frac{g(a) \, Df(a) - f(a) \, Dg(a)}{(g(a))^2} \, .$$

Proof. Problem. *Hint.* Write $\frac{f}{g}$ as $f \cdot (r \circ g)$ and use Propositions 8.4.16 and 8.4.19 and Example 8.4.18.

8.4.21. Proposition. *If $f \in \mathcal{D}_a$ and $Df(a) > 0$, then there exists $r > 0$ such that*

(a) $f(x) > f(a)$ *whenever* $a < x < a + r$, *and*

(b) $f(x) < f(a)$ *whenever* $a - r < x < a$.

Proof. Problem. *Hint.* Define $g(h) = h^{-1} \Delta f_a(h)$ if $h \neq 0$ and $g(0) = Df(a)$. Use Proposition 7.2.3 to show that g is continuous at 0. Then apply Proposition 3.3.21.

8.4.22. Proposition. *If $f \in \mathcal{D}_a$ and $Df(a) < 0$, then there exists $r > 0$ such that*

(a) $f(x) < f(a)$ *whenever* $a < x < a + r$, *and*

(b) $f(x) > f(a)$ *whenever* $a - r < x < a$.

Proof. Problem. *Hint.* Of course it is possible to obtain this result by doing Proposition 8.4.21 again with some inequalities reversed. That is the hard way.

8.4.23. Definition. Let $f \colon A \to \mathbb{R}$ where $A \subseteq \mathbb{R}$. The function f has a LOCAL (or RELATIVE) MAXIMUM at a point $a \in A$ if there exists $r > 0$ such that $f(a) \geq f(x)$ whenever $|x - a| < r$ and $x \in \operatorname{dom} f$. It has a LOCAL (or RELATIVE) MINIMUM at a point $a \in A$ if there exists $r > 0$ such that $f(a) \leq f(x)$ whenever $|x - a| < r$ and $x \in \operatorname{dom} f$.

Recall from Chapter 6 that $f \colon A \to \mathbb{R}$ is said to attain a MAXIMUM at a if $f(a) \geq f(x)$ for all $x \in \operatorname{dom} f$. This is often called a GLOBAL (or ABSOLUTE) MAXIMUM to help distinguish it from the local version just defined. It is clear that every global maximum is also a local maximum but not vice versa. (Of course a similar remark holds for minima.)

8.4.24. Proposition. *If $f \in \mathcal{D}_a$ and f has either a local maximum or a local minimum at a, then $Df(a) = 0$.*

Proof. Problem. *Hint.* Use Propositions 8.4.21 and 8.4.22.

8.4.25. Proposition (Rolle's theorem). *Let $a < b$. If $f \colon [a, b] \to \mathbb{R}$ is continuous, if it is differentiable on (a, b), and if $f(a) = f(b)$, then there exists a point c in (a, b) such that $Df(c) = 0$.*

Proof. Problem. *Hint.* Argue by contradiction. Use the extreme value theorem (Theorem 6.3.3) and Proposition 8.4.24.

8.4.26. Theorem (Mean value theorem). *Let $a < b$. If $f \colon [a, b] \to \mathbb{R}$ is continuous and if it is differentiable on (a, b), then there exists a point c in (a, b) such that*

$$Df(c) = \frac{f(b) - f(a)}{b - a}.$$

Proof. Problem. *Hint.* Let $y = g(x)$ be the equation of the line that passes through the points $(a, f(a))$ and $(b, f(b))$. Show that the function $f - g$ satisfies the hypotheses of Rolle's theorem (Proposition 8.4.25)

8.4.27. Proposition. *Let J be an open interval in $\mathbb{R}$. If $f \colon J \to \mathbb{R}$ is differentiable and $Df(x) = 0$ for every $x \in J$, then f is constant on J.*

Proof. Problem. *Hint.* Use the mean value theorem (Theorem 8.4.26).

Chapter 9

Metric spaces

Underlying the definition of the principal objects of study in calculus—derivatives, integrals, and infinite series—is the notion of "limit". What we mean when we write

$$\lim_{x \to a} f(x) = L$$

is that $f(x)$ can be made arbitrarily close to L by choosing x sufficiently close to a. To say what we mean by "closeness", we need the notion of the distance between two points. In this chapter we study "distance functions", also known as "metrics".

In the preceding chapters we have looked at such topics as limits, continuity, connectedness, and compactness from the point of view of a single example, the real line $\mathbb{R}$, where the distance between two points is the absolute value of their difference. There are other familiar distance functions (the Euclidean metric in the plane $\mathbb{R}^2$ or in three-space $\mathbb{R}^3$, for example, where the distance between points is usually taken to be the square root of the sum of the squares of the differences of their coordinates), and there are many less familiar ones that are also useful in analysis. Each of these has its own distinctive properties that merit investigation. But that would be a poor place to start. It is easier to study first those properties they have in common. We list four conditions we may reasonably expect any distance function to satisfy. If x and y are points, then the distance between x and y should be

(i) the same as the distance between y and x;

(ii) no larger than the sum of the distances produced by taking a detour through a point z;

(iii) 0 if and only if $x = y$; and

(iv) always greater than or equal to 0.

We formalize these conditions to define a "metric" on a set.

9.1. Definitions

9.1.1. Definition. Let M be a nonempty set. A function $d\colon M \times M \to \mathbb{R}$ is a METRIC (or DISTANCE FUNCTION) on M if for all x, y, $z \in M$,

(i) $d(x,y) = d(y,x)$,

(ii) $d(x,y) \leq d(x,z) + d(z,y)$,

(iii) $d(x,y) = 0$ if and only if $x = y$.

If d is a metric on a set M, then we say that the pair (M,d) is a METRIC SPACE.

It is standard practice—although certainly an abuse of language—to refer to "the metric space M" when in fact we mean "the metric space (M,d)". We will adopt this convention when it appears that no confusion will result. We must keep in mind, however, that there are situations where it is clearly inappropriate; if, for example, we are considering two different metrics on the same set M, a reference to "the metric space M" would be ambiguous.

In the chapter introduction, which precedes the formal definition of "metric", there was a condition (iv) requiring that a metric be nonnegative. What happened to that? It is an easy exercise to show that it is implied by the other conditions.

9.1.2. Proposition. *If d is a metric on a set M, then $d(x,y) \geq 0$ for all x, $y \in M$.*

Proof. Exercise.

9.1.3. Definition. For each point a in a metric space (M,d) and each number $r > 0$, we define $B_r(a)$, the OPEN BALL about a of radius r, to be the set of all those points whose distance from a is less than r. That is,

$$B_r(a) := \{x \in M : d(x,a) < r\}.$$

9.2. Examples

9.2.1. Example. The absolute value of the difference of two numbers is a metric on $\mathbb{R}$. We will call this the usual metric on $\mathbb{R}$. Notice that in this metric the open ball about a of radius r is just the open interval $(a-r, a+r)$. (*Proof.* $x \in B_r(a)$ if and only if $d(x,a) < r$ if and only if $|x - a| < r$ if and only if $a - r < x < a + r$.)

9.2.2. Problem. Define $d(x,y) = |\arctan x - \arctan y|$ for all x, $y \in \mathbb{R}$.

(a) Show that d is a metric on $\mathbb{R}$.

(b) Find $d(-1, \sqrt{3})$.

(c) Solve the equation $d(x,0) = d(x,\sqrt{3})$.

9.2.3. Problem. Let $f(x) = \frac{1}{1+x}$ for all $x \geq 0$. Define a metric d on $[0,\infty)$ by $d(x,y) = |f(x) - f(y)|$. Find a point $z \neq 1$ in this space whose distance from 2 is equal to the distance between 1 and 2.

9.2.4. Example. Define $d(x,y) = |x^2 - y^2|$ for all x, $y \in \mathbb{R}$. Then d is *not* a metric on $\mathbb{R}$.

Proof. Problem.

9.2.5. Example. Let $f(x) = \frac{x}{1+x^2}$ for $x \geq 0$. Define a function d on $[0, \infty) \times [0, \infty)$ by $d(x, y) = |f(x) - f(y)|$. Then d is *not* a metric on $[0, \infty)$.

Proof. Problem.

For our next example we make use of (a special case of) an important fact known as *Minkowski's inequality*. This we derive from another standard result, *Schwarz's inequality*.

9.2.6. Proposition (Schwarz's inequality). *Let* $u_1, \ldots, u_n, v_1, \ldots, v_n \in \mathbb{R}$. *Then*

$$\left(\sum_{k=1}^{n} u_k v_k \right)^2 \leq \left(\sum_{k=1}^{n} u_k^2 \right) \left(\sum_{k=1}^{n} v_k^2 \right).$$

Proof. To simplify notation, make some abbreviations: let $a = \sum_{k=1}^{n} u_k^2$, $b = \sum_{k=1}^{n} v_k^2$, and $c = \sum_{k=1}^{n} u_k v_k$. Then

$$0 \leq \sum_{k=1}^{n} \left(\sqrt{b}\, u_k - \frac{c}{\sqrt{b}} v_k \right)^2$$
$$= ab - 2c^2 + c^2$$
$$= ab - c^2. \qquad \square$$

9.2.7. Proposition (Minkowski's inequality). *Let* $u_1, \ldots, u_n, v_1, \ldots, v_n \in \mathbb{R}$. *Then*

$$\left(\sum_{k=1}^{n} (u_k + v_k)^2 \right)^{\frac{1}{2}} \leq \left(\sum_{k=1}^{n} u_k^2 \right)^{\frac{1}{2}} + \left(\sum_{k=1}^{n} v_k^2 \right)^{\frac{1}{2}}.$$

Proof. Let a, b, and c be as in Proposition 9.2.6. Then

$$\sum_{k=1}^{n} (u_k + v_k)^2 = a + 2c + b$$
$$\leq a + 2|c| + b$$
$$\leq a + 2\sqrt{ab} + b \qquad \text{(by Proposition 9.2.6)}$$
$$= \left(\sqrt{a} + \sqrt{b} \right)^2. \qquad \square$$

9.2.8. Example. For points $x = (x_1, \ldots, x_n)$ and $y = (y_1, \ldots, y_n)$ in $\mathbb{R}^n$, let

$$d(x, y) := \left(\sum_{k=1}^{n} (x_k - y_k)^2 \right)^{\frac{1}{2}}.$$

Then d is the USUAL (or EUCLIDEAN) METRIC on $\mathbb{R}^n$. The only nontrivial part of the proof that d is a metric is the verification of the *triangle inequality* (that is, condition (2) of the definition):

$$d(x, y) \leq d(x, z) + d(z, y).$$

To accomplish this, let $x = (x_1, \ldots, x_n)$, $y = (y_1, \ldots, y_n)$, and $z = (z_1, \ldots, z_n)$ be points in $\mathbb{R}^n$. Apply *Minkowski's inequality* (Proposition 9.2.7) with $u_k = x_k - z_k$ and $v_k = z_k - y_k$ for $1 \leq k \leq n$ to obtain

$$\begin{aligned}
d(x, y) &= \left(\sum_{k=1}^{n} (x_k - y_k)^2 \right)^{\frac{1}{2}} \\
&= \left(\sum_{k=1}^{n} ((x_k - z_k) + (z_k - y_k))^2 \right)^{\frac{1}{2}} \\
&\leq \left(\sum_{k=1}^{n} (x_k - z_k)^2 \right)^{\frac{1}{2}} + \left(\sum_{k=1}^{n} (z_k - y_k)^2 \right)^{\frac{1}{2}} \\
&= d(x, z) + d(z, y) \,.
\end{aligned}$$

9.2.9. Problem. Let d be the usual metric on $\mathbb{R}^2$.

(a) Find $d(x, y)$ when $x = (3, -2)$ and $y = (-3, 1)$.

(b) Let $x = (5, -1)$ and $y = (-3, -5)$. Find a point z in $\mathbb{R}^2$ such that $d(x, y) = d(y, z) = d(x, z)$.

(c) Sketch $B_r(a)$ when $a = (0, 0)$ and $r = 1$.

The Euclidean metric is by no means the only metric on $\mathbb{R}^n$ that is useful. Two more examples follow (Examples 9.2.10 and 9.2.12).

9.2.10. Example. For points $x = (x_1, \ldots, x_n)$ and $y = (y_1, \ldots, y_n)$ in $\mathbb{R}^n$, let

$$d_1(x, y) := \sum_{k=1}^{n} |x_k - y_k| \,.$$

It is easy to see that d_1 is a metric on $\mathbb{R}^n$. When $n = 2$, this is frequently called the TAXICAB METRIC. (Why?)

9.2.11. Problem. Let d_1 be the taxicab metric on $\mathbb{R}^2$ (see Example 9.2.10).

(a) Find $d_1(x, y)$ where $x = (3, -2)$ and $y = (-3, 1)$.

(b) Let $x = (5, -1)$ and $y = (-3, -5)$. Find a point z in $\mathbb{R}^2$ such that $d_1(x, y) = d_1(y, z) = d_1(x, z)$.

(c) Sketch $B_r(a)$ for the metric d_1 when $a = (0, 0)$ and $r = 1$.

9.2.12. Example. For $x = (x_1, \ldots, x_n)$ and $y = (y_1, \ldots, y_n)$ in $\mathbb{R}^n$, let

$$d_u(x, y) := \max\{|x_k - y_k| : 1 \leq k \leq n\} \,.$$

Then d_u is a metric on $\mathbb{R}^n$. The triangle inequality is verified as follows:

$$\begin{aligned}
|x_k - y_k| &\leq |x_k - z_k| + |z_k - y_k| \\
&\leq \max\{|x_i - z_i| : 1 \leq i \leq n\} + \max\{|z_i - y_i| : 1 \leq i \leq n\} \\
&= d_u(x, z) + d_u(z, y)
\end{aligned}$$

whenever $1 \leq k \leq n$. Thus

$$\begin{aligned}
d_u(x, y) &= \max\{|x_k - y_k| : 1 \leq k \leq n\} \\
&\leq d_u(x, z) + d_u(z, y) \,.
\end{aligned}$$

The metric d_u is called the "uniform metric". The reason for this name will become clear later.

Notice that on the real line the three immediately preceding metrics agree; the distance between points is just the absolute value of their difference. That is, when $n = 1$ the metrics given in Examples 9.2.8, 9.2.10, and 9.2.12 reduce to the one given in Example 9.2.1.

9.2.13. Problem. This problem concerns the metric d_u (defined in Example 9.2.12) on $\mathbb{R}^2$.

(a) Find $d_u(x, y)$ when $x = (3, -2)$ and $y = (-3, 1)$.

(b) Let $x = (5, -1)$ and $y = (-3, -5)$. Find a point z in $\mathbb{R}^2$ such that $d_u(x, y) = d_u(y, z) = d_u(x, z)$.

(c) Sketch $B_r(a)$ for the metric d_u when $a = (0, 0)$ and $r = 1$.

9.2.14. Example. Let M be any nonempty set. For x, $y \in M$, define

$$d(x, y) = \begin{cases} 1 & \text{if } x \neq y, \\ 0 & \text{if } x = y. \end{cases}$$

It is easy to see that d is a metric; this is the DISCRETE METRIC on M. Although the discrete metric is rather trivial, it proves quite useful in constructing counterexamples.

9.2.15. Example. Let d be the usual metric on $\mathbb{R}^2$ and 0 be the origin. Define a function ρ on $\mathbb{R}^2$ as

$$\rho(x, y) := \begin{cases} d(x, y) & \text{if } x \text{ and } y \text{ are collinear with } 0, \text{ and} \\ d(x, 0) + d(0, y) & \text{otherwise.} \end{cases}$$

The function ρ is a metric on $\mathbb{R}^2$. This metric is sometimes called the GREEK AIRLINE METRIC.

Proof. Problem.

9.2.16. Problem. Let ρ be the Greek airline metric on $\mathbb{R}^2$.

(a) Let $x = (-1, 2)$, $y = (-3, 6)$, and $z = (-3, 4)$. Find $\rho(x, y)$ and $\rho(x, z)$. Which point, y or z, is closer to x with respect to ρ?

(b) Let $r = 1$. Sketch $B_r(a)$ for the metric ρ when $a = (0, 0)$, $a = (\frac{1}{4}, 0)$, $a = (\frac{1}{2}, 0)$, $a = (\frac{3}{4}, 0)$, $a = (1, 0)$, and $a = (3, 0)$.

9.2.17. Proposition. *Let (M, d) be a metric space, and let x, y, $z \in M$. Then*

$$|d(x, z) - d(y, z)| \leq d(x, y).$$

Proof. Problem.

9.2.18. Proposition. *If a and b are distinct points in a metric space, then there exists a number $r > 0$ such that $B_r(a)$ and $B_r(b)$ are disjoint.*

Proof. Problem.

9.2.19. Proposition. *Let a and b be points in a metric space, and let r, $s > 0$. If c belongs to $B_r(a) \cap B_s(b)$, then there exists a number $t > 0$ such that $B_t(c) \subseteq B_r(a) \cap B_s(b)$.*

Proof. Problem.

9.2.20. Problem. Let $f(x) = \frac{1}{1+x^2}$ for all $x \geq 0$, and define a metric d on the interval $[0, \infty)$ by
$$d(x, y) = |f(x) - f(y)| \, .$$

(a) With respect to this metric, find the point halfway between 1 and 2.

(b) Find the open ball $B_{\frac{3}{10}}(1)$.

9.3. Equivalent metrics

Ahead of us lie many situations in which it will be possible to replace a computationally complicated metric on some space by a simpler one without affecting the fundamental properties of the space. As it turns out, a sufficient condition for this process to be legitimate is that the two metrics be "equivalent". For the moment we content ourselves with the *definition* of this term and an example; applications will be discussed later when we introduce the weaker notion of "topologically equivalent metrics" (see Definition 11.2.2).

9.3.1. Definition. Two metrics d_1 and d_2 on a set M are EQUIVALENT if there exist numbers α, $\beta > 0$ such that
$$d_1(x, y) \leq \alpha \, d_2(x, y) \quad \text{and}$$
$$d_2(x, y) \leq \beta \, d_1(x, y)$$
for all x and y in M.

9.3.2. Proposition. *On $\mathbb{R}^2$ the three metrics d, d_1, and d_u, defined in Examples 9.2.8, 9.2.10, and 9.2.12, are equivalent.*

Proof. Exercise. *Hint.* First prove that if a, $b \geq 0$, then
$$\max\{a, b\} \leq a + b \leq \sqrt{2} \, \sqrt{a^2 + b^2} \leq 2 \max\{a, b\} \, .$$

9.3.3. Problem. Let d and ρ be equivalent metrics on a set M. Then every open ball in the space (M, d) contains an open ball of the space (M, ρ) (and vice versa).

9.3.4. Problem. Let a, b, c, $d \in \mathbb{R}$. Establish each of the following:

(a) $(\frac{1}{3}a + \frac{2}{3}b)^2 \leq \frac{1}{3}a^2 + \frac{2}{3}b^2$.

(b) $(\frac{1}{2}a + \frac{1}{3}b + \frac{1}{6}c)^2 \leq \frac{1}{2}a^2 + \frac{1}{3}b^2 + \frac{1}{6}c^2$.

(c) $(\frac{5}{12}a + \frac{1}{3}b + \frac{1}{6}c + \frac{1}{12}d)^2 \leq \frac{5}{12}a^2 + \frac{1}{3}b^2 + \frac{1}{6}c^2 + \frac{1}{12}d^2$.

Hint. If (a), (b), and (c) are all special cases of some general result, it may be easier to give one proof (of the general theorem) rather than three proofs (of the special cases). In each case what can you say about the numbers multiplying a, b, c, and d? Notice that if $x > 0$, then $xy = \sqrt{x} \, (\sqrt{x} \, y)$. Use Schwarz's inequality, Proposition 9.2.6.

9.3.5. Proposition. *Let* (M, d) *be a metric space. The function* ρ *defined on* $M \times M$ *by*

$$\rho(x, y) = \frac{d(x, y)}{1 + d(x, y)}$$

is a metric on M.

Proof. Problem. *Hint.* Show first that $\frac{u}{1+u} \leq \frac{v}{1+v}$ whenever $0 \leq u \leq v$.

9.3.6. Problem. In Proposition 9.3.5 take M to be the real line $\mathbb{R}$ and d to be the usual metric on $\mathbb{R}$ (see Example 9.2.1).

(a) Find the open ball $B_{\frac{3}{5}}(1)$ in the metric space $(\mathbb{R}, \rho)$.

(b) Show that the metrics d and ρ are *not* equivalent on $\mathbb{R}$.

Chapter 10

Interiors, closures, and boundaries

10.1. Definitions and examples

10.1.1. Definition. Let (M, d) be a metric space, and let M_0 be a nonempty subset of M. If d_0 is the restriction of d to $M_0 \times M_0$, then, clearly, (M_0, d_0) is a metric space. It is a METRIC SUBSPACE of (M, d). In practice the restricted function (often called the INDUCED METRIC) is seldom given a name of its own; one usually writes "(M_0, d) is a (metric) subspace of (M, d)". When the metric on M is understood, this is further shortened to "M_0 is a subspace of M".

10.1.2. Example. Let M be $\mathbb{R}^2$ equipped with the usual Euclidean metric d and $M_0 = \mathbb{Q}^2$. The induced metric d_0 agrees with d where they are both defined

$$d(x, y) = d_0(x, y) = \sqrt{(x_1 - y_1)^2 + (x_2 - y_2)^2},$$

where $x = (x_1, x_2)$ and $y = (y_1, y_2)$. The only difference is that $d_0(x, y)$ is defined only when both x and y have rational coordinates.

10.1.3. Exercise. Regard $M_0 = \{-1\} \cup [0, 4)$ as a subspace of $\mathbb{R}$ under its usual metric. In this subspace find the open balls $B_1(-1)$, $B_1(0)$, and $B_2(0)$.

10.1.4. Definition. Let A be a subset of a metric space M. A point a is an INTERIOR POINT of A if some open ball about a lies entirely in A. The INTERIOR of A, denoted by A°, is the set of all interior points of A. That is,

$$A^\circ := \{x \in M : B_r(x) \subseteq A \text{ for some } r > 0\}.$$

10.1.5. Example. Let M be $\mathbb{R}$ with its usual metric and A be the closed interval $[0, 1]$. Then $A^\circ = (0, 1)$.

10.1.6. Example. Let M be $\mathbb{R}^2$ with its usual metric and A be the unit disk $\{(x, y): x^2 + y^2 \leq 1\}$. Then the interior of A is the open disk $\{(x, y): x^2 + y^2 < 1\}$.

10.1.7. Example. Let $M = \mathbb{R}^2$ with its usual metric and $A = \mathbb{Q}^2$. Then $A^\circ = \emptyset$.

Proof. No open ball in $\mathbb{R}^2$ contains only points both of whose coordinates are rational. ∎

10.1.8. Example. Consider the metrics d, d_1, and d_u on $\mathbb{R}^2$. Let $A = \{x \in \mathbb{R}^2 : d(x,0) \leq 1\}$, $A_1 = \{x \in \mathbb{R}^2 : d_1(x,0) \leq 1\}$, and $A_u = \{x \in \mathbb{R}^2 : d_u(x,0) \leq 1\}$. The point $\left(\frac{2}{3}, \frac{3}{8}\right)$ belongs to A° and A_u°, but not to A_1°.

Proof. Problem.

10.1.9. Definition. A point x in a metric space M is an ACCUMULATION POINT of a set $A \subseteq M$ if every open ball about x contains a point of A distinct from x. (We do *not* require that x belong to A.) We denote the set of all accumulation points of A by A'. This is sometimes called the DERIVED SET of A. The CLOSURE of the set A, denoted by $\overline{A}$, is $A \cup A'$.

10.1.10. Example. Let $\mathbb{R}^2$ have its usual metric, and let A be $\left[(0,1) \times (0,1)\right] \cup \{(2,3)\} \subseteq \mathbb{R}^2$. Then $A' = [0,1] \times [0,1]$ and $\overline{A} = \left([0,1] \times [0,1]\right) \cup \{(2,3)\}$.

10.1.11. Example. The set $\mathbb{Q}^2$ is a subset of the metric space $\mathbb{R}^2$. Every ordered pair of real numbers is an accumulation point of $\mathbb{Q}^2$ since every open ball in $\mathbb{R}^2$ contains (infinitely many) points with both coordinates rational. So the closure of $\mathbb{Q}^2$ is all of $\mathbb{R}^2$.

10.1.12. Definition. The BOUNDARY of a set A in a metric space is the intersection of the closures of A and its complement. We denote it by ∂A. In symbols,

$$\partial A := \overline{A} \cap \overline{A^c}.$$

10.1.13. Example. Take M to be $\mathbb{R}$ with its usual metric. If $A = (0,1)$, then $\overline{A} = A' = [0,1]$ and $\overline{A^c} = A^c = (-\infty, 0] \cup [1, \infty)$; so $\partial A = \{0,1\}$.

10.1.14. Exercise. In each of the following find A°, A', $\overline{A}$, and ∂A.

(a) Let $A = \{\frac{1}{n} : n \in \mathbb{N}\}$. Regard A as a subset of the metric space $\mathbb{R}$.

(b) Let $A = \mathbb{Q} \cap (0, \infty)$. Regard A as a subset of the metric space $\mathbb{R}$.

(c) Let $A = \mathbb{Q} \cap (0, \infty)$. Regard A as a subset of the metric space $\mathbb{Q}$ (where $\mathbb{Q}$ is a subspace of $\mathbb{R}$).

10.2. Interior points

10.2.1. Lemma. *Let M be a metric space, $a \in M$, and $r > 0$. If $c \in B_r(a)$, then there is a number $t > 0$ such that*

$$B_t(c) \subseteq B_r(a).$$

Proof. Exercise.

10.2.2. Proposition. *Let A and B be subsets of a metric space.*

(a) *If $A \subseteq B$, then $A^\circ \subseteq B^\circ$.*

(b) *$A^{\circ\circ} = A^\circ$. ($A^{\circ\circ}$ means $\left(A^\circ\right)^\circ$.)*

Proof. Exercise.

10.2.3. Proposition. *If A and B are subsets of a metric space, then*
$$(A \cap B)^\circ = A^\circ \cap B^\circ \,.$$

Proof. Problem.

10.2.4. Proposition. *Let $\mathfrak{A}$ be a family of subsets of a metric space. Then*

(a) $\bigcup \{A^\circ \colon A \in \mathfrak{A}\} \subseteq \left(\bigcup \mathfrak{A}\right)^\circ$.

(b) *Equality need not hold in* (a).

Proof. Exercise.

10.2.5. Proposition. *Let $\mathfrak{A}$ be a family of subsets of a metric space. Then*

(a) $\left(\bigcap \mathfrak{A}\right)^\circ \subseteq \bigcap \{A^\circ \colon A \in \mathfrak{A}\}$.

(b) *Equality need not hold in* (a).

Proof. Problem.

10.3. Accumulation points and closures

In Lemma 10.2.1 and Propositions 10.2.2–10.2.5 some of the fundamental properties of the interior operator $A \mapsto A^\circ$ were developed. In the next proposition we study accumulation points. Once their properties are understood, it is quite easy to derive the basic facts about the closure operator $A \mapsto \overline{A}$.

10.3.1. Proposition. *Let A and B be subsets of a metric space.*

(a) *If $A \subseteq B$, then $A' \subseteq B'$.*

(b) $A'' \subseteq A'$.

(c) *Equality need not hold in* (b).

(d) $(A \cup B)' = A' \cup B'$.

(e) $(A \cap B)' \subseteq A' \cap B'$.

(f) *Equality need not hold in* (e).

Proof.

(a) Let $x \in A'$. Then each open ball about x contains a point of A, hence of B, distinct from x. Thus $x \in B'$.

(b) Let $a \in A''$. If $r > 0$, then $B_r(a)$ contains a point, say b, of A' distinct from a. By Lemma 10.2.1 there exists $s > 0$ such that $B_s(b) \subseteq B_r(a)$. Let $t = \min\{s, d(a,b)\}$. Note that $t > 0$. Since $b \in A'$, there is a point $c \in B_t(b) \subseteq B_r(a)$ such that $c \in A$. Since $t \le d(a,b)$, it is clear that $c \ne a$. Thus every open ball $B_r(a)$ contains a point c of A distinct from a. This establishes that $a \in A'$.

(c) Problem.

(d) Problem.

(e) Since $A \cap B \subseteq A$, part (a) implies that $(A \cap B)' \subseteq A'$. Similarly, $(A \cap B)' \subseteq B'$. Conclusion: $(A \cap B)' \subseteq A' \cap B'$.

(f) In the metric space $\mathbb{R}$ let $A = \mathbb{Q}$ and $B = \mathbb{Q}^c$. Then $(A \cap B)' = \emptyset' = \emptyset$, while $A' \cap B' = \mathbb{R} \cap \mathbb{R} = \mathbb{R}$.

$\square$

10.3.2. Proposition. *Let A and B be subsets of a metric space with $A \subseteq B$. Then*

(a) $\overline{A} \subseteq \overline{B}$.

(b) $\overline{\overline{A}} = \overline{A}$.

Proof. Problem.

10.3.3. Proposition. *If A and B are subsets of a metric space, then*
$$\overline{A \cup B} = \overline{A} \cup \overline{B}.$$

Proof. Problem.

10.3.4. Proposition. *Let $\mathfrak{A}$ be a family of subsets of a metric space. Then*

(a) $\bigcup \{\overline{A} : A \in \mathfrak{A}\} \subseteq \overline{\bigcup \mathfrak{A}}$.

(b) *Equality need not hold in* (a).

Proof. Problem.

10.3.5. Proposition. *Let $\mathfrak{A}$ be a family of subsets of a metric space. Then*

(a) $\overline{\bigcap \mathfrak{A}} \subseteq \bigcap \{\overline{A} : A \in \mathfrak{A}\}$.

(b) *Equality need not hold in* (a).

Proof. Problem.

10.3.6. Proposition. *Let A be a subset of a metric space. Then*

(a) $\left(A^{\circ}\right)^c = \overline{A^c}$.

(b) $\left(A^c\right)^{\circ} = \left(\overline{A}\right)^c$.

Proof. Problem.

10.3.7. Problem. Use Proposition 10.3.6 and Proposition 10.2.2 (but not Proposition 10.3.1) to give another proof of Proposition 10.3.2.

10.3.8. Problem. Use Proposition 10.3.6 and Proposition 10.2.3 (but not Proposition 10.3.1) to give another proof of Proposition 10.3.3.

Chapter 11

The topology of metric spaces

11.1. Open and closed sets

11.1.1. Definition. A subset A of a metric space M is OPEN in M if $A° = A$. That is, a set is open if it contains an open ball about each of its points. To indicate that A is open in M we write $A \overset{\circ}{\subseteq} M$.

11.1.2. Example. Care must be taken to claim that a particular set is open (or not open) *only* when the metric space in which the set "lives" is clearly understood. For example, the assertion "the set $[0, 1)$ is open" is false if the metric space in question is $\mathbb{R}$. It is true, however, if the metric space being considered is $[0, \infty)$ (regarded as a subspace of $\mathbb{R}$). The reason: in the space $[0, \infty)$ the point 0 is an interior point of $[0, 1)$; in $\mathbb{R}$ it is not.

11.1.3. Example. In a metric space every open ball is an open set. Notice that this is exactly what Lemma 10.2.1 says: each point of an open ball is an interior point of that ball.

The fundamental properties of open sets may be deduced easily from information we already possess concerning interiors of sets. Three facts about open sets are given in Propositions 11.1.4–11.1.6. The first of these is very simple.

11.1.4. Proposition. *Every nonempty open set is a union of open balls.*

Proof. Let U be an open set. For each a in U there is an open ball $B(a)$ about a contained in U. Then clearly

$$U = \bigcup \{B_a : a \in U\}. \qquad \square$$

11.1.5. Proposition. *Let M be a metric space.*

(a) *The union of any family of open subsets of M is open.*

(b) *The intersection of any* finite *family of open subsets of M is open.*

Proof.

(a) Let $\mathfrak{U}$ be a family of open subsets of M. Since the interior of a set is always contained in the set, we need only show that $\bigcup \mathfrak{U} \subseteq \left(\bigcup \mathfrak{U} \right)^{\circ}$. By Proposition 10.2.4

$$\bigcup \mathfrak{U} = \bigcup \{ U : U \in \mathfrak{U} \}$$
$$= \bigcup \{ U^{\circ} : U \in \mathfrak{U} \}$$
$$\subseteq \left(\bigcup \mathfrak{U} \right)^{\circ}.$$

(b) It is enough to show that the intersection of *two* open sets is open. Let $U, V \overset{\circ}{\subseteq} M$. Then by Proposition 10.2.3

$$(U \cap V)^{\circ} = U^{\circ} \cap V^{\circ} = U \cap V. \qquad \square$$

11.1.6. Proposition. *The interior of a set A is the largest open set contained in A. (Precisely, A° is the union of all the open sets contained in A.)*

Proof. Exercise.

11.1.7. Definition. A subset A of a metric space is CLOSED if $\overline{A} = A$. That is, a set is closed if it contains all its accumulation points.

11.1.8. Example. As is the case with open sets, care must be taken when affirming or denying that a particular set is closed. It must be clearly understood in which metric space the set "lives". For example the interval $(0, 1]$ is not closed in the metric space $\mathbb{R}$, but it is a closed subset of the metric space $(0, \infty)$ (regarded as a subspace of $\mathbb{R}$).

REMINDER. Recall the remarks made after Definition 2.2.11: *sets are not like doors or windows; they are not necessarily either open or closed.* One cannot show that a set is closed, for example, by showing that it fails to be open.

11.1.9. Proposition. *A subset of a metric space is open if and only if its complement is closed.*

Proof. Exercise. *Hint.* Use Proposition 10.3.6.

Facts already proved concerning closures of sets give us one way of dealing with closed sets; the preceding proposition gives us another. To illustrate this, we give two proofs of the next proposition.

11.1.10. Proposition. *The intersection of an arbitrary family of closed subsets of a metric space is closed.*

First proof. Let $\mathfrak{A}$ be a family of closed subsets of a metric space. Then $\bigcap \mathfrak{A}$ is the complement of $\bigcup \{A^c \colon A \in \mathfrak{A}\}$. Since each set A^c is open (by Proposition 11.1.9), the union of $\{A^c \colon A \in \mathfrak{A}\}$ is open (by Proposition 11.1.5(a)), and its complement $\bigcap \mathfrak{A}$ is closed (Proposition 11.1.9 again). $\qquad\square$

Second proof. Let $\mathfrak{A}$ be a family of closed subsets of a metric space. Since a set is always contained in its closure, we need only show that $\overline{\bigcap \mathfrak{A}} \subseteq \bigcap \mathfrak{A}$. Use Proposition 10.3.5(a):

$$\overline{\bigcap \mathfrak{A}} \subseteq \bigcap \{\overline{A} \colon A \in \mathfrak{A}\}$$
$$= \bigcap \{A \colon A \in \mathfrak{A}\}$$
$$= \bigcap \mathfrak{A}.$$

$\qquad\square$

11.1.11. Problem. The union of a finite family of closed subsets of a metric space is closed.

(a) Prove this assertion using Propositions 11.1.5(b) and 11.1.9.

(b) Prove this assertion using Proposition 10.3.3.

11.1.12. Problem. Give an example to show that the intersection of an arbitrary family of open subsets of a metric space need not be open.

11.1.13. Problem. Give an example to show that the union of an arbitrary family of closed subsets of a metric space need not be closed.

11.1.14. Definition. Let M be a metric space, let $a \in M$, and let $r > 0$. The CLOSED BALL $C_r(a)$ about a of radius r is $\{x \in M \colon d(a,x) \leq r\}$. The SPHERE $S_r(a)$ about a of radius r is $\{x \in M \colon d(a,x) = r\}$.

11.1.15. Problem. Let M be a metric space, let $a \in M$, and let $r > 0$.

(a) Give an example to show that the closed ball about a of radius r need not be the same as the closure of the open ball about a of radius r. That is, the sets $C_r(a)$ and $\overline{B_r(a)}$ may differ.

(b) Show that every closed ball in M is a closed subset of M.

(c) Show that every sphere in M is a closed subset of M.

11.1.16. Proposition. *In a metric space the closure of a set A is the smallest closed set containing A. (Precisely, $\overline{A}$ is the intersection of the family of all closed sets that contain A.)*

Proof. Problem.

11.1.17. Proposition. *If A is a subset of a metric space, then its boundary ∂A is equal to $\overline{A} \setminus A^\circ$. Thus ∂A is closed.*

Proof. Problem.

11.1.18. Proposition. *Let A be a subset of a metric space M. If A is closed in M or if it is open in M, then $(\partial A)^\circ = \emptyset$.*

Proof. Problem.

11.1.19. Problem. Give an example of a subset A of the metric space $\mathbb{R}$ the interior of whose boundary is all of $\mathbb{R}$.

11.1.20. Definition. Let $A \subseteq B \subseteq M$ where M is a metric space. We say that A is DENSE in B if $\overline{A} \supseteq B$. Thus, in particular, A is dense in the space M if $\overline{A} = M$.

11.1.21. Example. The rational numbers are dense in the reals; so are the irrationals.

Proof. That $\overline{Q} = \mathbb{R}$ was proved in Example 2.2.4. The proof that $\overline{Q^c} = \mathbb{R}$ is similar. $\qquad\square$

The following proposition gives a useful and easily applied criterion for determining when a set is dense in a metric space.

11.1.22. Proposition. *A subset D of a metric space M is dense in M if and only if every open ball contains a point of D.*

Proof. Exercise.

11.1.23. Problem. Let M be a metric space. Prove the following.

(a) If $A \subseteq M$ and $U \overset{\circ}{\subseteq} M$, then $U \cap \overline{A} \subseteq \overline{U \cap A}$.

(b) If D is dense in M and $U \overset{\circ}{\subseteq} M$, then $U \subseteq \overline{U \cap D}$.

11.1.24. Proposition. *Let $A \subseteq B \subseteq C \subseteq M$ where M is a metric space. If A is dense in B and B is dense in C, then A is dense in C.*

Proof. Problem.

11.2. The relative topology

In Example 11.1.2 we considered the set $A = [0,1)$, which is contained in both the metric spaces $M = [0,\infty)$ and $N = \mathbb{R}$. We observed that the question, "Is A open?", is ambiguous; it depends on whether we mean "open in M" or "open in N". Similarly, the notation $B_r(a)$ is equivocal. In M the open ball $B_{\frac{1}{2}}(0)$ is the interval $[0, \frac{1}{2})$, while in N it is the interval $(-\frac{1}{2}, \frac{1}{2})$. When working with sets that are contained in two different spaces, considerable confusion can be created by ambiguous choices of notation or terminology. In the next proposition, where we examine the relationship between open subsets of a metric space N and open subsets of a subspace M, it is necessary, in the proof, to consider open balls in both M and N. To avoid confusion, we use the usual notation $B_r(a)$ for open balls in M and a different one $D_r(a)$ for those in N.

The point of the following proposition is that even if we are dealing with a complicated or badly scattered subspace of a metric space, its open sets are easily identified. When an open set in the larger space is intersected with the subspace M, what results is an open set in M; and, less obviously, *every* open set in M can be produced in this fashion.

11.2.1. Proposition. *Let M be a subspace of a metric space N. A set $U \subseteq M$ is open in M if and only if there exists a set V open in N such that $U = V \cap M$.*

Proof. Let us establish some notation. If $a \in M$ and $r > 0$, we write $B_r(a)$ for the open ball about a of radius r in the space M. If $a \in N$ and $r > 0$, we write $D_r(a)$ for the corresponding open ball in the space N. Notice that $B_r(a) = D_r(a) \cap M$. Define a mapping f from the set of all open balls in M into the set of open balls in N by

$$f(B_r(a)) = D_r(a).$$

Thus f is just the function that associates with each open ball in the space M the corresponding open ball (same center, same radius) in N; so $f(B) \cap M = B$ for each open ball B in M.

Now suppose U is open in M. By Proposition 11.1.4 there exists a family $\mathfrak{B}$ of open balls in M such that $U = \bigcup \mathfrak{B}$. Let $\mathfrak{D} = \{f(B) \colon B \in \mathfrak{B}\}$ and $V = \bigcup \mathfrak{D}$. Then V, being a union of open balls in N, is an open subset of N and

$$\begin{aligned}
V \cap M &= \left(\bigcup \mathfrak{D}\right) \cap M \\
&= \left(\bigcup \{f(B) : B \in \mathfrak{B}\}\right) \cap M \\
&= \bigcup \{f(B) \cap M : B \in \mathfrak{B}\} \\
&= \bigcup \mathfrak{B} \\
&= U.
\end{aligned}$$

(For the third equality in the preceding string, see Proposition F.2.10.)

The converse is even easier. Let V be an open subset of N, and let $a \in V \cap M$. In order to show that $V \cap M$ is open in M, it suffices to show that, in the space M, the point a is an interior point of the set $V \cap M$. Since V is open in N, there exists $r > 0$ such that $D_r(a) \subseteq V$. But then

$$B_r(a) \subseteq D_r(a) \cap M \subseteq V \cap M. \qquad \square$$

The family of all open subsets of a metric space is called the TOPOLOGY on the space. As was the case for the real numbers, the concepts of continuity, compactness, and connectedness can be characterized entirely in terms of the open subsets of the metric spaces involved and without any reference to the specific metrics that lead to these open sets. Thus we say that continuity, compactness, and connectedness are topological concepts. Proposition 11.2.3 tells us that equivalent metrics on a set produce identical topologies. Clearly, no topological property of a metric space is affected when we replace the given metric with another metric that generates the same topology.

11.2.2. Definition. Two metrics d_1 and d_2 on a set M are TOPOLOGICALLY EQUIV-ALENT if they induce the same topology on M.

We now prove that equivalent metrics are topologically equivalent.

11.2.3. Proposition. *Let d_1 and d_2 be metrics on a set M, and let $\mathfrak{T}_1$ and $\mathfrak{T}_2$ be the topologies on (M, d_1) and (M, d_2), respectively. If d_1 and d_2 are equivalent, then $\mathfrak{T}_1 = \mathfrak{T}_2$.*

Proof. Exercise.

11.2.4. Problem. Give an example to show that topologically equivalent metrics need not be equivalent.

11.2.5. Definition. If M is a subspace of the metric space (N, d), the family of open subsets of M induced by the metric d is called the RELATIVE TOPOLOGY on M. According to Proposition 11.2.1, the relative topology on M is $\{V \cap M : V \overset{\circ}{\subseteq} N\}$.

Chapter 12

Sequences in metric spaces

In Chapter 4 we were able to characterize several topological properties of the real line $\mathbb{R}$ by means of sequences. The same sort of thing works in general metric spaces. Early in this chapter we give sequential characterizations of open sets, closed sets, dense sets, closure, and interior. Later we discuss products of metric spaces.

12.1. Convergence of sequences

Recall that a SEQUENCE is any function whose domain is the set $\mathbb{N}$ of natural numbers. If S is a set and $x \colon \mathbb{N} \to S$, then we say that x is a sequence of members of S. A map $x \colon \mathbb{N} \to \mathbb{R}$, for example, is a sequence of real numbers. In dealing with sequences, one usually (but not always) writes x_n for $x(n)$. The element x_n in the range of a sequence is the nth TERM of the sequence. Frequently we use the notations $(x_n)_{n=1}^{\infty}$ or just (x_n) to denote the sequence x.

12.1.1. Definition. A NEIGHBORHOOD of a point in a metric space is any open set containing the point. Let x be a sequence in a set S, and let B be a subset of S. The sequence x is EVENTUALLY in the set B if there is a natural number n_0 such that $x_n \in B$ whenever $n \geq n_0$. A sequence x in a metric space M CONVERGES to a point a in M if x is eventually in every neighborhood of a (equivalently, if it is eventually in every open ball about a). The point a is the LIMIT of the sequence x. (In Proposition 12.2.4 we find that limits of sequences are unique, so references to "the" limit of a sequence are justified.)

If a sequence x converges to a point a in a metric space, we write

$$x_n \to a \quad \text{as} \quad n \to \infty$$

or

$$\lim_{n \to \infty} x_n = a \,.$$

It should be clear that the preceding definition may be rephrased as follows: The sequence x converges to the point a if for every $\epsilon > 0$ there exists $n_0 \in \mathbb{N}$ such that $d(x_n, a) < \epsilon$ whenever $n \geq n_0$. It follows immediately that $x_n \to a$ if and only if

$d(x_n, a) \to 0$ as $n \to \infty$. Notice that in the metric space $\mathbb{R}$, the current definition agrees with the one given in Chapter 4.

12.2. Sequential characterizations of topological properties

Now we proceed to characterize some metric space concepts in terms of sequences. The point of this is that sequences are often easier to work with than arbitrary open sets.

12.2.1. Proposition. *A subset U of a metric space M is open if and only if every sequence in M that converges to an element of U is eventually in U.*

Proof. Suppose that U is an open subset of M. Let (x_n) be a sequence in M that converges to a point a in U. Then U is a neighborhood of a. Since $x_n \to a$, the sequence (x_n) is eventually in U.

To obtain the converse, suppose that U is not open. Some point, say a, of U is not an interior point of U. Then for every n in $\mathbb{N}$ we may choose an element x_n in $B_{1/n}(a)$ such that $x_n \notin U$. Then the sequence (x_n) converges to $a \in U$, but it is certainly not true that (x_n) is eventually in U. $\qquad\square$

12.2.2. Proposition. *A subset A of a metric space is closed if and only if b belongs to A whenever (a_n) is a sequence in A that converges to b.*

Proof. Exercise.

12.2.3. Proposition. *A subset D of a metric space M is dense in M if and only if every point of M is the limit of a sequence of elements of D.*

Proof. Problem.

12.2.4. Proposition. *In metric spaces, limits of sequences are unique. (That is, if $a_n \to b$ and $a_n \to c$ in some metric space, then $b = c$.)*

Proof. Problem.

12.2.5. Problem. Show that a point p is in the closure of a subset A of a metric space if and only if there is a sequence of points in A that converges to p. Also, give a characterization of the interior of a set by means of sequences.

12.2.6. Problem. Since the rationals are dense in $\mathbb{R}$, it must be possible, according to Proposition 12.2.3, to find a sequence of rational numbers that converges to the number π. Identify one such sequence.

12.2.7. Proposition. *Let d_1 and d_2 be topologically equivalent metrics on a set M. If a sequence of points in M converges to a point b in the metric space (M, d_1), then it also converges to b in the space (M, d_2).*

Proof. Problem.

12.2.8. Problem. Use Proposition 12.2.7 to show that the Greek airline metric ρ defined in Example 9.2.15 is not equivalent on $\mathbb{R}^2$ to the usual Euclidean metric (Example 9.2.8). You gave a (presumably different) proof of this in Problem 9.3.6.

12.2.9. Problem. Let M be a metric space with the discrete metric. Give a simple characterization of the convergent sequences in M.

12.2.10. Definition. Let A and B be nonempty subsets of a metric space M. The DISTANCE between A and B, which we denote by $d(A, B)$, is defined to be $\inf\{d(a, b) : a \in A$ and $b \in B\}$. If $a \in M$, we write $d(a, B)$ for $d(\{a\}, B)$.

12.2.11. Problem. Let B be a nonempty subset of a metric space M.

(a) Show that if $x \in B$, then $d(x, B) = 0$.

(b) Give an example to show that the converse of (a) may fail.

(c) Show that if B is closed, the converse of (a) holds.

12.3. Products of metric spaces

12.3.1. Definition. Let (M_1, ρ_1) and (M_2, ρ_2) be metric spaces. We define three metrics, d, d_1, and d_u, on the product $M_1 \times M_2$. For $x = (x_1, x_2)$ and $y = (y_1, y_2)$ in $M_1 \times M_2$, let

$$d(x, y) = \left((\rho_1(x_1, y_1))^2 + (\rho_2(x_2, y_2))^2\right)^{\frac{1}{2}},$$

$$d_1(x, y) = \rho_1(x_1, y_1) + \rho_2(x_2, y_2), \quad \text{and}$$

$$d_u(x, y) = \max\{\rho_1(x_1, y_1), \rho_2(x_2, y_2)\}.$$

It is not difficult to show that these really are metrics. They are just generalizations, to arbitrary products, of the metrics on $\mathbb{R} \times \mathbb{R}$ defined in Examples 9.2.8, 9.2.10, and 9.2.12.

12.3.2. Proposition. *The three metrics on $M_1 \times M_2$ defined in Definition* 12.3.1 *are equivalent.*

Proof. Exercise. *Hint.* Review Proposition 9.3.2.

In light of the preceding result and Proposition 11.2.3, the three metrics defined on $M_1 \times M_2$ (in Definition 12.3.1) all give rise to exactly the same topology on $M_1 \times M_2$. Since we will be concerned primarily with *topological properties* of product spaces, it makes little difference which of these metrics we officially adopt as "the" product metric. We choose d_1 because it is arithmetically simple (no square roots of sums of squares).

12.3.3. Definition. If M_1 and M_2 are metric spaces, then we say that the metric space $(M_1 \times M_2, d_1)$, where d_1 is defined in Definition 12.3.1, is the PRODUCT (METRIC) SPACE of M_1 and M_2, and the metric d_1 is the PRODUCT METRIC. When we encounter a reference to "the metric space $M_1 \times M_2$", we assume, unless the contrary is explicitly stated, that this space is equipped with *the* product metric d_1.

A minor technical point, which is perhaps worth mentioning, is that the usual (Euclidean) metric on $\mathbb{R}^2$ is not (according to the definition just given) the product metric. Since these two metrics *are* equivalent and since most of the properties we consider are topological ones, this will cause little difficulty.

It is easy to work with sequences in product spaces. This is a consequence of the fact, which we prove next, that a necessary and sufficient condition for the convergence of a sequence in the product of two spaces is the convergence of its coordinates.

12.3.4. Proposition. *Let M_1 and M_2 be metric spaces. A sequence $\left((x_n, y_n) \right)_{n=1}^{\infty}$ in the product space converges to a point (a, b) in $M_1 \times M_2$ if and only if $x_n \to a$ and $y_n \to b$.*

Proof. For $k = 1, 2$, let ρ_k be the metric on the space M_k. The product metric d_1 on $M_1 \times M_2$ is defined in Definition 12.3.1.

Suppose $(x_n, y_n) \to (a, b)$. Then

$$\rho_1(x_n, a) \leq \rho_1(x_n, a) + \rho_2(y_n, b)$$
$$= d_1\big((x_n, y_n), (a, b)\big) \to 0;$$

so $x_n \to a$. Similarly, $y_n \to b$.

Conversely, suppose $x_n \to a$ in M_1 and $y_n \to b$ in M_2. Given $\epsilon > 0$, we may choose $n_1, n_2 \in \mathbb{N}$ such that $\rho_1(x_n, a) < \frac{1}{2}\epsilon$ when $n \geq n_1$ and $\rho_2(y_n, b) < \frac{1}{2}\epsilon$ when $n \geq n_2$. Thus, if $n \geq \max\{n_1, n_2\}$,

$$d_1\big((x_n, y_n), (a, b)\big) = \rho_1(x_n, a) + \rho_2(y_n, b) < \tfrac{1}{2}\epsilon + \tfrac{1}{2}\epsilon = \epsilon\,;$$

so $(x_n, y_n) \to (a, b)$ in $M_1 \times M_2$. $\qquad\square$

Remark. By virtue of Proposition 12.2.7, the truth of the preceding proposition would not have been affected had either d or d_u (as in Definition 12.3.1) been chosen as the product metric for $M_1 \times M_2$.

12.3.5. Problem. Generalize Definitions 12.3.1 and 12.3.3 to $\mathbb{R}^n$ where $n \in \mathbb{N}$. That is, write appropriate formulas for $d(x, y)$, $d_1(x, y)$, and $d_u(x, y)$ for $x, y \in \mathbb{R}^n$, and explain what we mean by the product metric on an arbitrary finite product $M_1 \times M_2 \times \cdots \times M_n$ of metric spaces.

Also state and prove generalizations of Propositions 12.3.2 and 12.3.4 to arbitrary finite products.

Chapter 13

Uniform convergence

13.1. The uniform metric on the space of bounded functions

13.1.1. Definition. Let S be a nonempty set. A function $f\colon S \to \mathbb{R}$ is BOUNDED if there exists a number $M \geq 0$ such that

$$|f(x)| \leq M \qquad \text{for all } x \in S.$$

We denote by $\mathcal{B}(S, \mathbb{R})$ (or just by $\mathcal{B}(S)$) the set of all bounded real valued functions on S.

13.1.2. Proposition. *If f and g are bounded real valued functions on a nonempty set S and α is a real number, then the functions $f + g$, αf, and fg are all bounded.*

Proof. There exist numbers M, $N \geq 0$ such that $|f(x)| \leq M$ and $|g(x)| \leq N$ for all $x \in S$. Then, for all $x \in S$,

$$|(f + g)(x)| \leq |f(x)| + |g(x)| \leq M + N,$$
$$|(\alpha f)(x)| = |\alpha|\,|f(x)| \leq |\alpha|\,M, \text{ and}$$
$$|(fg)(x)| = |f(x)|\,|g(x)| \leq MN. \qquad \square$$

13.1.3. Definition. Let S be a nonempty set. We define a metric d_u on the set $\mathcal{B}(S, \mathbb{R})$ by

$$d_u(f, g) \equiv \sup\{|f(x) - g(x)|\colon x \in S\}$$

whenever f, $g \in \mathcal{B}(S, \mathbb{R})$. The metric d_u is the UNIFORM METRIC on $\mathcal{B}(S, \mathbb{R})$.

13.1.4. Example. Let $S = [-1, 1]$, and for all $x \in S$ let $f(x) = |x|$ and $g(x) = \frac{1}{2}(x - 1)$. Then $d_u(f, g) = 2$.

Proof. It is clear from the graphs of f and g that the functions are farthest apart at $x = -1$. Thus

$$d_u(f, g) = \sup\{|f(x) - g(x)|\colon -1 \leq x \leq 1\}$$
$$= |f(-1) - g(-1)| = 2. \qquad \square$$

13.1.5. Example. Let $f(x) = x^2$ and $g(x) = x^3$ for $0 \leq x \leq 1$. Then $d_u(f, g) = 4/27$.

Proof. Let $h(x) = |f(x) - g(x)| = f(x) - g(x)$ for $0 \leq x \leq 1$. To maximize h on $[0, 1]$, use elementary calculus to find critical points. Since $h'(x) = 2x - 3x^2 = 0$ only if $x = 0$ or $x = \frac{2}{3}$, it is clear that the maximum value of h occurs at $x = \frac{2}{3}$. Thus,

$$d_u(f, g) = \sup\{h(x) \colon 0 \leq x \leq 1\} = h\left(\frac{2}{3}\right) = \frac{4}{27}. \qquad \square$$

13.1.6. Exercise. Suppose f is the constant function defined on $[0, 1]$ whose value is 1. When asked to describe those functions in $\mathcal{B}([0, 1])$ that lie in the open ball about f of radius 1, a student replies (somewhat incautiously) that $B_1(f)$ is the set of all real valued functions g on $[0, 1]$ satisfying $0 < g(x) < 2$ for all $x \in [0, 1]$. Why is this response wrong?

13.1.7. Problem. Let $f(x) = \sin x$, and let $g(x) = \cos x$ for $0 \leq x \leq \pi$. Find $d_u(f, g)$ in the set of functions $\mathcal{B}([0, \pi])$.

13.1.8. Problem. Let $f(x) = 3x - 3x^3$, and let $g(x) = 3x - 3x^2$ for $0 \leq x \leq 2$. Find $d_u(f, g)$ in the set of functions $\mathcal{B}([0, 2])$.

13.1.9. Problem. Explain why it is reasonable to use the same notation d_u (and the same name) for both the metric in Example 9.2.12 and the one defined in Definition 13.1.3.

The terminology in Definition 13.1.3 is somewhat optimistic. We have not yet verified that the "uniform metric" is indeed a metric on $\mathcal{B}(S, \mathbb{R})$. We now remedy this.

13.1.10. Proposition. *Let S be a nonempty set. The function d_u as in Definition 13.1.3 is a metric on the set of functions $\mathcal{B}(S, \mathbb{R})$.*

Proof. Exercise.

13.1.11. Problem. Let $f(x) = x$ and $g(x) = 0$ for all $x \in [0, 1]$. Find a function h in $\mathcal{B}([0, 1])$ such that

$$d_u(f, h) = d_u(f, g) = d_u(g, h).$$

13.1.12. Definition. Let (f_n) be a sequence of real valued functions on a nonempty set S. If there is a function g in $\mathcal{F}(S, \mathbb{R})$ such that

$$\sup\{|f_n(x) - g(x)| \colon x \in S\} \to 0 \qquad \text{as } n \to \infty,$$

we say that the sequence (f_n) CONVERGES UNIFORMLY to g, and we write

$$f_n \to g \text{ (unif)}.$$

The function g is the UNIFORM LIMIT of the sequence (f_n). Notice that if g and all the f_n's belong to $\mathcal{B}(S, \mathbb{R})$, then uniform convergence of (f_n) to g is the same thing as convergence of f_n to g in the uniform metric.

13.1.13. Example. For each $n \in \mathbb{N}$ and $x \in \mathbb{R}$, let

$$f_n(x) = \frac{1}{n}\sin(nx).$$

Then $f_n \to \mathbf{0}$ (unif). (Here $\mathbf{0}$ is the constant function zero.)

Proof.

$$d_u(f_n, \mathbf{0}) = \sup\left\{\tfrac{1}{n}|\sin nx| \colon x \in \mathbb{R}\right\}$$
$$= \frac{1}{n} \to 0 \qquad \text{as } n \to \infty. \qquad \square$$

13.1.14. Example. Let $g(x) = x$ and $f_n(x) = x + \frac{1}{n}$ for all $x \in \mathbb{R}$ and $n \in \mathbb{N}$. Then $f_n \to g$ (unif) since

$$\sup\{|f_n(x) - g(x)| \colon x \in \mathbb{R}\} = \frac{1}{n} \to 0 \qquad \text{as } n \to \infty.$$

It is not correct (although perhaps tempting) to write $d_u(f_n, g) \to 0$. This expression is meaningless since the functions f_n and g do not belong to the metric space $\mathcal{B}(\mathbb{R})$ on which d_u is defined.

13.2. Pointwise convergence

Sequences of functions may converge in many different and interesting ways. Another mode of convergence that is frequently encountered is "pointwise convergence".

13.2.1. Definition. Let (f_n) be a sequence of real valued functions on a nonempty set S. If there is a function g such that

$$f_n(x) \to g(x) \qquad \text{for all } x \in S,$$

then (f_n) CONVERGES POINTWISE to g. In this case we write

$$f_n \to g \,(\text{ptws}).$$

The function g is the POINTWISE LIMIT of the sequence f_n.

In the following proposition we make the important, if elementary, observation that uniform convergence is stronger than pointwise convergence.

13.2.2. Proposition. *Uniform convergence implies pointwise convergence, but not conversely.*

Proof. Exercise.

13.2.3. Problem. Find an example of a sequence (f_n) of functions in $\mathcal{B}([0,1])$ that converges pointwise to the zero function $\mathbf{0}$ but satisfies

$$d_u(f_n, \mathbf{0}) \to \infty \qquad \text{as } n \to \infty.$$

Next we show that the uniform limit of a sequence of bounded functions is itself bounded.

13.2.4. Proposition. *Let (f_n) be a sequence in $\mathcal{B}(S)$, and let g be a real valued function on S.*

(a) *If $f_n \to g$ (unif), then $g \in \mathcal{B}(S)$.*

(b) *The assertion in (a) does not hold if uniform convergence is replaced by pointwise convergence.*

Proof. Exercise.

13.2.5. Exercise. Let $f_n(x) = x^n - x^{2n}$ for $0 \le x \le 1$, and let $n \in \mathbb{N}$. Does the sequence (f_n) converge pointwise on $[0, 1]$? Is the convergence uniform?

13.2.6. Problem. We give in each of the following the nth term of a sequence of real valued functions defined on $[0, 1]$. Which of these converge pointwise on $[0, 1]$? For which is the convergence uniform?

(a) $x \mapsto x^n$.

(b) $x \mapsto nx$.

(c) $x \mapsto xe^{-nx}$.

13.2.7. Problem. We give in each of the following the nth term of a sequence of real valued functions defined on $[0, 1]$. Which of these converge pointwise on $[0, 1]$? For which is the convergence uniform?

(a) $x \mapsto \dfrac{1}{nx + 1}$.

(b) $x \mapsto \dfrac{x}{nx + 1}$.

(c) $x \mapsto \dfrac{x^2}{n} - \dfrac{x}{n^2}$.

13.2.8. Problem. Let $f_n(x) = \dfrac{(n-1)x + x^2}{n + x}$ for all $x \ge 1$ and $n \in \mathbb{N}$. Does the sequence (f_n) have a pointwise limit on $[1, \infty)$? A uniform limit?

Chapter 14

More on continuity and limits

14.1. Continuous functions

As is the case with real valued functions of a real variable, a function $f\colon M_1 \to M_2$ between two metric spaces is continuous at a point a in M_1 if $f(x)$ can be made arbitrarily close to $f(a)$ by insisting that x be sufficiently close to a.

14.1.1. Definition. Let (M_1, d_1) and (M_2, d_2) be metric spaces. A function $f\colon M_1 \to M_2$ is CONTINUOUS AT a point a in M_1 if every neighborhood of $f(a)$ contains the image under f of a neighborhood of a. Since every neighborhood of a point contains an open ball about the point and since every open ball about a point *is* a neighborhood of that point, we may restate the definition as follows. The function f is CONTINUOUS AT a if every open ball about $f(a)$ contains the image under f of an open ball about a; that is, if the following condition is satisfied: for every $\epsilon > 0$, there exists $\delta > 0$ such that

$$(11) \qquad\qquad f^{\to}(B_\delta(a)) \subseteq B_\epsilon(f(a)).$$

There are many equivalent ways of expressing (11). Here are three:

$$B_\delta(a) \subseteq f^{\leftarrow}(B_\epsilon(f(a))),$$
$$x \in B_\delta(a) \text{ implies } f(x) \in B_\epsilon(f(a)),$$
$$d_1(x, a) < \delta \text{ implies } d_2(f(x), f(a)) < \epsilon.$$

Notice that if f is a real valued function of a real variable, then the definition above agrees with the one given at the beginning of Chapter 3.

14.1.2. Definition. A function $f\colon M_1 \to M_2$ between two metric spaces is CONTINUOUS if it is continuous at each point of M_1.

In proving propositions concerning continuity, one should not slavishly insist on specifying the radii of open balls when these particular numbers are of no interest. As an illustration, the next proposition, concerning the composite of continuous functions, is given two proofs—one with the radii of open balls specified, and a smoother one in which they are suppressed.

14.1.3. Proposition. *Let $f \colon M_1 \to M_2$ and $g \colon M_2 \to M_3$ be functions between metric spaces. If f is continuous at a in M_1 and g is continuous at $f(a)$ in M_2, then the composite function $g \circ f$ is continuous at a.*

Proof 1. Let $\eta > 0$. We wish to show that there exists $\delta > 0$ such that

$$B_\delta(a) \subseteq (g \circ f)^{\leftarrow}(B_\eta(g(f(a)))).$$

Since g is continuous at $f(a)$, there exists $\epsilon > 0$ such that

$$B_\epsilon(f(a)) \subseteq g^{\leftarrow}(B_\eta(g(f(a)))).$$

Since f is continuous at a, there exists $\delta > 0$ such that

$$B_\delta(a) \subseteq f^{\leftarrow}(B_\epsilon(f(a))).$$

Thus we have

$$
\begin{aligned}
(g \circ f)^{\leftarrow}(B_\eta(g(f(a)))) &= f^{\leftarrow}(g^{\leftarrow}(B_\eta(g(f(a))))) \\
&\supseteq f^{\leftarrow}(B_\epsilon(f(a))) \\
&\supseteq B_\delta(a). \qquad \square
\end{aligned}
$$

Proof 2. Let B_3 be an arbitrary open ball about $g(f(a))$. We wish to show that the inverse image of B_3 under $g \circ f$ contains an open ball about a. Since g is continuous at $f(a)$, the set $g^{\leftarrow}(B_3)$ contains an open ball B_2 about $f(a)$. And since f is continuous at a, the set $f^{\leftarrow}(B_2)$ contains an open ball B_1 about a. Thus we have

$$
\begin{aligned}
(g \circ f)^{\leftarrow}(B_3) &= f^{\leftarrow}(g^{\leftarrow}(B_3)) \\
&\supseteq f^{\leftarrow}(B_2) \\
&\supseteq B_1. \qquad \square
\end{aligned}
$$

The two preceding proofs are essentially the same. The only difference is that the first proof suffers from a severe case of clutter. It certainly is not more rigorous; it is just harder to read. It is good practice to relieve proofs (and their readers) of extraneous detail. The following corollary is an obvious consequence of the proposition we have just proved.

14.1.4. Corollary. *The composite of two continuous functions is continuous.*

Next we prove a result emphasizing that continuity is a topological notion; that is, it can be expressed in terms of open sets. A necessary and sufficient condition for a function to be continuous is that the inverse image of every open set be open.

14.1.5. Proposition. *A function $f \colon M_1 \to M_2$ between metric spaces is continuous if and only if $f^{\leftarrow}(U)$ is an open subset of M_1 whenever U is open in M_2.*

Proof. Exercise.

14.1.6. Example. As an application of the preceding proposition, we show that the function

$$f \colon \mathbb{R}^2 \to \mathbb{R} \colon (x, y) \mapsto 2x - 5y$$

is continuous. One approach to this problem is to find, given a point (a, b) in $\mathbb{R}^2$ and $\epsilon > 0$, a number $\delta > 0$ sufficiently small that $\sqrt{(x-a)^2 + (y-b)^2} < \delta$ implies $|(2x - 5y) - (2a - 5b)| < \epsilon$. This is not excessively difficult, but it is made somewhat awkward by the appearance of squares and a square root in the definition of the usual metric on $\mathbb{R}^2$. A simpler approach is possible. We wish to prove continuity of f with respect to the usual metric d on $\mathbb{R}^2$ (defined in Example 9.2.8).

We know that the metric d_1 (defined in Example 9.2.10) on $\mathbb{R}^2$ is equivalent to d (Proposition 9.3.2) and that equivalent metrics produce identical topologies. Thus $(\mathbb{R}^2, d_1)$ and $(\mathbb{R}^2, d)$ have the same topologies. Since the continuity of a function is a topological concept (this was the point of Proposition 14.1.5), we know that f will be continuous with respect to d if and only if it is continuous with respect to d_1. Since the metric d_1 is algebraically simpler, we prove continuity with respect to d_1. To this end, let $(a, b) \in \mathbb{R}^2$ and $\epsilon > 0$. Choose $\delta = \epsilon/5$. If $d_1((x, y), (a, b)) = |x - a| + |y - b| < \delta$, then

$$
\begin{aligned}
|f(x, y) - f(a, b)| &= |2(x - a) - 5(y - b)| \\
&\leq 5(|x - a| + |y - b|) \\
&< 5\delta \\
&= \epsilon.
\end{aligned}
$$

14.1.7. Example. The function

$$f \colon \mathbb{R}^2 \to \mathbb{R} \colon (x, y) \mapsto 5x - 7y$$

is continuous.

Proof. Problem.

The principle used in Example 14.1.6 works generally: replacing a metric on the domain of a function by an equivalent metric does not affect the continuity of the function. The same assertion is true for the codomain of a function as well. We state this formally.

14.1.8. Proposition. *Let* $f \colon M_1 \to M_2$ *be a continuous function between two metric spaces* (M_1, d_1) *and* (M_2, d_2). *If* ρ_1 *is a metric on* M_1 *equivalent to* d_1 *and* ρ_2 *is equivalent to* d_2, *then* f *considered as a function from the space* (M_1, ρ_1) *to the space* (M_2, ρ_2) *is still continuous.*

Proof. This is an immediate consequence of Propositions 11.2.3 and 14.1.5. □

14.1.9. Example. Multiplication is a continuous function on $\mathbb{R}$. That is, if we define

$$M \colon \mathbb{R}^2 \to \mathbb{R} \colon (x, y) \mapsto xy \,,$$

then the function M is continuous.

Proof. Exercise. *Hint.* Use Proposition 14.1.8.

14.1.10. Example. Addition is a continuous function on $\mathbb{R}$. That is, if we define
$$A\colon \mathbb{R}^2 \to \mathbb{R}\colon (x,y) \mapsto x+y\,,$$
then the function A is continuous.

Proof. Problem.

14.1.11. Problem. Let d be the usual metric on $\mathbb{R}^2$, let ρ be the metric on $\mathbb{R}^2$ defined in Example 9.2.15, and let $f\colon (\mathbb{R}^2, d) \to (\mathbb{R}^2, \rho)$ be the identity function on $\mathbb{R}^2$. Show that f is *not* continuous.

14.1.12. Problem. Let $\mathbb{R}_d$ be the set of real numbers with the discrete metric, and let M be an arbitrary metric space. Describe the family of all continuous functions $f\colon \mathbb{R}_d \to M$.

14.1.13. Proposition. *Let $f\colon M_1 \to M_2$ where M_1 and M_2 are metric spaces. Then f is continuous if and only if $f^{\leftarrow}(C)$ is a closed subset of M_1 whenever C is a closed subset of M_2.*

Proof. Problem.

14.1.14. Proposition. *Let $f\colon M_1 \to M_2$ be a function between metric spaces. Then f is continuous if and only if*
$$f^{\leftarrow}(B^{\circ}) \subseteq \left(f^{\leftarrow}(B)\right)^{\circ}$$
for all $B \subseteq M_2$.

Proof. Problem.

14.1.15. Proposition. *Let $f\colon M_1 \to M_2$ be a function between metric spaces. Then f is continuous if and only if*
$$f^{\rightarrow}\left(\overline{A}\right) \subseteq \overline{f^{\rightarrow}(A)}$$
for every $A \subseteq M_1$.

Proof. Problem.

14.1.16. Proposition. *Let $f\colon M_1 \to M_2$ be a function between metric spaces. Then f is continuous if and only if*
$$\overline{f^{\leftarrow}(B)} \subseteq f^{\leftarrow}\left(\overline{B}\right)$$
for every $B \subseteq M_2$.

Proof. Problem.

14.1.17. Proposition. *Let f be a real valued function on a metric space M, and let $a \in M$. If f is continuous at a and $f(a) > 0$, then there exists a neighborhood B of a such that $f(x) > \frac{1}{2}f(a)$ for all $x \in B$.*

Proof. Problem.

14.1.18. Proposition. *Let N be a metric space and M be a subspace of N.*

(a) *The inclusion map $\iota\colon M \to N\colon x \mapsto x$ is continuous.*

(b) *Restrictions of continuous functions are continuous. That is, if $f: M_1 \to M_2$ is a continuous mapping between metric spaces and $A \subseteq M_1$, then $f|_A$ is continuous.*

Proof. Problem.

14.1.19. Problem. Show that alteration of the codomain of a continuous function does not affect its continuity. Precisely, if $f: M_0 \to M$ and $g: M_0 \to N$ are functions between metric spaces such that $f(x) = g(x)$ for all $x \in M_0$ and if their common image is a metric subspace of both M and N, then f is continuous if and only if g is.

In the next proposition we show that if two continuous functions agree on a dense subset of a metric space, they agree everywhere on that space.

14.1.20. Proposition. *If f, $g: M \to N$ are continuous functions between metric spaces, if D is a dense subset of M, and if $f|_D = g|_D$, then $f = g$.*

Proof. Problem. *Hint.* Suppose that there is a point a where f and g differ. Consider the inverse images under f and g, respectively, of disjoint neighborhoods of $f(a)$ and $g(a)$. Use Proposition 11.1.22.

14.1.21. Problem. Suppose that M is a metric space and that $f: M \to M$ is continuous but is *not* the identity map. Show that there exists a proper closed set $C \subseteq M$ such that
$$C \cup f^{\leftarrow}(C) = M.$$
Hint. Choose x so that $x \neq f(x)$. Look at the complement of $U \cap f^{\leftarrow}(V)$ where U and V are disjoint neighborhoods of x and $f(x)$, respectively.

There are two ways in which metric spaces may be regarded as essentially the same: they may be isometric (having essentially the same distance function), or they may be topologically equivalent (having essentially the same open sets).

14.1.22. Definition. Let (M, d) and (N, ρ) be metric spaces. A bijection $f: M \to N$ is an ISOMETRY if
$$\rho(f(x), f(y)) = d(x, y)$$
for all x, $y \in M$. If an isometry exists between two metric spaces, the spaces are said to be ISOMETRIC.

14.1.23. Definition. A bijection $g: M \to N$ between metric spaces is a HOMEO-MORPHISM if both g and g^{-1} are continuous. Notice that if g is a homeomorphism, then $g^{\rightarrow}$ establishes a one-to-one correspondence between the family of open subsets of M and the family of open subsets of N. For this reason two metric spaces are said to be TOPOLOGICALLY EQUIVALENT or HOMEOMORPHIC if there exists a homeomorphism between them. Since the open sets of a space are determined by its metric, it is clear that every isometry is automatically a homeomorphism. The converse, however, is not correct; see Example 14.1.25 below.

14.1.24. Problem. Give an example of a bijection between metric spaces that is continuous but is not a homeomorphism.

14.1.25. Example. The open interval $(0,1)$ and the real line $\mathbb{R}$ (with their usual metrics) are homeomorphic but not isometric.

Proof. Problem.

We have seen (in Chapter 12) that certain properties of sets in metric spaces can be characterized by means of sequences. Continuity of functions between metric spaces also has a simple and useful sequential characterization.

14.1.26. Proposition. *Let* $f\colon M_1 \to M_2$ *be a function between metric spaces, and let* a *be a point in* M_1. *The function* f *is continuous at* a *if and only if* $f(x_n) \to f(a)$ *whenever* $x_n \to a$.

Proof. Exercise.

14.1.27. Problem. Give a second solution to Proposition 14.1.20, this time making use of Propositions 12.2.3 and 14.1.26.

14.1.28. Problem. Use Examples 14.1.10 and 14.1.9 to show that if $x_n \to a$ and $y_n \to b$ in $\mathbb{R}$, then $x_n + y_n \to a + b$ and $x_n y_n \to ab$. (Do not give an "ϵ-δ proof".)

14.1.29. Problem. Let c be a point in a metric space M. Show that the function
$$f\colon M \to \mathbb{R}\colon x \mapsto d(x,c)$$
is continuous. *Hint.* Use Proposition 9.2.17.

14.1.30. Problem. Let C be a nonempty subset of a metric space. Then the function
$$g\colon M \to \mathbb{R}\colon x \mapsto d(x,C)$$
is continuous. (See Definition 12.2.10 for the definition of $d(x,C)$.)

14.1.31. Proposition (Urysohn's lemma). *Let* A *and* B *be nonempty disjoint closed subsets of a metric space* M. *Then there exists a continuous function* $f\colon M \to \mathbb{R}$ *such that* ran $f \subseteq [0,1]$, $f^\to(A) = \{0\}$, *and* $f^\to(B) = \{1\}$.

Proof. Problem. *Hint.* Consider $\dfrac{d(x,A)}{d(x,A) + d(x,B)}$. Use Problems 12.2.11(c) and 14.1.30.

14.1.32. Problem. (*Definition.* Disjoint sets A and B in a metric space M are said to be SEPARATED BY OPEN SETS if there exist $U, V \overset{\circ}{\subseteq} M$ such that $U \cap V = \emptyset$, $A \subseteq U$, and $B \subseteq V$.) Show that in a metric space every pair of disjoint closed sets can be separated by open sets.

14.1.33. Problem. If f and g are continuous real valued functions on a metric space M, then $\{x \in M\colon f(x) \neq g(x)\}$ is an open subset of M.

14.1.34. Problem. Show that if f is a continuous real valued function on a metric space, then $|f|$ is continuous. (We denote by $|f|$ the function $x \mapsto |f(x)|$.)

14.1.35. Problem. Show that metrics are continuous functions. That is, show that if M is a set and $d\colon M \times M \to \mathbb{R}$ is a metric, then d is continuous. Conclude from this that if $x_n \to a$ and $y_n \to b$ in a metric space, then $d(x_n, y_n) \to d(a,b)$.

14.1.36. Problem. Show that if f and g are continuous real valued functions on a metric space M and $f(a) = g(a)$ at some point where $a \in M$, then for every $\epsilon > 0$ there exists a neighborhood U of a such that $f(x) < g(x) + \epsilon$ for all $x \in U$.

14.2. Maps into and from products

Let (M_1, ρ_1) and (M_2, ρ_2) be metric spaces. Define the COORDINATE PROJECTIONS π_1 and π_2 on the product $M_1 \times M_2$ by

$$\pi_k \colon M_1 \times M_2 \to M_k \colon (x_1, x_2) \mapsto x_k \qquad \text{for } k = 1, 2.$$

If $M_1 \times M_2$ has the product metric d_1 (see Definitions 12.3.1 and 12.3.3), then the coordinate projections turn out to be continuous functions.

14.2.1. Exercise. Prove the assertion made in the preceding sentence.

14.2.2. Notation. Let S_1, S_2, and T be sets. If $f \colon T \to S_1 \times S_2$, then we define functions $f^1 := \pi_1 \circ f$ and $f^2 := \pi_2 \circ f$. These are the COMPONENTS of f.

If, on the other hand, functions $g \colon T \to S_1$ and $h \colon T \to S_2$ are given, we define the function (g, h) by

$$(g, h) \colon T \to S_1 \times S_2 \colon x \mapsto (g(x), h(x)) \,.$$

Thus it is clear that whenever $f \colon T \to S_1 \times S_2$, we have

$$f = (f^1, f^2) \,.$$

14.2.3. Proposition. *Let M_1, M_2, and N be metric spaces, and let $f \colon N \to M_1 \times M_2$. The function f is continuous if and only if its components f^1 and f^2 are.*

Proof. Exercise.

14.2.4. Proposition. *Let f and g be continuous real valued functions on a metric space.*

(a) *The product fg is a continuous function.*

(b) *For every real number α the function $\alpha g \colon x \mapsto \alpha g(x)$ is continuous.*

Proof. Exercise.

14.2.5. Problem. Let f and g be continuous real valued functions on a metric space and suppose that g is never zero. Show that the function f/g is continuous.

14.2.6. Proposition. *If f and g are continuous real valued functions on a metric space, then $f + g$ is continuous.*

Proof. Problem.

14.2.7. Problem. Show that every polynomial function on $\mathbb{R}$ is continuous. *Hint.* An induction on the degree of the polynomial works nicely.

14.2.8. Definition. Let S be a set, and let $f, g\colon S \to \mathbb{R}$. Then $f \vee g$, the SUPREMUM (or MAXIMUM) of f and g, is defined by

$$(f \vee g)(x) := \max\{f(x), g(x)\}$$

for every $x \in S$. Similarly, $f \wedge g$, the INFIMUM (or MINIMUM) of f and g, is defined by

$$(f \wedge g)(x) := \min\{f(x), g(x)\}$$

for every $x \in S$.

14.2.9. Problem. Let $f(x) = \sin x$ and $g(x) = \cos x$ for $0 \le x \le 2\pi$. Make a careful sketch of $f \vee g$ and $f \wedge g$.

14.2.10. Problem. Show that if f and g are continuous real valued functions on a metric space, then $f \vee g$ and $f \wedge g$ are continuous. *Hint.* Consider things like $f + g + |f - g|$.

In Proposition 14.2.3 we have dealt with the continuity of functions that map *into* products of metric spaces. We now turn to functions that map *from* products; that is, to functions of several variables.

14.2.11. Notation. Let S_1, S_2, and T be sets, and let $f\colon S_1 \times S_2 \to T$. For each $a \in S_1$, we define the function

$$f(a, \cdot)\colon S_2 \to T\colon y \mapsto f(a, y),$$

and for each $b \in S_2$, we define the function

$$f(\cdot, b)\colon S_1 \to T\colon x \mapsto f(x, b).$$

Loosely speaking, $f(a, \cdot)$ is the result of regarding f as a function of only its second variable; $f(\cdot, b)$ results from thinking of f as depending on only its first variable.

14.2.12. Proposition. *Let M_1, M_2, and N be metric spaces and $f\colon M_1 \times M_2 \to N$. If f is continuous, then so are $f(a, \cdot)$ and $f(\cdot, b)$ for all $a \in M_1$ and $b \in M_2$.*

Proof. Exercise.

This proposition is sometimes paraphrased as follows: joint continuity implies separate continuity. The converse is *not* true. (See Problem 14.2.13.)

Remark. It should be clear how to extend the results of Propositions 14.2.3 and 14.2.12 to products of any finite number of metric spaces.

14.2.13. Problem. Let $f\colon \mathbb{R}^2 \to \mathbb{R}$ be defined by

$$f(x, y) = \begin{cases} xy(x^2 + y^2)^{-1} & \text{for } (x, y) \ne (0, 0), \\ 0 & \text{for } x = y = 0. \end{cases}$$

(a) Show that f is continuous at each point of $\mathbb{R}^2$ except at $(0, 0)$, where it is *not* continuous.

(b) Show that the converse of Proposition 14.2.12 is not true.

14.2.14. Notation. Let M and N be metric spaces. We denote by $\mathcal{C}(M, N)$ the family of all continuous functions f taking M into N.

In Proposition 13.2.4 we showed that the uniform limit of a sequence of bounded real valued functions is bounded. We now prove an analogous result for continuous real valued functions: the uniform limit of a sequence of continuous real valued functions is continuous.

14.2.15. Proposition. *If (f_n) is a sequence of continuous real valued functions on a metric space M that converges uniformly to a real valued function g on M, then g is continuous.*

Proof. Exercise.

14.2.16. Problem. If the word "pointwise" is substituted for "uniformly" in Proposition 14.2.15, the conclusion no longer follows. In particular, find an example of a sequence (f_n) of continuous functions on $[0, 1]$ that converges pointwise to a function g on $[0, 1]$ that is *not* continuous.

14.3. Limits

We now generalize to metric spaces the results of Chapter 7 on limits of real valued functions. Most of this generalization is accomplished quite simply: just replace open intervals on the real line with open balls in metric spaces.

14.3.1. Definition. If $B_r(a)$ is the open ball of radius r about a point a in a metric space M, then $B_r^*(a)$, the DELETED OPEN BALL of radius r about a, is just $B_r(a)$ with the point a deleted. That is, $B_r^*(a) = \{x \in M : 0 < d(x, a) < r\}$.

14.3.2. Definition. Let (M, d) and (N, ρ) be metric spaces, let $A \subseteq M$, let $f : A \to N$, let a be an accumulation point of A, and let $l \in N$. We say that l is the LIMIT OF f AS x APPROACHES a (of the LIMIT OF f AT a) if, for every $\epsilon > 0$, there exists $\delta > 0$ such that $f(x) \in B_\epsilon(l)$ whenever $x \in A \cap B_\delta^*(a)$. In slightly different notation,

$$(\forall \epsilon > 0)(\exists \delta > 0)(\forall x \in A)\ 0 < d(x, a) < \delta \implies \rho(f(x), l) < \epsilon.$$

When this condition is satisfied, we write

$$f(x) \to l \quad \text{as } x \to a$$

or

$$\lim_{x \to a} f(x) = l.$$

As in Chapter 7 this notation is optimistic. We will show in the next proposition that limits, if they exist, are unique.

14.3.3. Proposition. *Let $f : A \to N$ where M and N are metric spaces and $A \subseteq M$, and let $a \in A'$. If $f(x) \to b$ as $x \to a$ and $f(x) \to c$ as $x \to a$, then $b = c$.*

Proof. Exercise.

For a function between metric spaces, the relationship between its continuity at a point and its limit there is exactly the same as in the case of real valued functions. (See Proposition 7.2.3 and the two examples that precede it.)

14.3.4. Proposition. *Let M and N be metric spaces, let $f\colon A \to N$ where $A \subseteq M$, and let $a \in A \cap A'$. Then f is continuous at a if and only if*

$$\lim_{x \to a} f(x) = f(a).$$

Proof. Problem. *Hint.* Modify the proof of Proposition 7.2.3.

14.3.5. Proposition. *If M is a metric space, $f\colon A \to M$ where $A \subseteq \mathbb{R}$, and $a \in A'$, then*

$$\lim_{h \to 0} f(a+h) = \lim_{x \to a} f(x)$$

in the sense that if either limit exists, then so does the other and the two limits are equal.

Proof. Problem. *Hint.* Modify the proof of Proposition 7.2.4.

We conclude this chapter by examining the relationship between "double" and "iterated" limits of real valued functions of two real variables. A limit of the form

$$\lim_{(x,y) \to (a,b)} f(x,y)$$

is a DOUBLE LIMIT; limits of the form

$$\lim_{x \to a} \left(\lim_{y \to b} f(x,y) \right) \quad \text{and} \quad \lim_{y \to b} \left(\lim_{x \to a} f(x,y) \right)$$

are ITERATED LIMITS. The meaning of the expression $\lim_{x \to a} \left(\lim_{y \to b} f(x,y) \right)$ should be clear: it is $\lim_{x \to a} h(x)$ where h is the function defined by $h(x) = \lim_{y \to b} f(x,y)$.

14.3.6. Example. Let $f(x,y) = x \sin(1 + x^2 y^2)$ for all $x, y \in \mathbb{R}$. Then

$$\lim_{(x,y) \to (0,0)} f(x,y) = 0.$$

Proof. The function f maps $\mathbb{R}^2$ into $\mathbb{R}$. We take the usual Euclidean metric on both of these spaces. Given $\epsilon > 0$, choose $\delta = \epsilon$. If $(x,y) \in B_\delta^*(0,0)$, then

$$|f(x,y) - 0| = |x||\sin(1 + x^2 y^2)| \le |x| \le \sqrt{x^2 + y^2} = d\big((x,y),(0,0)\big) < \delta = \epsilon. \quad \square$$

14.3.7. Example. Let $f(x,y) = x \sin(1 + x^2 y^2)$ for all $x, y \in \mathbb{R}$. Then

$$\lim_{x \to 0} \left(\lim_{y \to 0} f(x,y) \right) = 0.$$

Proof. Compute the inner limit first:

$$\lim_{x \to 0} \left(\lim_{y \to 0} \big(x \sin(1 + x^2 y^2) \big) \right) = \lim_{x \to 0} (x \sin 1) = 0. \quad \square$$

Because of the intimate relationship between continuity and limits (Proposition 14.3.4) and because of the fact that joint continuity implies separate continuity (Proposition 14.2.12), many persons wrongly conclude that the existence of a double limit implies the existence of the corresponding iterated limits. One of the last problems in this chapter will provide you with an example of a function having a double limit at the origin but failing to have one of the corresponding iterated limits. In the next proposition we prove that if in addition to the existence of the double limit we assume that $\lim_{x \to a} f(x,y)$ and $\lim_{y \to b} f(x,y)$ always exist, then both iterated limits exist and are equal.

14.3.8. Proposition. *Let f be a real valued function of two real variables. If the limit*

$$l = \lim_{(x,y) \to (a,b)} f(x,y)$$

exists and if $\lim_{x \to a} f(x,y)$ and $\lim_{y \to b} f(x,y)$ exist for all y and x, respectively, then the iterated limits

$$\lim_{x \to a} \left(\lim_{y \to b} f(x,y) \right) \quad and \quad \lim_{y \to b} \left(\lim_{x \to a} f(x,y) \right)$$

both exist and are equal to l.

Proof. Exercise. *Hint.* Let $g(x) = \lim_{y \to b} f(x,y)$ for all $x \in \mathbb{R}$. We wish to show that $\lim_{x \to a} g(x) = l$. The quantity $|g(x) - l|$ is small whenever both $|g(x) - f(x,y)|$ and $|f(x,y) - l|$ are. Since $\lim_{(x,y) \to (a,b)} f(x,y) = l$ it is easy to make $|f(x,y) - l|$ small: insist that (x,y) lie in some sufficiently small open ball about (a,b) of radius, say, δ. This can be accomplished by requiring, for example, that

(12) $$|x - a| < \delta/2$$

and that

(13) $$|y - b| < \delta/2.$$

Since $g(x) = \lim_{y \to b} f(x,y)$ for every x, we can make $|g(x) - f(x,y)|$ small (for fixed x) by supposing that

(14) $$|y - b| < \eta$$

for some sufficiently small η. So the proof is straightforward: require x to satisfy (12), and for such x require y to satisfy (13) and (14).

It is sometimes necessary to show that certain limits do *not* exist. There is a rather simple technique that is frequently useful for showing that the limit of a given real valued function does not exist at a point a. Suppose we can find two numbers $\alpha \neq \beta$ such that in *every* neighborhood of a the function f assumes (at points other than a) both the values α and β. (That is, for every $\delta > 0$ there exist points u and v in $B_\delta(a)$ distinct from a such that $f(u) = \alpha$ and $f(v) = \beta$.) Then it is easy to see that f cannot have a limit as x approaches a. Argue by contradiction: Suppose $\lim_{x \to a} f(x) = l$. Let $\epsilon = |\alpha - \beta|$. Then $\epsilon > 0$; so there exists $\delta > 0$ such that $|f(x) - l| < \epsilon/2$ whenever $0 < d(x,a) < \delta$. Let u and v be points in $B_\delta^*(a)$ satisfying $f(u) = \alpha$ and $f(v) = \beta$. Since $|f(u) - l| < \epsilon/2$ and $|f(v) - l| < \epsilon/2$, it follows that

$$\begin{aligned}
\epsilon &= |\alpha - \beta| \\
&= |f(u) - f(v)| \\
&\leq |f(u) - l| + |l - f(v)| \\
&< \epsilon,
\end{aligned}$$

which is an obvious contradiction.

14.3.9. Example. Let $f(x, y) = \dfrac{x^2 y^2}{x^2 y^2 + (x+y)^4}$ provided that $(x, y) \neq (0,0)$.
Then $\lim_{(x,y) \to (0,0)} f(x, y)$ does not exist.

Proof. Exercise.

14.3.10. Example. The limit as $(x, y) \to (0,0)$ of $\dfrac{x^3 y^3}{x^{12} + y^4}$ does not exist.

Proof. Problem.

14.3.11. Problem. Let $f \colon A \to \mathbb{R}$ where $A \subseteq M$ and M is a metric space, and
let $a \in A'$. Show that
$$\lim_{x \to a} f(x) = 0 \quad \text{if and only if} \quad \lim_{x \to a} |f(x)| = 0 \,.$$

14.3.12. Problem. Let f, g, $h \colon A \to \mathbb{R}$ where $A \subseteq M$ and M is a metric space,
and let $a \in A'$. Show that if $f \leq g \leq h$ and
$$\lim_{x \to a} f(x) = \lim_{x \to a} h(x) = l \,,$$
then $\lim_{x \to a} g(x) = l$.

14.3.13. Problem. Let M, N, and P be metric spaces, and let $a \in A \subseteq M$,
$f \colon A \to N$, and $g \colon N \to P$.

(a) Show that if $l = \lim_{x \to a} f(x)$ exists and g is continuous at l, then
$\lim_{x \to a}(g \circ f)(x)$ exists and is equal to $g(l)$.

(b) Show by example that the following assertion need not be true: If $l = \lim_{x \to a} f(x)$ exists and $\lim_{y \to l} g(y)$ exists, then $\lim_{x \to a}(g \circ f)(x)$ exists.

14.3.14. Problem. Let a be a point in a metric space. Show that
$$\lim_{x \to a} d(x, a) = 0 \,.$$

14.3.15. Problem. In this problem $\mathbb{R}^n$ has its usual metric; in particular,
$$d(x, 0) = \left(\sum_{k=1}^{n} x_k^2 \right)^{1/2}$$
for all $x = (x_1, \ldots, x_n) \in \mathbb{R}^n$.

(a) Show that
$$\lim_{x \to 0} \frac{x_j x_k}{d(x, 0)} = 0$$
whenever $1 \leq j \leq n$ and $1 \leq k \leq n$.

(b) For $1 \leq k \leq n$ show that $\lim_{x \to 0} \dfrac{x_k}{d(x, 0)}$ does not exist.

14.3.16. Problem. Let $f(x, y) = \dfrac{x^2 - y^2}{x^2 + y^2}$ for $(x, y) \neq (0,0)$. Find the following
limits, if they exist.

(a) $\lim_{(x,y) \to (0,0)} f(x, y)$

(b) $\lim_{x \to 0}\left(\lim_{y \to 0} f(x, y)\right)$

(c) $\lim_{y \to 0}\left(\lim_{x \to 0} f(x, y)\right)$

14.3.17. Problem. Same as Problem 14.3.16, but $f(x,y) = \dfrac{xy}{x^2 + y^2}$.

14.3.18. Problem. Same as Problem 14.3.16, but $f(x,y) = \dfrac{x^2 y^2}{x^2 + y^2}$.

14.3.19. Problem. Same as Problem 14.3.16, but $f(x,y) = y\sin(1/x)$ if $x \neq 0$ and $f(x,y) = 0$ if $x = 0$.

Chapter 15

Compact metric spaces

15.1. Definition and elementary properties

15.1.1. Definition. Recall that a family $\mathfrak{U}$ of sets is said to COVER a set S if $\bigcup \mathfrak{U} \supseteq S$. The phrases "$\mathfrak{U}$ covers S", "$\mathfrak{U}$ is a cover for S", and "$\mathfrak{U}$ is a covering of S" are used interchangeably. If a cover $\mathfrak{U}$ for a metric space M consists entirely of open subsets of M, then $\mathfrak{U}$ is an OPEN COVER for M. If $\mathfrak{U}$ is a family of sets that covers a set S and $\mathfrak{V}$ is a subfamily of $\mathfrak{U}$ that also covers S, then $\mathfrak{V}$ is a SUBCOVER of $\mathfrak{U}$ for S. A metric space M is COMPACT if *every* open cover of M has a finite subcover.

We have just defined what we mean by a compact *space*. It will be convenient also to speak of a compact *subset* of a metric space M. If $A \subseteq M$, we say that A is a COMPACT SUBSET of M if, regarded as a subspace of M, it is a compact metric space.

Notice that the definition of compactness is identical to the one given for subsets of the real line in Definition 6.1.3. Recall also that we have proved that every closed bounded subset of $\mathbb{R}$ is compact (see Example 6.3.6).

Remark (A matter of technical convenience). Suppose we wish to show that some particular subset A of a metric space M is compact. Is it really necessary that we work with coverings made up of open subsets of A (as the definition demands), or can we just as well use coverings whose members are open subsets of M? Fortunately, either will do. This is an almost obvious consequence of Proposition 11.2.1. Nevertheless, providing a careful verification takes a few lines, and it is good practice to attempt it.

15.1.2. Proposition. *A subset A of a metric space M is compact if and only if every cover of A by open subsets of M has a finite subcover.*

Proof. Exercise.

We give an obvious corollary of the preceding proposition: If M_1 is a subspace of a metric space M_2 and $K \subseteq M_1$, then K is a compact subset of M_1 if and only if it is a compact subset of M_2.

15.1.3. Problem. Generalize the result of Proposition 6.2.2 to metric spaces. That is, show that every closed subset of a compact metric space is compact.

15.1.4. Definition. A subset of a metric space is BOUNDED if it is contained in some open ball.

15.1.5. Problem. Generalize the result of Proposition 6.2.3 to metric spaces. That is, show that every compact subset of a metric space is closed and bounded.

As we will see, the converse of the preceding theorem holds for subsets of $\mathbb{R}^n$ under its usual metric; this is the Heine–Borel theorem. It is most important to know, however, that this converse is not true in arbitrary metric spaces, where sets that are closed and bounded may fail to be compact.

15.1.6. Example. Consider an infinite set M under the discrete metric. Regarded as a subset of itself, M is clearly closed and bounded. But since the family $\mathfrak{U} = \{\{x\} : x \in M\}$ is a cover for M that has no proper subcover, the space M is certainly not compact.

15.1.7. Example. With its usual metric $\mathbb{R}^2$ is not compact.

Proof. Problem.

15.1.8. Example. The open unit ball $\{(x, y, z) : x^2 + y^2 + z^2 < 1\}$ in $\mathbb{R}^3$ is not compact.

Proof. Problem.

15.1.9. Example. The strip $\{(x, y) : 2 \leq y \leq 5\}$ in $\mathbb{R}^2$ is not compact.

Proof. Problem.

15.1.10. Example. The closed first quadrant $\{(x, y) : x \geq 0 \text{ and } y \geq 0\}$ in $\mathbb{R}^2$ is not compact.

Proof. Problem.

15.1.11. Problem. Show that the intersection of an arbitrary nonempty collection of compact subsets of a metric space is itself compact.

15.1.12. Problem. Show that the union of a finite collection of compact subsets of a metric space is itself compact. *Hint.* What about *two* compact sets? Give an example to show that the union of an arbitrary collection of compact subsets of a metric space need not be compact.

15.2. The extreme value theorem

15.2.1. Definition. A real valued function f on a metric space M is said to have a (GLOBAL) MAXIMUM at a point a in M if $f(a) \geq f(x)$ for every x in M; the number $f(a)$ is the MAXIMUM VALUE of f. The function has a (GLOBAL) MINIMUM at a if $f(a) \leq f(x)$ for every x in M; and in this case $f(a)$ is the MINIMUM VALUE of f. A number is an EXTREME VALUE of f if it is either a maximum or a minimum value. It is clear that a function may fail to have maximum or minimum values. For example, on the open interval $(0, 1)$ the function $f \colon x \mapsto x$ assumes neither a maximum nor a minimum.

Our next goal is to show that every continuous function on a compact metric space attains both a maximum and a minimum. This turns out to be an easy consequence of the fact that the continuous image of a compact set is compact. All this works exactly as it did for $\mathbb{R}$.

15.2.2. Problem. Generalize the result of Theorem 6.3.2 to metric spaces. That is, show that if M and N are metric spaces, if M is compact, and if $f \colon M \to N$ is continuous, then $f^{\to}(M)$ is compact.

15.2.3. Problem (The extreme value theorem). Generalize the result of Theorem 6.3.3 to metric spaces. That is, show that if M is a compact metric space and $f \colon M \to \mathbb{R}$ is continuous, then f assumes both a maximum and a minimum value on M.

In Chapter 13 we defined the uniform metric on the family $\mathcal{B}(S, \mathbb{R})$ of all bounded real valued functions on S and agreed to call convergence of sequences in this space "uniform convergence". Since S was an arbitrary set (not a metric space), the question of continuity of members of $\mathcal{B}(S, \mathbb{R})$ is meaningless. For the moment we restrict our attention to functions defined on a compact metric space, where the issue of continuity is both meaningful and interesting.

A trivial, but crucial, observation is that if M is a compact metric space, then the family $\mathcal{C}(M) = \mathcal{C}(M, \mathbb{R})$ of continuous real valued functions on M is a subspace of the metric space $\mathcal{B}(M) = \mathcal{B}(M, \mathbb{R})$. This is an obvious consequence of the extreme value theorem (Problem 15.2.3), which says, in particular, that every continuous real valued function on M is bounded. Furthermore, since the uniform limit of continuous functions is continuous (see Proposition 14.2.15), it is clear that $\mathcal{C}(M, \mathbb{R})$ is a closed subset of $\mathcal{B}(M, \mathbb{R})$ (see Proposition 12.2.2 for the sequential characterization of closed sets). For future reference we record this formally.

15.2.4. Proposition. *If M is a compact metric space, then $\mathcal{C}(M)$ is a closed subset of $\mathcal{B}(M)$.*

15.2.5. Example. The circle $x^2 + y^2 = 1$ is a compact subset of $\mathbb{R}^2$.

Proof. Problem. *Hint.* Parametrize.

15.2.6. Example. The ellipse $\frac{x^2}{16} + \frac{y^2}{9} = 1$ is a compact subset of $\mathbb{R}^2$.

Proof. Problem.

15.2.7. Example. Regard $\mathbb{R}^2$ as a metric space under the uniform metric (see Example 9.2.12). Then the boundary of the unit ball in this space is compact.

Proof. Problem.

15.2.8. Problem. Let $f\colon M \to N$ be a continuous bijection between metric spaces.

(a) Show by example that f need not be a homeomorphism.

(b) Show that if M is compact, then f must be a homeomorphism.

15.3. Dini's theorem

15.3.1. Definition. A family $\mathfrak{F}$ of sets is said to have the FINITE INTERSECTION PROPERTY if every finite subfamily of $\mathfrak{F}$ has nonempty intersection.

15.3.2. Problem. Show that a metric space M is compact if and only if every family of closed subsets of M having the finite intersection property has nonempty intersection.

15.3.3. Proposition (Dini's theorem). *Let M be a compact metric space, and let (f_n) be a sequence of continuous real valued functions on M such that $f_n(x) \geq f_{n+1}(x)$ for all x in M and all n in $\mathbb{N}$. If the sequence (f_n) converges pointwise on M to a continuous function g, then it converges uniformly to g.*

Proof. Problem. *Hint.* First establish the correctness of the assertion for the special case where $g = 0$. For $\epsilon > 0$ consider the sets $A_n = f_n^{\leftarrow}([\epsilon, \infty))$. Argue by contradiction to show that $\bigcap_{n=1}^{\infty} A_n$ is empty. Then use Problem 15.3.2.

15.3.4. Example. *Dini's theorem* (Proposition 15.3.3) is no longer true if we remove the hypothesis that

(a) the sequence (f_n) is decreasing,

(b) the function g is continuous, or

(c) the space M is compact.

Proof. Problem.

15.3.5. Example. On the interval $[0,1]$ the square root function $x \mapsto \sqrt{x}$ is the uniform limit of a sequence of polynomials.

Proof. Problem. *Hint.* Let p_0 be the zero function on $[0,1]$, and for $n \geq 0$ define p_{n+1} on $[0,1]$ by

$$p_{n+1}(t) = p_n(t) + \tfrac{1}{2}\left(t - (p_n(t))^2\right),$$

and verify that

$$0 \leq \sqrt{t} - p_n(t) \leq \frac{2\sqrt{t}}{2 + n\sqrt{t}} \leq \frac{2}{n}$$

for $0 \leq t \leq 1$ and $n \in \mathbb{N}$.

Chapter 16

Sequential characterization of compactness

We have previously characterized open sets, closed sets, closure, and continuity by means of sequences. Our next goal is to produce a characterization of compactness in terms of sequences. This is achieved in Theorem 16.2.1 where it is shown that compactness in metric spaces is equivalent to something called "sequential compactness". For this concept we need to speak of "subsequences" of sequences of points in a metric space. In Definition 4.4.1 we defined subsequence for sequences of real numbers. There is certainly no reason why we cannot speak of subsequences of arbitrary sequences.

16.1. Sequential compactness

16.1.1. Definition. If a is a sequence of elements of a set S and $n \colon \mathbb{N} \to \mathbb{N}$ is strictly increasing, then the composite function $a \circ n$ is a SUBSEQUENCE of the sequence a. (Notice that $a \circ n$ is itself a sequence of elements of S since it maps $\mathbb{N}$ into S.) The kth term of $a \circ n$ is frequently written as a_{n_k}; other acceptable notations are $a_{n(k)}$ and $a(n(k))$.

16.1.2. Example. Let $a = (a_k)$ be a sequence of points in a metric space. To get, say, the subsequence containing only the even terms of this sequence, use the increasing map $n \colon k \mapsto 2k$ on $\mathbb{N}$. Then $a \circ n = (a_{n(k)}) = (a_{2k})$ is the desired subsequence.

16.1.3. Example. Notice that it is possible for a sequence that fails to converge to have subsequences that do converge. For example, if $a_n = (-1)^n + (1/n)$ for each n, then the subsequence (a_{2n}) converges while the sequence (a_n) itself does not.

16.1.4. Definition. A metric space M is SEQUENTIALLY COMPACT if every sequence in M has a convergent subsequence.

16.1.5. Example. It is important to understand that for a space to be sequentially compact, the preceding definition requires that every sequence in the space have a subsequence that converges to a point *in that space*. It is not enough to find a subsequence that converges in some larger space. For example, we know that the metric space $(0,1]$ regarded as a subspace of $\mathbb{R}$ is not sequentially compact because the sequence $\left(\frac{1}{n}\right)$ has no subsequence that converges to something in $(0,1]$. That $\left(\frac{1}{n}\right)$ happens to converge to 0 in $\mathbb{R}$ is completely irrelevant.

A major goal of this chapter is to demonstrate that in metric spaces compactness and sequential compactness are the same thing. This is done in Theorem 16.2.1. Be aware however that in general topological spaces this is *not* true. An essential ingredient of the proof is the following chain of implications:

$$\text{sequentially compact} \implies \text{totally bounded} \implies \text{separable}.$$

So the next order of business is to define the last two concepts and to prove that the preceding implications do hold.

16.1.6. Definition. A metric space M is TOTALLY BOUNDED if for every $\epsilon > 0$ there exists a finite subset F of M such that for every $a \in M$ there is a point $x \in F$ such that $d(x,a) < \epsilon$. This definition has a more or less standard paraphrase: a space is totally bounded if it can be kept under surveillance by a finite number of arbitrarily near-sighted policemen.

16.1.7. Proposition. *Every totally bounded metric space is bounded.*

Proof. Problem.

16.1.8. Example. The converse of the preceding proposition is false. Any infinite set with the discrete metric is an example of a bounded metric space that is not totally bounded. (Why?)

16.1.9. Proposition. *Every sequentially compact metric space is totally bounded.*

Proof. Exercise. *Hint.* Assume that a metric space M is not totally bounded. Inductively, construct a sequence in M no two terms of which are closer together than some fixed distance $\epsilon > 0$.

16.1.10. Definition. A metric space is SEPARABLE if it possesses a countable dense subset.

16.1.11. Example. The space $\mathbb{R}^n$ is separable. The set of points $(q_1, \ldots, q_n)$ such that each coordinate q_k is rational is a countable dense subset of $\mathbb{R}^n$.

16.1.12. Example. It follows easily from Proposition 11.1.22 that the real line (or indeed any uncountable set) with the discrete metric is *not* separable. (Consider the open balls of radius 1 about each point.)

16.1.13. Proposition. *Every totally bounded metric space is separable.*

Proof. Problem. *Hint.* Let M be a totally bounded metric space. For each n in $\mathbb{N}$ choose a finite set F_n such that for each $a \in M$ the set $F_n \cap B_{\frac{1}{n}}(a)$ is nonempty.

16.1.14. Problem. The metric space $\mathbb{R}_d$ comprising the real numbers under the discrete metric is not separable.

16.1.15. Corollary. *Every sequentially compact metric space is separable.*

Proof. Propositions 16.1.9 and 16.1.13. $\square$

16.2. Conditions equivalent to compactness

We are now ready for a major result of this chapter—a sequential characterization of compactness. We show that a space M is compact if and only if every sequence in M has a convergent subsequence. In other words a space is compact if and only if it is sequentially compact. We will see later in the section how useful this result is when we use it to prove that a finite product of compact spaces is compact.

The following theorem also provides a second characterization of compactness: a space M is compact if and only if every infinite subset of M has a point of accumulation in M. The proof of the theorem is fairly straightforward except for one complicated bit. It is not easy to prove that every sequentially compact space is compact. This part of the proof comes equipped with a lengthy hint.

16.2.1. Theorem. *If M is a metric space, then the following are equivalent:*

(a) *M is compact;*

(b) *every infinite subset of M has an accumulation point in M;*

(c) *M is sequentially compact.*

Proof. Exercise. *Hint.* Showing that (c) implies (a) is not so easy. To show that a sequentially compact metric space M is compact, start with an arbitrary open cover $\mathfrak{U}$ for M and show first that

(A) $\qquad\qquad$ $\mathfrak{U}$ has a *countable* subfamily $\mathfrak{V}$ that covers M.

Then show that

(B) $\qquad\qquad$ there is a finite subfamily of $\mathfrak{V}$ that covers M.

The hard part is (A). According to Corollary 16.1.15, we may choose a countable dense subset A of M. Let $\mathfrak{B}$ be the family of all open balls $B_r(a)$ such that

(i) $a \in A$,

(ii) $r \in \mathbb{Q}$, and

(iii) $B_r(a) \subseteq U$ for some $U \in \mathfrak{U}$.

For each B in $\mathfrak{B}$ choose a set U_B in $\mathfrak{U}$ such that $B \subseteq U_B$, and let

$$\mathfrak{V} = \{U_B \colon B \in \mathfrak{B}\}.$$

Then verify that $\mathfrak{V}$ is a countable subfamily of $\mathfrak{U}$ that covers M.

To show that $\mathfrak{V}$ covers M, start with an arbitrary point $x \in M$ and a set $U \in \mathfrak{U}$ containing x. All that is needed is to find an open ball $B_s(a)$ in $\mathfrak{B}$ such that $x \in B_s(a) \subseteq U$. In order to do this, the point $a \in A$ must be taken sufficiently close to x so that it is possible to choose a rational number s that is both

(1) small enough for $B_s(a)$ to be a subset of U, and

(2) large enough for $B_s(a)$ to contain x.

To establish (B), let $(V_1, V_2, V_3, \ldots)$ be an enumeration of $\mathfrak{V}$. and let $W_n = \bigcup_{k=1}^{n} V_k$ for each $n \in \mathbb{N}$. If no one set W_n covers M, then for every $k \in \mathbb{N}$ there is a point $x_k \in W_k{}^c$.

16.2.2. Problem. Use Theorem 16.2.1 to give three different proofs that the metric space $[0, 1)$ (with the usual metric inherited from $\mathbb{R}$) is not compact.

CAUTION. It is a common (and usually helpful) mnemonic device to reduce statements of complicated theorems in analysis to brief paraphrases. In doing this, considerable care should be exercised so that crucial information is not lost. Here is an example of the kind of thing that can go wrong.

Consider the two statements:

(1) In the metric space $\mathbb{R}$ every infinite subset of the open unit interval has a point of accumulation; and

(2) A metric space is compact if every infinite subset has a point of accumulation.

Assertion (1) is a special case of Proposition 4.4.9, and (2) is just part of Theorem 16.2.1. The unwary tourist might be tempted to conclude from (1) and (2) that the open unit interval $(0, 1)$ is compact, which, of course, it is not. The problem here is that (1) is a correct assertion about $(0, 1)$ regarded as a subset of the space $\mathbb{R}$; every infinite subset of $(0, 1)$ does have a point of accumulation *lying in* $\mathbb{R}$.

If, however, the metric space under consideration is $(0, 1)$ itself, then (1) is no longer true. For example, the set of all numbers of the form $1/n$ for $n \geq 2$ has no accumulation point in $(0, 1)$. When we use (2) to establish the compactness of a metric space M, what we must verify is that every infinite subset of M has a point of accumulation *that lies in* M. Showing that these points of accumulation exist in some space that contains M just does not do the job. The complete statements of Proposition 4.4.9 and Theorem 16.2.1 make this distinction clear; the paraphrases (1) and (2) above do not.

16.3. Products of compact spaces

It is possible using just the definition of compactness to prove that the product of two compact metric spaces is compact. It is a pleasant reward for the effort put into proving Theorem 16.2.1 that it can be used to give a genuinely simple proof of this important result.

16.3.1. Theorem. *If M_1 and M_2 are compact metric spaces, then so is $M_1 \times M_2$.*

Proof. Problem.

16.3.2. Corollary. *If $M_1, \ldots, M_n$ are compact metric spaces, then the product space $M_1 \times \cdots \times M_n$ is compact.*

Proof. Induction.   □

16.3.3. Problem. Let A and B be subsets of a metric space. Recall from Definition 12.2.10 that the distance between A and B is defined by

$$d(A, B) := \inf\{d(a, b) \colon a \in A \text{ and } b \in B\}.$$

Prove or disprove the following.

(a) If $A \cap B = \emptyset$, then $d(A, B) > 0$.

(b) If A and B are closed and $A \cap B = \emptyset$, then $d(A, B) > 0$.

(c) If A is closed, B is compact, and $A \cap B = \emptyset$, then $d(A, B) > 0$. *Hint.* If $d(A, B) = 0$, then there exist sequences a in A and b in B such that $d(a_n, b_n) \to 0$.

(d) If A and B are compact and $A \cap B = \emptyset$, then $d(A, B) > 0$.

16.3.4. Problem. Let a, $b > 0$. The elliptic disk

$$D := \left\{ (x, y) \colon \frac{x^2}{a^2} + \frac{y^2}{b^2} \le 1 \right\}$$

is a compact subset of $\mathbb{R}^2$. *Hint.* Write the disk as a continuous image of the unit square $[0, 1] \times [0, 1]$.

16.3.5. Problem. The unit sphere

$$S^2 \equiv \{ (x, y, z) \colon x^2 + y^2 + z^2 = 1 \}$$

is a compact subset of $\mathbb{R}^3$. *Hint.* Spherical coordinates.

16.3.6. Problem. Show that the interval $[0, \infty)$ is not a compact subset of $\mathbb{R}$ using each of the following.

(a) The definition of compactness.

(b) Proposition 15.1.2.

(c) Proposition 15.1.5.

(d) The *extreme value theorem* (Problem 15.2.3).

(e) Theorem 16.2.1, condition (2).

(f) Theorem 16.2.1, condition (3).

(g) The *finite intersection property* (see Problem 15.3.2).

(h) *Dini's theorem* (see Proposition 15.3.3).

16.4. The Heine–Borel theorem

We have seen in Problem 15.1.5 that in an arbitrary metric space, compact sets are always closed and bounded. Recall also that the converse of this is not true in general (see Problem 6.3.9). In $\mathbb{R}^n$, however, the converse does indeed hold. This assertion is the *Heine–Borel theorem*. Notice that in Example 6.3.6 we have already established its correctness for the case $n = 1$. The proof of the general case is now just as easy—we have done all the hard work in proving that the product of finitely many compact spaces is compact (Corollary 16.3.2).

16.4.1. Theorem (Heine–Borel theorem). *A subset of $\mathbb{R}^n$ is compact if and only if it is closed and bounded.*

Proof. Exercise.

16.4.2. Example. The triangular region T whose vertices are $(0,0)$, $(1,0)$, and $(0,1)$ is a compact subset of $\mathbb{R}^2$.

Proof. Define functions $f, g, h : \mathbb{R}^2 \to \mathbb{R}$ by $f(x, y) = x$, $g(x, y) = y$, and $h(x, y) = x + y$. Each of these functions is continuous. Thus the sets

$$A := f^{\leftarrow}[0, \infty) = \{(x, y) : x \geq 0\},$$
$$B := g^{\leftarrow}[0, \infty) = \{(x, y) : y \geq 0\}, \quad \text{and}$$
$$C := h^{\leftarrow}(-\infty, 1] = \{(x, y) : x + y \leq 1\}$$

are all closed sets. Thus $T = A \cap B \cap C$ is closed. It is bounded since it is contained in the open ball about the origin with radius 2. Thus by the Heine–Borel theorem (Theorem 16.4.1), T is compact. $\qquad\square$

16.4.3. Problem. Do Problem 16.3.4 again, this time using the Heine–Borel theorem.

16.4.4. Problem (Bolzano–Weierstrass theorem). Every bounded infinite subset of $\mathbb{R}^n$ has at least one point of accumulation in $\mathbb{R}^n$ (compare Proposition 4.4.9).

16.4.5. Problem (Cantor intersection theorem, Version I). If (A_n) is a nested sequence of nonempty closed bounded subsets of $\mathbb{R}^n$, then $\bigcap_{n=1}^{\infty} A_n$ is nonempty. Furthermore, if diam $A_n \to 0$, then $\bigcap_{n=1}^{\infty} A_n$ is a single point.

16.4.6. Problem. Use the Cantor intersection theorem (Problem 16.4.5) to show that the medians of a triangle are concurrent.

16.4.7. Problem. Let m be the set of all bounded sequences of real numbers under the uniform metric

$$d_u(a, b) = \sup\{|a_n - b_n| : n \in \mathbb{N}\}$$

whenever $a, b \in m$.

(a) Show by example that the Heine–Borel theorem (Theorem 16.4.1) does not hold for subsets of m.

(b) Show by example that the Bolzano–Weierstrass theorem does not hold for subsets of m (see Problem 16.4.4).

(c) Show by example that the Cantor intersection theorem does not hold for subsets of m (see Problem 16.4.5).

16.4.8. Problem. Find a metric space M with the property that every infinite subset of M is closed and bounded but not compact.

16.4.9. Problem. Prove or disprove: If both the interior and the boundary of a set $A \subseteq \mathbb{R}$ are compact, then so is A.

Chapter 17

Connectedness

In Chapter 5 we discussed connected subsets of the real line. Although they are easily characterized (they are just the intervals), they possess important properties, most notably the *intermediate value property*. Connected subsets of arbitrary metric spaces can be somewhat more complicated, but they are no less important.

17.1. Connected spaces

17.1.1. Definition. A metric space M is DISCONNECTED if there exist disjoint nonempty open sets U and V whose union is M. In this case we say that the sets U and V DISCONNECT M. A metric space is CONNECTED if it is not disconnected. A subset of a metric space M is CONNECTED (respectively, DISCONNECTED) if it is connected (respectively, disconnected) as a subspace of M. Thus a subset N of M is disconnected if there exist nonempty disjoint sets U and V open in the *relative* topology on N whose union is N.

17.1.2. Example. Every discrete metric space with more than one point is disconnected.

17.1.3. Example. The set $\mathbb{Q}^2$ of points in $\mathbb{R}^2$ both of whose coordinates are rational is a disconnected subset of $\mathbb{R}^2$.

Proof. The subspace $\mathbb{Q}^2$ is disconnected by the sets $\{(x,y) \in \mathbb{Q}^2 \colon x^2 < 2\}$ and $\{(x,y) \in \mathbb{Q}^2 \colon x^2 > 2\}$. (Why are these sets open in the relative topology on $\mathbb{Q}^2$?) $\square$

17.1.4. Example. The following subset of $\mathbb{R}^2$ is not connected:
$$\{(x, x^{-1}) \colon x > 0\} \cup \{(0, y) \colon y \in \mathbb{R}\}$$

Proof. Problem.

17.1.5. Proposition. *A metric space M is disconnected if and only if there exists a continuous surjection from M onto a two-element discrete space, say $\{0, 1\}$. A* *metric space M is connected if and only if every continuous function f from M into a two-element discrete space is constant.*

Proof. Problem.

Proposition 5.1.2 remains true for arbitrary metric spaces, and the same proof works.

17.1.6. Proposition. *A metric space M is connected if and only if the only subsets of M that are both open and closed are the null set and M itself.*

Proof. Exercise.

Just as in Chapter 5, dealing with the relative topology on a subset of a metric space can sometimes be a nuisance. The remedy used there is available here: work with mutually separated sets.

17.1.7. Definition. Two nonempty subsets C and D of a metric space M are said to be MUTUALLY SEPARATED if

$$C \cap \overline{D} = \overline{C} \cap D = \emptyset \, .$$

17.1.8. Proposition. *A subset N of a metric space M is disconnected if and only if it is the union of two nonempty sets mutually separated in M.*

Proof. Exercise. *Hint.* Make a few changes in the proof of Proposition 5.1.7.

17.1.9. Proposition. *If A is a connected subset of a metric space, then any set B satisfying $A \subseteq B \subseteq \overline{A}$ is also connected.*

Proof. Problem. *Hint.* Use Proposition 17.1.8.

If a metric space is disconnected, it is often a rather simple job to demonstrate this fact. All one has to do is track down two subsets that disconnect the space. If the space is connected, however, one is confronted with the unenviable task of showing that *every* pair of subsets fails for some reason to disconnect the space. How, for example, does one go about showing that the unit square $[0, 1] \times [0, 1]$ is connected? Or the unit circle? Or the curve $y = \sin x$? In Theorem 17.1.10, Proposition 17.1.13, and Proposition 17.2.2 we give sufficient conditions for a metric space to be connected. The first of these is that the space be the continuous image of a connected space.

17.1.10. Theorem. *A metric space N is connected if there exist a connected metric space M and a continuous surjection from M onto N.*

Proof. In the hint to the proof of Theorem 5.2.1, change the first "$\mathbb{R}$" to "N" and the second to "M". Then try the same proof.   □

17.1.11. Example. The graph of the curve $y = \sin x$ is a connected subset of $\mathbb{R}^2$.

Proof. Exercise.

17.1.12. Example. The unit circle $\{(x, y) \in \mathbb{R}^2 : x^2 + y^2 = 1\}$ is a connected subset of the plane.

Proof. Problem.

17.1.13. Proposition. *A metric space is connected if it is the union of a family of connected subsets with nonempty intersection.*

Proof. Exercise. *Hint.* Argue by contradiction. Use Definition 17.1.1.

17.1.14. Problem. Use Proposition 17.1.5 to give a second proof of Proposition 17.1.13.

17.1.15. Example. The closed unit square $[0,1] \times [0,1]$ in $\mathbb{R}^2$ is connected.

Proof. Exercise.

17.2. Arcwise connected spaces

A concept closely related to (but stronger than) connectedness is *arcwise connectedness*.

17.2.1. Definition. A metric space is ARCWISE CONNECTED (or PATH CONNECTED) if for every $x, y \in M$ there exists a continuous map $f \colon [0,1] \to M$ such that $f(0) = x$ and $f(1) = y$. Such a function f is an ARC (or PATH or CURVE) connecting x to y. It is very easy to prove that arcwise connected spaces are connected (Proposition 17.2.2). The converse is false (Example 17.2.7). If, however, we restrict our attention to open subsets of $\mathbb{R}^n$, then the converse does hold (Proposition 17.2.8).

17.2.2. Proposition. *If a metric space is arcwise connected, then it is connected.*

Proof. Problem.

17.2.3. Example. The following subset of $\mathbb{R}^2$ is not connected:
$$\{(x,y) \colon (x-1)^2 + (y-1)^2 < 4\} \cup \{(x,y) \colon x < 0\}$$
$$\cup \{(x,y) \colon (x-10)^2 + (y-1)^2 < 49\}.$$

Proof. Problem.

17.2.4. Example. The following subset of $\mathbb{R}^2$ is connected:
$$\{(x,y) \colon (x-1)^2 + (y-1)^2 < 4\} \cup \{(x,y) \colon y < 0\}$$
$$\cup \{(x,y) \colon (x-10)^2 + (y-1)^2 < 49\}.$$

Proof. Problem.

17.2.5. Example. The following subset of $\mathbb{R}^2$ is connected:
$$\{(x, x^3 + 2x) \colon x \in \mathbb{R}\} \cup \{(x, x^2 + 56) \colon x \in \mathbb{R}\}.$$

Proof. Problem.

17.2.6. Example. Every open ball in $\mathbb{R}^n$ is connected. So is every closed ball.

Proof. Problem.

17.2.7. Example. Let $B = \{(x, \sin x^{-1}) \colon 0 < x \le 1\}$. Then $\overline{B}$ is a connected subset of $\mathbb{R}^2$ but is not arcwise connected.

Proof. Exercise. *Hint.* Let $M = \overline{B}$. To show that M is not arcwise connected, argue by contradiction. Assume there exists a continuous function $f \colon [0,1] \to M$ such that $f(1) \in B$ and $f(0) \notin B$. Prove that $f^2 = \pi_2 \circ f$ is discontinuous at the point $t_0 = \sup f^{\leftarrow}(M \setminus B)$. To this end show that $t_0 \in f^{\leftarrow}(M \setminus B)$. Then, given $\delta > 0$, choose t_1 in $[0,1]$ so that $t_0 < t_1 < t_0 + \delta$. Without loss of generality one may suppose that $f^2(t_0) \leq 0$. Show that $\left(f^1\right)^{\to}[t_0, t_1]$ is an interval containing 0 and $f^1(t_1)$ (where $f^1 = \pi_1 \circ f$). Find a point t in $[t_0, t_1]$ such that $0 < f^1(t) < f^1(t_1)$ and $f^2(t) = 1$.

17.2.8. Proposition. *Every connected open subset of $\mathbb{R}^n$ is arcwise connected.*

Proof. Exercise. *Hint.* Let A be a connected open subset of $\mathbb{R}^n$, and let $a \in A$. Let U be the set of all points in A that can be joined to a by an arc in A. Show that $A \setminus U$ is empty by showing that U and $A \setminus U$ disconnect A.

17.2.9. Problem. Does there exist a continuous bijection from a closed disk in $\mathbb{R}^2$ to its circumference? Does there exist a continuous bijection from the interval $[0,1]$ to the circumference of a disk in $\mathbb{R}^2$? *Hint.* For the first question suppose that there is such a bijection and consider what will happen if you remove two points from the domain of the function. Try something similar for the second question.

17.2.10. Problem. Let x be a point in a metric space M. Define the COMPONENT of M containing x to be the largest connected subset of M that contains x. Discover as much about components of metric spaces as you can. First, of course, you must make sure that the definition just given makes sense. (How do we know that there really *is* a "largest" connected set containing x?)

Here are some more things to think about.

(a) The components of a metric space are a disjoint family whose union is the whole space.

(b) It is fairly clear that the components of a discrete metric space are the points of the space. If the components are points, must the space be homeomorphic to a discrete metric space?

(c) Components of a metric space are closed sets; must they be open?

(d) Distinct components of a metric space are mutually separated.

(e) If a metric space M is the union of two mutually separated sets C and D and if points x and y belong to the same component of M, then both points are in C or both are in D. What about the converse? Suppose x and y are points in M such that whenever M is written as the union of two mutually separated sets C and D, both points lie in C or both lie in D. Must x and y lie in the same component?

17.2.11. Problem. A function f in $\mathcal{C}(M, \mathbb{R})$, where M is a metric space, is IDEM-POTENT if $(f(x))^2 = f(x)$ for all $x \in M$. The constant functions 0 and 1 are the TRIVIAL idempotents of $\mathcal{C}(M, \mathbb{R})$. Show that $\mathcal{C}(M, \mathbb{R})$ possesses a nontrivial idempotent if and only if the underlying metric space is disconnected. (This is one of a large number of results that link algebraic properties of $\mathcal{C}(M, \mathbb{R})$ to topological properties of the underlying space M.)

Chapter 18

Complete spaces

18.1. Cauchy sequences

18.1.1. Definition. A sequence (x_n) in a metric space is a CAUCHY SEQUENCE if for every $\epsilon > 0$ there exists $n_0 \in \mathbb{N}$ such that $d(x_m, x_n) < \epsilon$ whenever m, $n \geq n_0$. This condition is often abbreviated as $d(x_m, x_n) \to 0$ as m, $n \to \infty$ (or $\lim_{m,n\to\infty} d(x_m, x_n) = 0$).

18.1.2. Example. In the metric space $\mathbb{R}$ the sequence $(1/n)$ is Cauchy.

Proof. Given $\epsilon > 0$, choose $n_0 > 2/\epsilon$. If m, $n > n_0$, then $d(1/m, 1/n) = |(1/m) - (1/n)| \leq (1/m) + (1/n) \leq 2/n_0 < \epsilon$. Notice that in $\mathbb{R}$ this sequence is also convergent. □

18.1.3. Example. In the metric space $\mathbb{R} \setminus \{0\}$ the sequence $(1/n)$ is Cauchy. (The proof is exactly the same as in the preceding example.) Notice, however, that this sequence does not converge in $\mathbb{R} \setminus \{0\}$.

18.1.4. Proposition. *In a metric space every convergent sequence is Cauchy.*

Proof. Exercise.

18.1.5. Proposition. *Every Cauchy sequence that has a convergent subsequence is itself convergent (and to the same limit as the subsequence).*

Proof. Exercise.

18.1.6. Proposition. *Every Cauchy sequence is bounded.*

Proof. Exercise.

Although every convergent sequence is Cauchy (Proposition 18.1.4), the converse need not be true (Example 18.1.3). Those spaces for which the converse is true are said to be *complete*.

18.2. Completeness

18.2.1. Definition. A metric space M is COMPLETE if every Cauchy sequence in M converges to a point of M.

18.2.2. Example. The metric space $\mathbb{R}$ is complete.

Proof. Let (x_n) be a Cauchy sequence in $\mathbb{R}$. By Proposition 18.1.6 the sequence (x_n) is bounded; by Corollary 4.4.4 it has a convergent subsequence; and so by Proposition 18.1.5 it converges. $\qquad\square$

18.2.3. Example. If (x_n) and (y_n) are Cauchy sequences in a metric space, then $\bigl(d(x_n, y_n)\bigr)_{n=1}^{\infty}$ is a Cauchy sequence in $\mathbb{R}$.

Proof. Problem. *Hint.* Proposition 9.2.17.

18.2.4. Example. The set $\mathbb{Q}$ of rational numbers (regarded as a subspace of $\mathbb{R}$) is not complete.

Proof. Problem.

18.2.5. Proposition. *Every compact metric space is complete.*

Proof. Problem. *Hint.* Theorem 16.2.1 and Proposition 18.1.5.

18.2.6. Problem. Let M be a metric space with the discrete metric.

(a) Which sequences in M are Cauchy?

(b) Show that M is complete.

18.2.7. Problem. Show that completeness is not a topological property.

18.2.8. Proposition. *Let M be a complete metric space, and let M_0 be a subspace of M. Then M_0 is complete if and only if it is a closed subset of M.*

Proof. Problem.

18.2.9. Proposition. *The product of two complete metric spaces is complete.*

Proof. Exercise.

18.2.10. Proposition. *If d and ρ are equivalent metrics on a set M, then the space (M, d) is complete if and only if (M, ρ) is.*

Proof. Exercise.

18.2.11. Example. With its usual metric the space $\mathbb{R}^n$ is complete.

Proof. Since $\mathbb{R}$ is complete (Example 18.2.2), Proposition 18.2.9 and induction show that $\mathbb{R}^n$ is complete under the metric d_1 (defined in Example 9.2.10). Since the usual metric is equivalent to d_1, we may conclude from Proposition 18.2.10 that $\mathbb{R}^n$ is complete under its usual metric. $\qquad\square$

Here is one more example of a complete metric space.

18.2.12. Example. If S is a set, then the metric space $\mathcal{B}(S, \mathbb{R})$ is complete.

Proof. Exercise.

18.2.13. Example. If M is a compact metric space, then $\mathcal{C}(M, \mathbb{R})$ is a complete metric space.

Proof. Problem.

18.2.14. Problem. Give examples of metric spaces M and N, a homeomorphism $f \colon M \to N$, and a Cauchy sequence (x_n) in M such that the sequence $\big(f(x_n)\big)$ is *not* Cauchy in N.

18.2.15. Problem. Show that if D is a dense subset of a metric space M and every Cauchy sequence in D converges to a point of M, then M is complete.

18.3. Completeness vs. compactness

In Proposition 18.2.5 we saw that every compact metric space is complete. The converse of this is not true without additional assumptions. (Think of the reals.) In the remainder of this chapter we show that adding total boundedness to completeness will suffice. For the next problem we require the notion of the "diameter" of a set in a metric space.

18.3.1. Definition. The DIAMETER of a subset A of a metric space is defined by

$$\operatorname{diam} A := \sup\{d(x, y) \colon x, y \in A\}$$

if this supremum exists. Otherwise $\operatorname{diam} A := \infty$.

18.3.2. Problem. Show that $\operatorname{diam} A = \operatorname{diam} \overline{A}$ for every subset A of a metric space.

18.3.3. Proposition (Cantor intersection theorem, Version II). *In a metric space M the following are equivalent:*

(a) *M is complete.*

(b) *Every nested sequence of nonempty closed sets in M whose diameters approach 0 has nonempty intersection.*

Proof. Problem. *Hint.* For the definition of "nested" see Definition 4.4.5. To show that (a) implies (b), let (F_k) be a nested sequence of nonempty closed subsets of M. For each k choose $x_k \in F_k$. Show that the sequence (x_k) is Cauchy. To show that (b) implies (a), let (x_k) be a Cauchy sequence in M. Define $A_n = \{x_k \colon k \geq n\}$ and $F_n = \overline{A_n}$. Show that (F_n) is a nested sequence of closed sets whose diameters approach 0 (see the preceding problem). Choose a point a in $\bigcap F_n$. Find a subsequence (x_{n_k}) of (x_n) such that $d(a, x_{n_k}) < 2^{-k}$.

18.3.4. Problem. The preceding proposition tells us that in a complete metric space every nested sequence of nonempty closed subsets of $\mathbb{R}$ whose diameters approach 0 has nonempty intersection.

(a) Show that this assertion is no longer correct if the phrase "whose diameters approach 0" is simply deleted.

(b) Show that the assertion is no longer correct if the word "closed" is deleted.

18.3.5. Proposition. *In a totally bounded metric space every sequence has a Cauchy subsequence.*

Proof. Problem. *Hint.* Let (x_n) be a sequence in a totally bounded metric space M. For every $n \in \mathbb{N}$ the space M can be covered by a finite collection of open balls of radius $1/n$. Thus, in particular, there is an open ball of radius 1 that contains infinitely many of the terms of the sequence (x_n). Show that it is possible inductively to choose subsets $N_1, N_2, \ldots$ of $\mathbb{N}$ such that for every $m, n \in \mathbb{N}$

(i) $n > m$ implies $N_n \subseteq N_m$,

(ii) N_n is infinite, and

(iii) $\{x_k : k \in N_n\}$ is contained in some open ball of radius $1/n$.

Then show that we may choose (again, inductively) $n_1, n_2, \ldots$ in $\mathbb{N}$ such that for every $j, k \in \mathbb{N}$

(iv) $k > j$ implies $n_k > n_j$, and

(v) $n_k \in N_k$.

Finally, show that the sequence (x_n) is Cauchy.

18.3.6. Proposition. *A metric space is compact if and only if it is complete and totally bounded.*

Proof. Problem.

18.3.7. Problem. Let (x_n) be a sequence of real numbers with the property that each term of the sequence (from the third term on) is the average of the two preceding terms. Show that the sequence converges and find its limit. *Hint.* Proceed as follows.

(a) Compute the distance between x_{n+1} and x_n in terms of the distance between x_n and x_{n-1}.

(b) Show (inductively) that
$$|x_{n+1} - x_n| = 2^{1-n}|x_2 - x_1|.$$

(c) Prove that (x_n) has a limit by showing that for $m < n$
$$|x_n - x_m| \le 2^{2-m}|x_2 - x_1|.$$

(d) Show (again inductively) that $2x_{n+1} + x_n = 2x_2 + x_1$.

18.3.8. Problem. Show that if (x_n) is a sequence lying in the interval $[-1, 1]$ that satisfies
$$|x_{n+1} - x_n| \le \tfrac{1}{4}|x_n^2 - x_{n-1}^2| \qquad \text{for } n \ge 2,$$
then (x_n) converges.

Chapter 19

A fixed point theorem

We now have enough information about metric spaces to consider an interesting application. We will prove a result known as the *contractive mapping theorem* and then use it to find solutions to some integral equations. Since this chapter contains mostly examples, we will make liberal use of computations from beginning calculus. Although it would perhaps be more logical to defer these matters until we have developed the necessary facts concerning integrals, derivatives, and power series, there is nevertheless much to be said for presenting a nontrivial application relatively early.

19.1. The contractive mapping theorem

19.1.1. Definition. A mapping $f \colon M \to N$ between metric spaces is CONTRACTIVE if there exists a constant c such that $0 < c < 1$ and

$$d(f(x), f(y)) \leq c \, d(x, y)$$

for all x, $y \in M$. Such a number c is a CONTRACTION CONSTANT for f. A contractive map is also called a CONTRACTION.

19.1.2. Exercise. Show that every contractive map is continuous.

19.1.3. Example. The map $f \colon \mathbb{R}^2 \to \mathbb{R}^3$, defined by

$$f(x, y) = \left(1 - \tfrac{1}{3}x, 1 + \tfrac{1}{3}y, 2 + \tfrac{1}{3}x - \tfrac{1}{3}y\right),$$

is a contraction.

Proof. Exercise.

19.1.4. Example. The map

$$f \colon \mathbb{R}^2 \to \mathbb{R}^2 : (x, y) \mapsto \left(\tfrac{1}{2}(1 + y), \tfrac{1}{2}(3 - x)\right)$$

is a contraction on $\mathbb{R}^2$, where $\mathbb{R}^2$ has its usual (Euclidean) metric.

Proof. Problem.

The next theorem is the basis for a number of interesting applications. Also it will turn out to be a crucial ingredient of the extremely important inverse function theorem (in Chapter 29). Although the statement of Theorem 19.1.5 is important in applications, its proof is even more so. The theorem guarantees the existence (and uniqueness) of solutions to certain kinds of equations; its proof allows us to approximate these solutions as closely as our computational machinery permits. Recall from Chapter 5 that a FIXED POINT of a mapping f from a set S into itself is a point $p \in S$ such that $f(p) = p$.

19.1.5. Theorem (Contractive mapping theorem). *Every contraction from a complete metric space into itself has a unique fixed point.*

Proof. *Exercise. Hint.* Let M be a complete metric space, and let $f\colon M \to M$ be a contractive mapping with contractive constant c. Start with an arbitrary point x_0 in M. Obtain a sequence $(x_n)_{n=0}^{\infty}$ of points in M by letting $x_1 = f(x_0)$, $x_2 = f(x_1)$, and so on. Show that whenever $m < n$,

$$(15) \qquad d(x_m, x_n) \le d(x_0, x_1) \frac{c^m}{1-c}$$

from which it follows that the sequence (x_n) is Cauchy.

19.1.6. Example. We use the contractive mapping theorem to solve the system of equations

$$(16) \qquad \begin{cases} 9x - 2y = 7, \\ 3x + 8y = 11 \,. \end{cases}$$

Define

$$S\colon \mathbb{R}^2 \to \mathbb{R}^2\colon (x,y) \mapsto (9x - 2y, 3x + 8y)\,.$$

The system (16) may be written as a single equation

$$S(x,y) = (7,11)\,,$$

or equivalently as

$$(17) \qquad (x,y) - S(x,y) + (7,11) = (x,y)\,.$$

(*Definition.* Addition and subtraction on $\mathbb{R}^2$ are defined coordinatewise. That is, if (x,y) and (u,v) are points in $\mathbb{R}^2$, then $(x,y) + (u,v) := (x+u, y+v)$ and $(x,y) - (u,v) := (x - u, y - v)$. Similar definitions hold for $\mathbb{R}^n$ with $n > 2$.) Let $T(x,y)$ be the left-hand side of (17); that is, define

$$T\colon \mathbb{R}^2 \to \mathbb{R}^2\colon (x,y) \mapsto (x,y) - S(x,y) + (7,11)\,.$$

With this notation (17) becomes

$$T(x,y) = (x,y)\,.$$

Thus, and this is the crucial point, to solve (16) we need only find a fixed point of the mapping T. If T is contractive, then the preceding theorem guarantees that T has a unique fixed point and, therefore, that the system of equations (16) has a unique solution.

Unfortunately, as things stand, T is *not* contractive with respect to the product metric on $\mathbb{R}^2$. (It is for convenience that we use the product metric d_1 on $\mathbb{R}^2$ rather than the usual Euclidean metric. Square roots are a nuisance.) To see that T

is not contractive, notice that $d_1\big((1,0),(0,0)\big) = 1$ whereas $d_1\big(T(1,0),T(0,0)\big) = d_1\big((-1,8),(7,11)\big) = 11$. All is not lost however. One simple-minded remedy is to divide everything in (16) by a large constant c. A little experimentation shows that $c = 10$ works. Instead of working with the system (16) of equations, consider the system

$$(18) \qquad \begin{cases} 0.9x - 0.2y = 0.7, \\ 0.3x + 0.8y = 1.1, \end{cases}$$

which obviously has the same solutions as (16). Redefine S and T in the obvious fashion. Let

$$S\colon \mathbb{R}^2 \to \mathbb{R}^2\colon (x,y) \mapsto (0.9x - 0.2y, 0.3x + 0.8y)$$

and

$$T\colon \mathbb{R}^2 \to \mathbb{R}^2\colon (x,y) \mapsto (x,y) - S(x,y) + (0.7, 1.1)\,.$$

Thus redefined, T is contractive with respect to the product metric. *Proof.* Since

$$(19) \qquad T(x,y) = (0.1x + 0.2y + 0.7, -0.3x + 0.2y + 1.1)$$

for all $(x,y) \in \mathbb{R}^2$, we see that

$$\begin{aligned}
d_1\big(T(x,y), T(u,v)\big) &= 10^{-1}\big(|(x+2y)-(u+2v)| + |(-3x+2y)-(-3u+2v)|\big) \\
&\leq 10^{-1}\big(|x-u| + 2|y-v| + 3|x-u| + 2|y-v|\big) \\
&= 0.4\,(|x-u| + |y-v|) \\
(20) \qquad &= 0.4\,d_1\big((x,y),(u,v)\big)
\end{aligned}$$

for all points (x,y) and (u,v) in $\mathbb{R}^2$.

Now since T is contractive and $\mathbb{R}^2$ is complete (with respect to the product metric—see Proposition 18.2.9), the contractive mapping theorem (Theorem 19.1.5) tells us that T has a unique fixed point. But a fixed point of T is a solution for the system (18) and consequently for (16).

The construction used in the proof of Theorem 19.1.5 allows us to approximate the fixed point of T to any desired degree of accuracy. As in that proof choose x_0 to be any point whatever in $\mathbb{R}^2$. Then the points $x_0, x_1, x_2, \ldots$ (where $x_n = T\big(x_{n-1}\big)$ for each n) converge to the fixed point of T. This is a technique of SUCCESSIVE APPROXIMATION.

For the present example let $x_0 = (0,0)$. (The origin is chosen just for convenience.) Now use (19) and compute.

$$\begin{aligned}
x_0 &= (0,0) \\
x_1 &= T(0,0) = (0.7, 1.1) \\
x_2 &= T(x_1) = (1.021, 1.025) \\
x_3 &= T(x_2) = (1.0071, 0.9987)
\end{aligned}$$

$$\vdots$$

It is reasonable to conjecture that the system (16) has a solution consisting of rational numbers and then to guess that the points $x_0, x_1, x_2, \ldots$, as computed above, are converging to the point $(1,1)$ in $\mathbb{R}^2$. Putting $x = 1$ and $y = 1$ in (19), we see that the point $(1,1)$ is indeed the fixed point of T and therefore the solution to (16).

I hope it is not necessary to point out that the preceding example is not offered as a useful application of the contractive mapping theorem, or as a suggestion that it provides a reasonable way to solve simple algebraic equations. Its intended purpose is to display, as clearly as possible, the workings of the theorem in a fashion that is in no way compromised by complexities that may arise from the problem itself. In it we achieved an exact solution to a system of equations. In general, of course, we cannot hope that any successive approximation technique will yield exact answers. In those cases in which it does not, it is most important to have some idea how accurate our approximations are. After n iterations, how close to the true solution are we? How many iterations must be computed in order to achieve a desired degree of accuracy? The answer to these questions in an easy consequence of the proof of Theorem 19.1.5.

19.1.7. Corollary. *Let the space M, the mapping f, the constant c, and the sequence (x_n) be as in the hint for the proof of Theorem 19.1.5. Also let p be the limit of the sequence (x_n). Then for every $m \geq 0$,*

$$d(x_m, p) \leq d(x_0, x_1) \frac{c^m}{(1-c)} \, .$$

Proof. Take the limit of inequality (15) in the hint for the proof of Theorem 19.1.5 as $n \to \infty$.   □

19.1.8. Definition. Use the notation in the preceding corollary. If we think of the point x_n as being the nth approximation to p, then the distance $d(x_n, p)$ between x_n and p is the ERROR associated with the nth approximation.

Notice that because the product metric d_1 was chosen for $\mathbb{R}^2$ in Example 19.1.6, the word "error" there means the *sum* of the errors in x and y. Had we wished for "error" to mean the *maximum* of the errors in x and y, we would have used the uniform metric d_u on $\mathbb{R}^2$. Similarly, if a *root-mean-square* "error" were desired (that is, the square root of the sum of the squares of the errors in x and y), then we would have used the usual Euclidean metric on $\mathbb{R}^2$.

19.1.9. Exercise. Let (x_n) be the sequence of points in $\mathbb{R}^2$ considered in Example 19.1.6. We showed that (x_n) converges to the point $p = (1,1)$.

(a) Use Corollary 19.1.7 to find an upper bound for the error associated with the approximation x_4.

(b) What is the actual error associated with x_4?

(c) According to Corollary 19.1.7, how many terms of the sequence (x_n) should we compute to be sure of obtaining an approximation that is correct to within 10^{-4}?

19.1.10. Problem. Show by example that the conclusion of the contractive mapping theorem fails if

(a) the contraction constant is allowed to have the value 1; or

(b) the space is not complete.

19.1.11. Problem. Show that the map

$$g \colon [0, \infty) \to [0, \infty) \colon x \mapsto \frac{1}{x+1}$$

is not a contraction even though

$$d\big(g(x), g(y)\big) < d(x, y)$$

for all x, $y \geq 0$ with $x \neq y$.

19.1.12. Problem. Let $f(x) = (x/2) + (1/x)$ for $x \geq 1$.

(a) Show that f maps the interval $[1, \infty)$ into itself.

(b) Show that f is contractive.

(c) Let $x_0 = 1$, and for $n \geq 0$ let

$$x_{n+1} = \frac{x_n}{2} + \frac{1}{x_n}.$$

Show that the sequence (x_n) converges.

(d) Find $\lim_{n \to \infty} x_n$.

(e) Show that the distance between x_n and the limit found in (d) is no greater than 2^{-n}.

19.1.13. Problem. Solve the system of equations

$$\begin{aligned}
9x \quad -y \;\; +2z &= 37, \\
x+10y \;\; -3z &= -69, \\
-2x \;\; +3y+11z &= 58,
\end{aligned}$$

following the procedure of Example 19.1.6. *Hint.* As in Example 19.1.6 divide by 10 to obtain a contractive mapping. Before guessing at a rational solution, compute 10 or 11 successive approximations. Since this involves a lot of arithmetic, it will be helpful to have some computational assistance—a programmable calculator, for example.

19.1.14. Problem. Consider the following system of equations:

$$\begin{aligned}
75x+16y-20z &= 40, \\
33x+80y+30z &= -48, \\
-27x+32y+80z &= 36.
\end{aligned}$$

(a) Solve the system following the method of Example 19.1.6. *Hint.* Because the contraction constant is close to 1, the approximations converge slowly. It may take 20 or 30 iterations before it is clear what the exact (rational) solutions should be. So as in the preceding example, it will be desirable to use computational assistance.

(b) Let (x_n) be the sequence of approximations in $\mathbb{R}^3$ converging to the solution of the system in (a). Use Corollary 19.1.7 to find an upper bound for the error associated with the approximation x_{25}.

(c) To four decimal places, what is the actual error associated with x_{25}?

(d) According to Corollary 19.1.7, how many terms must be computed to be sure that the error in our approximation is no greater than 10^{-3}?

19.1.15. Problem. Let $f : \mathbb{R}^2 \to \mathbb{R}^2$ be the rotation of the plane about the point $(0, 1)$ through an angle of π radians. Let $g \colon \mathbb{R}^2 \to \mathbb{R}^2$ be the map that takes the point (x, y) to the midpoint of the line segment connecting (x, y) and $(1, 0)$.

(a) Prove that $g \circ f$ is a contraction on $\mathbb{R}^2$ (with its usual metric).

(b) Find the unique fixed point of $g \circ f$.

(c) Let $(x_0, y_0) = (0, 1)$, and define (as in the proof of Theorem 19.1.5)

$$(x_{n+1}, y_{n+1}) = (g \circ f)(x_n, y_n)$$

for all $n \geq 0$. For each n compute the exact Euclidean distance between (x_n, y_n) and the fixed point of $g \circ f$.

19.2. Application to integral equations

19.2.1. Exercise. Use Theorem 19.1.5 to solve the integral equation

$$(21) \qquad\qquad f(x) = \tfrac{1}{3}x^3 + \int_0^x t^2 f(t)\, dt\,.$$

Hint. We wish to find a continuous function f that satisfies (21) for all $x \in \mathbb{R}$. Consider the mapping T that takes each continuous function f into the function Tf whose value at x is given by

$$Tf(x) = \tfrac{1}{3}x^3 + \int_0^x t^2 f(t)\, dt\,.$$

(It is important to keep in mind that T acts on functions, not on numbers. Thus $Tf(x)$ is to be interpreted as $(T(f))(x)$ and not $T(f(x))$.) In order to make use of Theorem 19.1.5, the mapping T must be contractive. One way to achieve this is to restrict our attention to continuous functions on the interval $[0, 1]$ and to use the uniform metric on $\mathcal{C}([0, 1], \mathbb{R})$. Once a continuous function f is found such that (21) is satisfied for all x in $[0, 1]$, it is a simple matter to check whether (21) holds for all x in $\mathbb{R}$. Consider then the map

$$T \colon \mathcal{C}([0, 1], \mathbb{R}) \to \mathcal{C}([0, 1], \mathbb{R}) \colon f \mapsto Tf,$$

where

$$Tf(x) := \tfrac{1}{3}x^3 + \int_0^x t^2 f(t)\, dt$$

for all $x \in [0, 1]$. The space $\mathcal{C}([0, 1], \mathbb{R})$ is complete. (Why?) Show that T is contractive by estimating $|Tf(x) - Tg(x)|$, where f and g are continuous functions on $[0, 1]$, and by taking the supremum over all x in $[0, 1]$. What can be concluded from Theorem 19.1.5 about (21)?

To actually find a solution to (21), use the proof of Theorem 19.1.5 (that is, successive approximations). For simplicity start with the zero function in $\mathcal{C}([0, 1], \mathbb{R})$: let $g_0(x) = 0$ for $0 \leq x \leq 1$. For $n \geq 0$, let $g_{n+1}(x) = Tg_n(x)$ for $0 \leq x \leq 1$. Compute g_1, g_2, g_3, and g_4. You should be able to guess what g_n will be. (It is easy to verify the correctness of your guess by induction, but it is not necessary to do this.) Next, let f be the function that is the uniform limit of the sequence (g_n).

That is, f is the function whose power series expansion has g_n as its nth partial sum. This power series expansion should be one with which you are familiar from beginning calculus; what elementary function does it represent?

Finally, show by direct computation that this elementary function does in fact satisfy (21) for all x in $\mathbb{R}$.

19.2.2. Problem. Give a careful proof that there exists a unique continuous real valued function f on $[0, 1]$ that satisfies the integral equation

$$f(x) = x^2 + \int_0^x t^2 f(t)\, dt\,.$$

(You are not asked to *find* the solution.)

19.2.3. Problem. Use Theorem 19.1.5 to solve the integral equation

$$f(x) = x + \int_0^x f(t)\, dt\,.$$

Hint. Follow the procedure of Exercise 19.2.1. Keep in mind that the only reason for choosing the particular interval $[0, 1]$ in Exercise 19.2.1 was to make the map T contractive.

19.2.4. Problem. For every f in $\mathcal{C}([0, \pi/4], \mathbb{R})$ define

$$Tf(x) = x^2 - 2 - \int_0^x f(t)\, dt,$$

where $0 \le x \le \pi/4$. Show that T is a contraction. Find the fixed point of T. What integral equation have you solved?

Chapter 20

Vector spaces

Most introductory calculus texts, for pedagogical reasons, do calculus twice: once for a single variable, and then for either two or three variables, leaving the general finite-dimensional and infinite-dimensional cases for future courses. It is our goal eventually (in Chapter 25) to develop differential calculus in a manner that is valid for any number of variables (even infinitely many).

A certain amount of algebra always underlies analysis. Before one studies the calculus of a single variable, a knowledge of arithmetic in $\mathbb{R}$ is required. For the calculus of a finite number of variables, it is necessary to know something about $\mathbb{R}^n$. In this chapter and the next we lay the algebraic foundations for the differential calculus of an arbitrary number of variables; we study vector spaces and the operation preserving maps between them, called linear transformations.

20.1. Definitions and examples

20.1.1. Definition. A (REAL) VECTOR SPACE is a set V together with a binary operation $(x, y) \mapsto x + y$ (called ADDITION) from $V \times V$ into V and a mapping $(\alpha, x) \mapsto \alpha x$ (called SCALAR MULTIPLICATION) from $\mathbb{R} \times V$ into V satisfying the following conditions.

(1) Addition is associative. That is,

$$x + (y + z) = (x + y) + z \qquad \text{for all } x,\, y,\, z \in V \,.$$

(2) In V there is an element $\mathbf{0}$ (called the ZERO VECTOR) such that

$$x + \mathbf{0} = x \qquad \text{for all } x \in V \,.$$

(3) For each x in V there is a corresponding element $-x$ (the ADDITIVE INVERSE of x) such that

$$x + (-x) = \mathbf{0} \,.$$

(4) Addition is commutative. That is,

$$x + y = y + x \qquad \text{for all } x,\, y \in V \,.$$

(5) If $\alpha \in \mathbb{R}$ and x, $y \in V$, then
$$\alpha(x + y) = (\alpha x) + (\alpha y).$$
(6) If α, $\beta \in \mathbb{R}$ and $x \in V$, then
$$(\alpha + \beta)x = (\alpha x) + (\beta x).$$
(7) If α, $\beta \in \mathbb{R}$ and $x \in V$, then
$$\alpha(\beta x) = (\alpha \beta)x.$$
(8) If $x \in V$, then
$$1 \cdot x = x.$$

An element of V is a VECTOR; an element of $\mathbb{R}$ is, in this context, often called a SCALAR. Concerning the order of performing operations, we agree that scalar multiplication takes precedence over addition. Thus, for example, condition (5) above may be unambiguously written as
$$\alpha(x + y) = \alpha x + \alpha y.$$
(Notice that the parentheses on the left-hand side may not be omitted.)

If x and y are vectors, we define $x - y$ to be $x + (-y)$. If A and B are subsets of a vector space, we define
$$A + B := \{a + b \colon a \in A, b \in B\};$$
and if $\alpha \in \mathbb{R}$,
$$\alpha A := \{\alpha a \colon a \in A\}.$$

Condition (3) above is somewhat optimistic. No uniqueness is asserted in (2) for the zero vector $\mathbf{0}$, so one may well wonder whether (3) is supposed to hold for *some* zero vector $\mathbf{0}$ or for *all* such vectors. Fortunately, the problem evaporates since we can easily show that the zero vector is in fact unique.

20.1.2. Exercise. A vector space has exactly one zero vector. That is, if $\mathbf{0}$ and $\mathbf{0}'$ are members of a vector space V that satisfy $x + \mathbf{0} = x$ and $x + \mathbf{0}' = x$ for all $x \in V$, then $\mathbf{0} = \mathbf{0}'$.

In a vector space, not only is the zero vector unique but so are additive inverses.

20.1.3. Problem. For every vector x in a vector space V there exists only one vector $-x$ such that
$$x + (-x) = \mathbf{0}.$$

In Exercises 20.1.4–20.1.6 and Problem 20.1.7 we state four useful, if elementary, facts concerning the arithmetic of vectors.

20.1.4. Exercise. If x is a vector (in some vector space) and $x + x = x$, then $x = \mathbf{0}$.

20.1.5. Exercise. Let x be a vector (in a some vector space), and let α be a real number. Then $\alpha x = \mathbf{0}$ if and only if $x = \mathbf{0}$ or $\alpha = 0$. *Hint.* Show three things:

(a) $\alpha \mathbf{0} = \mathbf{0}$,

(b) $0\, x = \mathbf{0}$, and

(c) If $\alpha \neq 0$ and $\alpha x = \mathbf{0}$, then $x = \mathbf{0}$.

(If it is not clear to you that proving (a), (b), and (c) is the same thing as proving Exercise 20.1.5, see the remark following this hint.) To prove (a), write $\mathbf{0} + \mathbf{0} = \mathbf{0}$, multiply by α, and use Exercise 20.1.4. For (c) use the fact that if $\alpha \in \mathbb{R}$ is not zero, it has a reciprocal. What happens if we multiply the vector αx by the scalar $1/\alpha$?

Remark. It should be clear that proving (a) and (b) of the preceding hint proves that

$$\text{if } x = \mathbf{0} \text{ or } \alpha = 0, \text{ then } \alpha x = \mathbf{0}.$$

What may not be clear is that proving (c) is enough to establish

(22) $\qquad\qquad$ if $\alpha x = \mathbf{0}$, then either $x = \mathbf{0}$ or $\alpha = 0$.

Some students feel that in addition to proving (c) it is also necessary to prove that

$$\text{if } x \neq \mathbf{0} \text{ and } \alpha x = \mathbf{0}, \text{ then } \alpha = 0.$$

To see that this is unnecessary, recognize that there are just two possible cases: either α is equal to zero, or it is not. In case α *is* equal to zero, then the conclusion of (22) is certainly true. The other case, where α is *not* zero, is dealt with by (c).

20.1.6. Exercise. If x is a vector, then $-(-x) = x$. *Hint.* Show that $(-x)+x = \mathbf{0}$. What does Problem 20.1.3 say about x?

20.1.7. Problem. If x is a vector, then $(-1)x = -x$. *Hint.* Show that $x+(-1)x = \mathbf{0}$. Use Problem 20.1.3.

20.1.8. Problem. Using nothing about vector spaces other than the definitions, prove that if x is a vector, then $3x - x = x + x$. Write your proof using at most one vector space axiom (or definition) at each step.

We now give some examples of vector spaces.

20.1.9. Example. Let $V = \mathbb{R}^n$. We make a standard notational convention. If x belongs to $\mathbb{R}^n$, then x is an n-tuple whose coordinates are $x_1, x_2, \ldots, x_n$; that is,

$$x = (x_1, x_2, \ldots, x_n).$$

It must be confessed that we do not *always* use this convention. For example, the temptation to denote a member of $\mathbb{R}^3$ by (x, y, z), rather than by (x_1, x_2, x_3), is often just too strong to resist. For n-tuples $x = (x_1, x_2, \ldots, x_n)$ and $y = (y_1, y_2, \ldots, y_n)$ in V, define

$$x + y := (x_1 + y_1, x_2 + y_2, \ldots, x_n + y_n).$$

Accordingly, we say that addition in $\mathbb{R}^n$ is defined coordinatewise. Scalar multiplication is also defined in a coordinatewise fashion. That is, if $x = (x_1, x_2, \ldots, x_n) \in V$ and $\alpha \in \mathbb{R}$, then we define

$$\alpha x := (\alpha x_1, \alpha x_2, \ldots, \alpha x_n).$$

Under these operations $\mathbb{R}^n$ becomes a vector space.

Proof. Problem. *Hint.* Just verify conditions (1)–(8) of Definition 20.1.1.

20.1.10. Problem. Let s be the family of all sequences of real numbers. Explain how to define addition and scalar multiplication on s in such a way that it becomes a vector space.

20.1.11. Example. Here is one of the ways in which we construct new vector spaces from old ones. Let V be an arbitrary vector space, and let S be a nonempty set. Let $\mathcal{F}(S, V)$ be the family of all V valued functions defined on S. That is, $\mathcal{F}(S, V)$ is the set of all functions f such that $f \colon S \to V$. We make $\mathcal{F}(S, V)$ into a vector space by defining operations in a pointwise fashion. For functions f, $g \in \mathcal{F}(S, V)$, define

$$(f + g)(x) := f(x) + g(x) \qquad \text{for all } x \in S.$$

It should be clear that the two "+" signs in the preceding equation denote operations in different spaces. The one on the left-hand side (that is being defined) represents addition in the space $\mathcal{F}(S, V)$; the one on the right-hand side is addition in V. Because we specify the value of $f + g$ at each point x by adding the values of f and g at that point, we say that we add f and g POINTWISE.

We also define scalar multiplication to be a pointwise operation. That is, if $f \in \mathcal{F}(S, V)$ and $\alpha \in \mathbb{R}$, then we define the function αf by

$$(\alpha f)(x) := \alpha(f(x)) \qquad \text{for every } x \in S.$$

Notice that according to the definitions above, both $f + g$ and αf belong to $\mathcal{F}(S, V)$. Under these pointwise operations, $\mathcal{F}(S, V)$ is a vector space. (Notice that the family of real valued functions on a set S is a special case of the preceding. Just let $V = \mathbb{R}$.)

Proof. Problem

Most of the vector spaces we encounter in the rest of the book are subspaces of $\mathcal{F}(S, V)$ for some appropriate set S and vector space V: so we now take up the topic of subspaces of vector spaces.

20.1.12. Definition. A subset W of a vector space V is a SUBSPACE of V if it is itself a vector space under the operations it inherits from V. In subsequent chapters we will regularly encounter objects that are simultaneously vector spaces and metric spaces. (One obvious example is $\mathbb{R}^n$.) We will often use the term VECTOR SUBSPACE (or LINEAR SUBSPACE) to distinguish a subspace of a vector space from a metric subspace. (*Example.* The unit circle $\{(x, y) \colon x^2 + y^2 = 1\}$ is a metric subspace of $\mathbb{R}^2$ but not a vector subspace thereof.)

Given a subset W of a vector space V, we do not actually need to check all eight vector space axioms to establish that it is a subspace of V. We need only know that W is nonempty and that it is closed under addition and scalar multiplication.

20.1.13. Proposition. *Let W be a subset of a vector space V. Then W is a subspace of V provided that*

(a) $W \neq \emptyset$,

(b) $x + y \in W$ *whenever* $x \in W$ *and* $y \in W$, *and*

(c) $\alpha x \in W$ *whenever* $x \in W$ *and* $\alpha \in \mathbb{R}$.

Proof. Exercise.

20.1.14. Example. Let S be a nonempty set. Then the family $\mathcal{B}(S, \mathbb{R})$ of all bounded real valued functions on S is a vector space because it is a subspace of $\mathcal{F}(S, \mathbb{R})$.

Proof. That it *is* a subspace of $\mathcal{F}(S, \mathbb{R})$ is clear from the preceding proposition: every constant function is bounded, so the set $\mathcal{B}(S, \mathbb{R})$ is nonempty; that it is closed under addition and scalar multiplication was proved in Proposition 13.1.2. ☐

20.1.15. Example. The x-axis (that is, $\{(x, 0, 0) \colon x \in \mathbb{R}\}$) is a subspace of $\mathbb{R}^3$. So is the xy-plane (that is, $\{(x, y, 0) \colon x, y \in \mathbb{R}\}$). In both cases it is clear that the set in question is nonempty and is closed under addition and scalar multiplication.

20.1.16. Example. Let M be a metric space. Then the set $\mathcal{C}(M, \mathbb{R})$ of all continuous real valued functions on M is a vector space.

Proof. Problem.

20.1.17. Problem. Let $a < b$ and $\mathcal{F}$ be the vector space of all real valued functions on the interval $[a, b]$. Consider the following subsets of $\mathcal{F}$.

$$\mathcal{K} = \{f \in \mathcal{F} \colon f \text{ is constant}\}$$
$$\mathcal{D} = \{f \in \mathcal{F} \colon f \text{ is differentiable}\}$$
$$\mathcal{B} = \{f \in \mathcal{F} \colon f \text{ is bounded}\}$$
$$\mathcal{P}_3 = \{f \in \mathcal{F} \colon f \text{ is a polynomial of degree 3}\}$$
$$\mathcal{Q}_3 = \{f \in \mathcal{F} \colon f \text{ is a polynomial of degree less than or equal to 3}\}$$
$$\mathcal{P} = \{f \in \mathcal{F} \colon f \text{ is a polynomial}\}$$
$$\mathcal{C} = \{f \in \mathcal{F} \colon f \text{ is continuous}\}$$

Which of these are subspaces of which? *Hint.* There is a ringer in the list.

20.1.18. Example. The family of all solutions of the differential equation

$$xy'' + y' + xy = 0$$

is a subspace of $\mathcal{C}(\mathbb{R}, \mathbb{R})$.

Proof. Problem.

Let A be a subset of a vector space V. *Question.* What is meant by the phrase "the smallest subspace of V that contains A"? *Answer.* The intersection of all the subspaces of V that contain A. It is important to realize that in order for this answer to make sense, it must be known that the intersection of the family of subspaces containing A is itself a subspace of V. This is an obvious consequence of the fact (proved below) that the intersection of any family of subspaces is itself a subspace. Once that is known, then it is clear that the intersection of all subspaces containing A is the "smallest" subspace containing A in the sense that it is contained in *every* subspace containing A. This subspace is sometimes called the subspace of V GENERATED by A.

20.1.19. Proposition. *Let $\mathfrak{S}$ be a nonempty family of subspaces of a vector space V. Then $\bigcap \mathfrak{S}$ is a subspace of V.*

Proof. Exercise. *Hint.* Use Proposition 20.1.13.

20.1.20. Example. Let V and W be vector spaces. If addition and scalar multiplication are defined on $V \times W$ by

$$(v, w) + (x, y) := (v + x, w + y)$$

and

$$\alpha(v, w) := (\alpha v, \alpha w)$$

for all v, $x \in V$, all w, $y \in W$, and all $\alpha \in \mathbb{R}$, then $V \times W$ becomes a vector space. (This is called the PRODUCT or (EXTERNAL) DIRECT SUM of V and W. It is frequently denoted by $V \oplus W$.)

Proof. Problem.

20.2. Linear combinations

20.2.1. Definition. Let V be a vector space. A LINEAR COMBINATION of a finite set $\{x_1, \ldots, x_n\}$ of vectors in V is a vector of the form $\sum_{k=1}^{n} \alpha_k x_k$ where $\alpha_1, \ldots, \alpha_n \in \mathbb{R}$. If $\alpha_1 = \alpha_2 = \cdots = \alpha_n = 0$, then the linear combination is TRIVIAL; if at least one α_k is different from zero, the linear combination is NONTRIVIAL.

CAUTION. There is a notational conflict here that may cause confusion. In the preceding definition, $\{x_1, \ldots, x_n\}$ is used to denote a list of n vectors in an arbitrary vector space. In Example 20.1.9, however, $(x_1, \ldots, x_n)$ was the notation for a single vector in $\mathbb{R}^n$. So, one may ask, what exactly does x_3 denote? The third vector in a list? The third coordinate of a vector in n-space? The best answer I can give is that *it should be clear from context.*

Even so, how does one denote, say, the fourth component of the third vector in a list of vectors in $\mathbb{R}^7$? One may find notations such as $(x_3)_4$ rather uninviting. Perhaps x_4^3 or $(x^3)_4$ looks better. In these notes I occasionally use superscripts for lists (see Example 20.2.6 below).

20.2.2. Exercise. Find a nontrivial linear combination of the following vectors in $\mathbb{R}^3$ that equals zero: $(1, 0, 0)$, $(1, 0, 1)$, $(1, 1, 1)$, and $(1, 1, 0)$.

20.2.3. Problem. Find, if possible, a nontrivial linear combination of the following vectors in $\mathbb{R}^3$ that equals zero: $(4, 1, 3)$, $(-1, 1, -7)$, and $(1, 2, -8)$.

20.2.4. Problem. Find, if possible, a nontrivial linear combination of the following vectors in $\mathbb{R}^3$ that equals zero: $(1, 2, -3)$, $(1, -1, 4)$, and $(5, 4, -1)$.

20.2.5. Problem. Find a nontrivial linear combination of the polynomials p^1, p^2, p^3, and p^4 that equal zero, where

$$p^1(x) = x + 1,$$
$$p^2(x) = x^3 - 1,$$
$$p^3(x) = 3x^3 + 2x - 1,$$
$$p^4(x) = -x^3 + x.$$

20.2.6. Example. Define vectors $e^1, \ldots, e^n$ in $\mathbb{R}^n$ by

$$e^1 := (1, 0, 0, \ldots, 0)$$

$$e^2 := (0, 1, 0, \ldots, 0)$$

$$\vdots$$

$$e^n := (0, 0, \ldots, 0, 1).$$

In other words, for $1 \leq j \leq n$ and $1 \leq k \leq n$, the kth coordinate of the vector e^j (denote it by $(e^j)_k$ or e^j_k) is 1 if $j = k$ and 0 if $j \neq k$. The vectors $e^1, \ldots, e^n$ are the STANDARD BASIS VECTORS in $\mathbb{R}^n$. (Note that the superscripts here have nothing to do with powers.) In $\mathbb{R}^3$ the three standard basis vectors are often denoted by $\mathbf{i}$, $\mathbf{j}$, and $\mathbf{k}$ rather than e^1, e^2, and e^3, respectively.

Every vector in $\mathbb{R}^n$ is a linear combination of the standard basis vectors in that space. In fact, if $x = (x_1, \ldots, x_n) \in \mathbb{R}^n$, then

$$x = \sum_{k=1}^{n} x_k e^k.$$

Proof. The proof is quite easy:

$$\begin{aligned}
x &= (x_1, x_2, \ldots, x_n) \\
&= (x_1, 0, \ldots, 0) + (0, x_2, \ldots, 0) + \cdots + (0, 0, \ldots, x_n) \\
&= x_1 e^1 + x_2 e^2 + \cdots + x_n e^n \\
&= \sum_{k=1}^{n} x_k e^k.
\end{aligned}$$ $\square$

20.2.7. Definition. A subset A (finite or not) of a vector space is LINEARLY DEPENDENT if the zero vector $\mathbf{0}$ can be written as a nontrivial linear combination of elements of A; that is, if there exist n vectors $\mathbf{x}_1, \ldots, \mathbf{x}_n \in A$ and scalars $\alpha_1, \ldots, \alpha_n$, **not all zero,** such that $\sum_{k=1}^{n} \alpha_k \mathbf{x}_k = \mathbf{0}$. A subset of a vector space is LINEARLY INDEPENDENT if it is not linearly dependent.

20.2.8. Definition. Let A be a set of vectors in a vector space V, and let M be a subspace of V which contains A. We say that A SPANS M (or that M is the SPAN of A) if every member of M can be written as a linear combination of members of A.

20.2.9. Problem. Let A be a nonempty subset of a vector space V. Show that the span of A is the smallest subspace of V containing A. (See the paragraph preceding Proposition 20.1.19.)

20.2.10. Problem. Let $e^1, e^2, \ldots, e^n$ be the standard basis vectors in $\mathbb{R}^n$ (see Example 20.2.6).

(a) Show that the set of standard basis vectors in $\mathbb{R}^n$ is a linearly independent set.

(b) Show that the standard basis vectors span $\mathbb{R}^n$.

(c) Show that in part (b) the representation of a vector in $\mathbb{R}^n$ as a linear combination of standard basis vectors is unique. (That is, show that if $x = \sum_{k=1}^{n} \alpha_k e^k = \sum_{k=1}^{n} \beta_k e^k$, then $\alpha_k = \beta_k$ for each k.)

20.3. Convex combinations

20.3.1. Definition. A linear combination $\sum_{k=1}^n \alpha_k x_k$ of the vectors $x_1, \ldots, x_n$ is a CONVEX COMBINATION if $\alpha_k \geq 0$ for each k $(1 \leq k \leq n)$ and if $\sum_{k=1}^n \alpha_k = 1$.

20.3.2. Exercise. Write the vector $(2, 1/4)$ in $\mathbb{R}^2$ as a convex combination of the vectors $(1,0)$, $(0,1)$, and $(3,0)$.

20.3.3. Problem. Write the vector $(1,1)$ as a convex combination of the vectors $(-2,2)$, $(2,2)$, and $(3,-3)$ in $\mathbb{R}^2$.

20.3.4. Definition. If x and y are vectors in the vector space V, then the CLOSED SEGMENT between x and y, denoted by $[x,y]$, is $\{(1-t)x + ty \colon 0 \leq t \leq 1\}$. (*Note.* In the vector space $\mathbb{R}$ this is the same as the closed interval $[x,y]$ provided that $x \leq y$. If $x > y$, however, the closed segment $[x,y]$ contains all numbers z such that $y \leq z \leq x$, whereas the closed interval $[x,y]$ is empty.)

A set $C \subseteq V$ is CONVEX if the closed segment $[x,y]$ is contained in C whenever $x, y \in C$.

20.3.5. Example. A disk is a convex subset of $\mathbb{R}^2$. The set $\{(x,y)\colon 1 \leq x^2 + y^2 \leq 2\}$ is not a convex subset of $\mathbb{R}^2$.

Proof. Problem.

20.3.6. Example. Every subspace of a vector space is convex.

Proof. Problem.

20.3.7. Example. The set
$$\{(x,y) \in \mathbb{R}^2 \colon x \geq 0, \ y \geq 0, \text{ and } x + y \leq 1\}$$
is a convex subset of $\mathbb{R}^2$.

Proof. Problem.

20.3.8. Example. Every convex subset of $\mathbb{R}^n$ is connected.

Proof. Problem.

20.3.9. Definition. Let A be a subset of a vector space V. The CONVEX HULL of A is the smallest convex set containing A; that is, it is the intersection of the family of all convex subsets of V that contain A.

20.3.10. Exercise. What fact must we know about convex sets in order for the preceding definition to make sense? Prove this fact. *Hint.* Review Proposition 20.1.19 and the discussion that precedes it.

20.3.11. Problem. Consider the following subsets of $\mathbb{R}^2$:
$$A = \{(x,y)\colon x \geq 0\},$$
$$B = \{(x,y)\colon 0 \leq y \leq 2\},$$
$$C = \{(x,y)\colon x + y \leq 4\},$$
$$D = A \cap B \cap C.$$

The set D can be described as the convex hull of four points. Which four?

20.3.12. Problem. The concepts of "convex combination" and "convex hull" have been introduced. The point of this problem is to explain the way in which these two ideas are related. Start with a set A. Let C be the set of all convex combinations of elements of A. Let H be the convex hull of A. What relation can you find between C and H? It might not be a bad idea to experiment a little; see what happens in some very simple concrete cases. For example, take A to be a pair of points in $\mathbb{R}^3$. Eventually you will have to consider the following question: Are convex sets closed under the taking of convex combinations?

Chapter 21

Linearity

21.1. Linear transformations

Linear transformations are central to our study of calculus. Functions are differentiable, for example, if they are smooth enough to admit decent approximation by (translates of) linear transformations. Thus, before tackling differentiation (in Chapter 25), we familiarize ourselves with some elementary facts about linearity.

21.1.1. Definition. A function $T\colon V \to W$ between vector spaces is LINEAR if

$$(23) \qquad T(x+y) = Tx + Ty \qquad \text{for all } x,y \in V$$

and

$$(24) \qquad T(\alpha x) = \alpha Tx \qquad \text{for all } x \in V \text{ and } \alpha \in \mathbb{R}.$$

   A linear function is most commonly called a LINEAR TRANSFORMATION, sometimes a LINEAR MAPPING. If the domain and codomain of a linear transformation are the same vector space, then it is often called a LINEAR OPERATOR, and occasionally a VECTOR SPACE ENDOMORPHISM. The family of all linear transformations from V into W is denoted by $\mathfrak{L}(V,W)$. Two oddities of notation concerning linear transformations deserve comment. First, the value of T at x is usually written Tx rather than $T(x)$. Naturally, the parentheses are used whenever their omission would create ambiguity. For example, in (23) above $Tx + y$ is not an acceptable substitute for $T(x+y)$. Second, the symbol for composition of two linear transformations is ordinarily omitted. If $S \in \mathfrak{L}(U,V)$ and $T \in \mathfrak{L}(V,W)$, then the composite of T and S is denoted by TS (rather than by $T \circ S$). This will cause no confusion since we will define no other "multiplication" of linear maps. As a consequence of this convention if T is a linear operator, then $T \circ T$ is written as T^2, $T \circ T \circ T$ as T^3, and so on. One may think of condition (23) in the definition of linearity in the following fashion. Let $T \times T$ be the mapping from $V \times V$ into $W \times W$ defined by

$$(T \times T)(x,y) = (Tx, Ty).$$

Then condition (23) holds if and only if the diagram

$$
\begin{array}{ccc}
V \times V & \xrightarrow{\;T \times T\;} & W \times W \\
{\scriptstyle +}\big\downarrow & & \big\downarrow{\scriptstyle +} \\
V & \xrightarrow{\quad T \quad} & W
\end{array}
$$

commutes. (The vertical maps are addition in V and in W.)

Condition (24) of the definition can similarly be thought of in terms of a diagram. For each scalar α define the function M_α, MULTIPLICATION BY α, from a vector space into itself by

$$M_\alpha(x) = \alpha x \,.$$

(We use the same symbol for multiplication by α in both of the spaces V and W.) Then condition (24) holds if and only if for every scalar α the following diagram commutes.

$$
\begin{array}{ccc}
V & \xrightarrow{\;T\;} & W \\
{\scriptstyle M_\alpha}\big\downarrow & & \big\downarrow{\scriptstyle M_\alpha} \\
V & \xrightarrow{\;T\;} & W
\end{array}
$$

21.1.2. Example. If $T \colon \mathbb{R}^3 \to \mathbb{R}^2 \colon x \mapsto (x_1 + x_3, x_1 - 2x_2)$, then T is linear.

Proof. Exercise.

21.1.3. Problem. Let $T \colon \mathbb{R}^2 \to \mathbb{R}^4$ be defined by

$$T(x, y) = (x + 2y, 3x - y, -2x, -x + y).$$

Show that T is linear.

21.1.4. Exercise. Let $T \colon \mathbb{R}^3 \to \mathbb{R}^3$ be a linear transformation that satisfies $T(e^1) = (1, 0, 1)$, $T(e^2) = (0, 2, -1)$, and $T(e^3) = (-4, -1, 3)$ (where e^1, e^2, and e^3 are the standard basis vectors for $\mathbb{R}^3$ defined in Example 20.2.6). Find $T(2, 1, 5)$. *Hint.* Use Problem 20.2.10.

21.1.5. Problem. Suppose that $T \in \mathfrak{L}(\mathbb{R}^3, \mathbb{R}^3)$ satisfies

$$
\begin{aligned}
Te^1 &= (1, 2, -3), \\
Te^2 &= (1, -1, 0), \\
Te^3 &= (-2, 0, 1).
\end{aligned}
$$

Find $T(3, -2, 1)$.

21.1.6. Proposition. *Let $T \colon V \to W$ be a linear transformation between two vector spaces. Then*

(a) $T(0) = 0$.

(b) $T(x - y) = Tx - Ty$ *for all* x, $y \in V$.

Proof. Exercise.

21.1.7. Example. The identity map from a vector space into itself is linear.

Proof. Obvious. □

21.1.8. Example. Each coordinate projection defined on $\mathbb{R}^n$,

$$\pi_k \colon \mathbb{R}^n \to \mathbb{R} \colon x \mapsto x_k,$$

is linear.

Proof. For $1 \leq k \leq n$, we have $\pi_k(x+y) = \pi_k(x_1 + y_1, \ldots, x_n + y_n) = x_k + y_k = \pi_k(x) + \pi_k(y)$ and $\pi_k(\alpha x) = \pi_k(\alpha x_1, \ldots, \alpha x_n) = \alpha x_k = \alpha \pi_k(x)$. □

21.1.9. Example. Let $\mathcal{F} = \mathcal{F}((a,b), \mathbb{R})$ be the family of all real valued functions defined on the open interval (a,b), and let $\mathcal{D} = \mathcal{D}((a,b), \mathbb{R})$ be the set of all members of $\mathcal{F}$ that are differentiable at each point of (a,b). Then $\mathcal{D}$ is a vector subspace of $\mathcal{F}$ and the differentiation operator

$$D \colon \mathcal{D} \to \mathcal{F} \colon f \mapsto f'$$

(where f' is the derivative of f) is linear.

Proof. We know from Example 20.1.11 that $\mathcal{F}$ is a vector space. To show that $\mathcal{D}$ is a vector subspace of $\mathcal{F}$, use Proposition 20.1.13. That $\mathcal{D}$ is nonempty is clear since constant functions are differentiable. That the space $\mathcal{D}$ of differentiable functions is closed under addition and scalar multiplication and that the operation D of differentiation is linear (that is, $D(\alpha f) = \alpha Df$ and $D(f+g) = Df + Dg$) are immediate consequences of Propositions 8.4.10 and 8.4.11. □

21.1.10. Example. We have not yet discussed integration of continuous functions, but recalling a few basic facts about integration from beginning calculus shows us that this is another example of a linear transformation. Let $\mathcal{C} = \mathcal{C}([a,b], \mathbb{R})$ be the family of all members of $\mathcal{F} = \mathcal{F}([a,b], \mathbb{R})$ that are continuous. We know from beginning calculus that any continuous function on a closed and bounded interval is (Riemann) integrable, that

$$\int_a^b \alpha f(x)\,dx = \alpha \int_a^b f(x)\,dx,$$

where $\alpha \in \mathbb{R}$ and $f \in \mathcal{C}$, and that

$$\int_a^b (f(x) + g(x))\,dx = \int_a^b f(x)\,dx + \int_a^b g(x)\,dx.$$

where $f, g \in \mathcal{C}$. It then follows easily (again from Proposition 20.1.13) that $\mathcal{C}$ is a vector subspace of $\mathcal{F}$ and that the function $K \colon \mathcal{C} \to \mathbb{R} \colon f \mapsto \int_a^b f(x)\,dx$ is linear.

An important observation is that the composite of two linear transformations is linear.

21.1.11. Proposition. *Let* U, V, *and* W *be vector spaces. If* $S \in \mathfrak{L}(U, V)$ *and* $T \in \mathfrak{L}(V, W)$, *then* $TS \in \mathfrak{L}(U, W)$.

Proof. Problem.

21.1.12. Definition. If $T: V \to W$ is a linear transformation between vector spaces, then the KERNEL (or NULL SPACE) of T, denoted by $\ker T$, is $T^{\leftarrow}\{\mathbf{0}\}$. That is,

$$\ker T := \{x \in V: Tx = \mathbf{0}\}.$$

21.1.13. Exercise. Let $T \in \mathfrak{L}(\mathbb{R}^3, \mathbb{R}^3)$ satisfy

$$Te^1 = (1, -2, 3),$$
$$Te^2 = (0, 0, 0),$$
$$Te^3 = (-2, 4, -6),$$

where e^1, e^2, and e^3 are the standard basis vectors in $\mathbb{R}^3$. Find and describe geometrically both the kernel of T and the range of T.

21.1.14. Problem. Let T be the linear transformation of Example 21.1.2. Find and describe geometrically the kernel and the range of T.

21.1.15. Problem. Let D be the linear transformation defined in Example 21.1.9. What is the kernel of D?

It is useful to know that the kernel of a linear transformation is always a vector subspace of its domain, that its range is a vector subspace of its codomain, and that a necessary and sufficient condition for a linear transformation to be injective is that its kernel contain only the zero vector.

21.1.16. Proposition. *If $T: V \to W$ is a linear transformation between vector spaces, then $\ker T$ is a subspace of V.*

Proof. Problem. *Hint.* Use Proposition 20.1.13.

21.1.17. Proposition. *If $T: V \to W$ is a linear transformation between vector spaces, then $\operatorname{ran} T$ is a subspace of W.*

Proof. Exercise. *Hint.* Use Proposition 20.1.13.

21.1.18. Proposition. *A linear mapping T is injective if and only if $\ker T = \{\mathbf{0}\}$.*

Proof. Problem. *Hint.* First show that if T is injective and $x \in \ker T$, then $x = 0$. For the converse, suppose that $\ker T = \{\mathbf{0}\}$ and that $Tx = Ty$. Show that $x = y$.

21.1.19. Problem. Let W be a vector space.

(a) Show that a linear transformation $T: \mathbb{R}^n \to W$ is completely determined by its values on the standard basis vectors $e^1, \dots, e^n$ of $\mathbb{R}^n$. *Hint.* Use Problem 20.2.10(b).

(b) Show that if $S, T \in \mathfrak{L}(\mathbb{R}^n, W)$ and $Se^k = Te^k$ for $1 \le k \le n$, then $S = T$.

(c) Let $w^1, \dots, w^n \in W$. Show that there exists a unique $T \in \mathfrak{L}(\mathbb{R}^n)$ such that $Te^k = w^k$ for $1 \le k \le n$.

Injective linear mappings take linearly independent sets to linearly independent sets.

21.1.20. Proposition. *If* $T \in \mathfrak{L}(V,W)$ *is injective and* A *is a linearly independent subset of* V, *then* $T^{\rightarrow}(A)$ *is a linearly independent set in* W.

Proof. Problem. *Hint.* Start with vectors $y^1, \ldots, y^n$ in $T^{\rightarrow}(A)$, and suppose that some linear combination of them $\sum_{k=1}^{n} \alpha_k x^k$ is 0. Show that all the scalars α_k are 0. Use Proposition 21.1.18.

Linear mappings take convex sets to convex sets.

21.1.21. Proposition. *If* V *and* W *are vector spaces,* C *is a convex subset of* V, *and* $T\colon V \to W$ *is linear, then* $T^{\rightarrow}(C)$ *is a convex subset of* W.

Proof. Problem. *Hint.* Let $u, v \in T^{\rightarrow}(C)$ and $0 \leq t \leq 1$. Show that $(1-t)u+tv \in T^{\rightarrow}(C)$.

21.1.22. Problem. Let $T \in \mathfrak{L}(\mathbb{R}^3, \mathbb{R}^3)$ satisfy

$$Te^1 = (0, 1, 0),$$
$$Te^2 = (0, 0, 1),$$
$$Te^3 = (3, -2, 0).$$

Show that T is bijective. *Hint.* To show that T is injective, use Proposition 21.1.18.

21.1.23. Problem. Let $\mathcal{C}^1 = \mathcal{C}^1([a,b], \mathbb{R})$ be the set of all functions f in $\mathcal{F} = \mathcal{F}([a,b], \mathbb{R})$ such that f' exists on $[a,b]$ in the usual sense of having one-sided derivatives at a and b and is continuous. (A function belonging to $\mathcal{C}^1$ is said to be CONTINUOUSLY DIFFERENTIABLE.) It is easy to see that the set of all $\mathcal{C}^1$ functions is a vector subspace of $\mathcal{F}$. Let $\mathcal{C} = \mathcal{C}([a,b], \mathbb{R})$ be the family of all continuous members of $\mathcal{F}$. For every $f \in \mathcal{C}$ and every $x \in [a,b]$, let

$$(Jf)(x) := \int_a^x f(t)\, dt.$$

(a) Why does Jf belong to $\mathcal{C}^1$?

(b) Show that the map $J\colon \mathcal{C} \to \mathcal{C}^1\colon f \mapsto Jf$ is linear.

21.1.24. Problem (Products). Recall (see Appendix N) that for every pair of functions $f^1\colon T \to S_1$ and $f^2\colon T \to S_2$ having the same domain, there exists a unique map, namely $f = (f^1, f^2)$, mapping T into the product space $S^1 \times S^2$ that satisfies $\pi_1 \circ f = f^1$ and $\pi_2 \circ f = f^2$. (See, in particular, Exercise N.1.4.) Now suppose that T, S_1, and S_2 are vector spaces and that f^1 and f^2 are linear. Then $S_1 \times S_2$ is a vector space (see Example 20.1.20). Show that the function $f = (f^1, f^2)$ is linear.

21.2. The algebra of linear transformations

The set $\mathfrak{L}(V,W)$ of linear transformations between two vector spaces is contained in the vector space $\mathcal{F}(V,W)$ of *all* W-valued functions whose domain is V. (That $\mathcal{F}$ is a vector space was proved in Example 20.1.11.) It is easy to show that $\mathfrak{L}(V,W)$ is a vector space; just show that it is a subspace of $\mathcal{F}(V,W)$.

21.2.1. Proposition. *Let V and W be vector spaces. Then $\mathfrak{L}(V,W)$ with pointwise operations of addition and scalar multiplication is a vector space.*

Proof. Exercise. *Hint.* Use Example 20.1.11 and Proposition 20.1.13.

Let $T\colon V \to W$ be a function between two sets. We say that T is INVERTIBLE if there exists a function T^{-1} mapping W to V such that $T \circ T^{-1}$ is the identity function on W and $T^{-1}\circ T$ is the identity function on V. (For details about this see Appendix M.) Now, suppose that V and W are vector spaces and that $T\colon V \to W$ is a linear transformation. In this context what do we mean when we say that T is "invertible"? For a linear transformation to be invertible, we will require two things: the transformation must possess an inverse function, and this function must itself be linear. It is a pleasant fact about linear transformations that the second condition is automatically satisfied whenever the first is.

21.2.2. Proposition. *If $T \in \mathfrak{L}(V,W)$ is bijective, then its inverse $T^{-1}\colon W \to V$ is linear.*

Proof. Exercise

21.2.3. Definition. A linear transformation $T\colon V \to W$ is INVERTIBLE (or is an ISOMORPHISM) if there exists a linear transformation T^{-1} such that $T^{-1} \circ T = I_V$ and $T \circ T^{-1} = I_W$.

The point of the preceding proposition is that this definition is somewhat redundant. In particular, the following are just different ways of saying the same thing about a linear transformation T:

(a) T is invertible.

(b) T is an isomorphism.

(c) As a function T has an inverse.

(d) T is bijective.

21.2.4. Definition. If there exists an isomorphism between vector spaces V and W, we say that the spaces are ISOMORPHIC and write $V \cong W$.

21.2.5. Problem. Let s be the set of all sequences of real numbers. Regard s as a vector space under pointwise operations. That is,

$$x + y := (x_1 + y_1, x_2 + y_2, \dots)$$
$$\alpha x := (\alpha x_1, \alpha x_2, \dots)$$

whenever $x = (x_1, x_2, \dots)$ and $y = (y_1, y_2, \dots)$ belong to s and α is a scalar. Define the UNILATERAL SHIFT OPERATOR $U\colon s \to s$ by

$$U(x_1, x_2, x_3, \dots) := (0, x_1, x_2, \dots).$$

(a) Show that $U \in \mathfrak{L}(s, s)$.

(b) Does U have a right inverse? If so, what is it?

(c) Does U have a left inverse? If so, what is it?

21.2.6. Definition. Suppose that a vector space V is equipped with an additional operation $(x, y) \mapsto xy$ from $V \times V$ into V (we will call it "multiplication") that satisfies

(a) $x(y + z) = xy + xz$,

(b) $(x + y)z = xz + yz$,

(c) $(xy)z = x(yz)$, and

(d) $\alpha(xy) = x(\alpha y)$,

whenever $x, y, z \in V$ and $\alpha \in \mathbb{R}$. Then V is an ALGEBRA. (Sometimes it is called a LINEAR ASSOCIATIVE ALGEBRA.) If an algebra possesses a multiplicative identity (that is, a vector $\mathbf{1}$ such that $\mathbf{1}\, x = x\, \mathbf{1} = x$ for all $x \in V$), then it is a UNITAL ALGEBRA. A subset of an algebra A that is closed under the operations of addition, multiplication, and scalar multiplication is a SUBALGEBRA of A. If A is a unital algebra and B is a subalgebra of A that contains the multiplicative identity of A, then B is a UNITAL SUBALGEBRA of A.

21.2.7. Example. If M is a compact metric space, then the vector space $\mathcal{B}(M, \mathbb{R})$ of bounded functions on M is a unital algebra under pointwise operations. (The constant function 1 is its multiplicative identity.) We have already seen in Chapter 15 that the space $\mathcal{C}(M, \mathbb{R})$ of continuous functions on M is a vector subspace of $\mathcal{B}(M, \mathbb{R})$. Since the product of continuous functions is continuous (and constant functions are continuous) $\mathcal{C}(M, \mathbb{R})$ is a unital subalgebra of $\mathcal{B}(M, \mathbb{R})$.

21.2.8. Problem. It has already been shown (in Proposition 21.2.1) that if V is a vector space, then so is $\mathfrak{L}(V, V)$. Show that with the additional operation of composition serving as multiplication $\mathfrak{L}(V, V)$ is a unital algebra.

21.2.9. Problem. If $T \in \mathfrak{L}(V, W)$ is invertible, then so is T^{-1} and $\left(T^{-1}\right)^{-1} = T$.

21.2.10. Problem. If $S \in \mathfrak{L}(U, V)$ and $T \in \mathfrak{L}(V, W)$ are both invertible, then so is TS and $(TS)^{-1} = S^{-1}T^{-1}$.

21.2.11. Problem. If $T \in \mathfrak{L}(V, V)$ satisfies the equation

$$T^2 - T + I = \mathbf{0},$$

then it is invertible. What is T^{-1}?

21.2.12. Problem. Let V be a vector space, let W be a set that is provided with operations $(u, v) \mapsto u + v$ from $W \times W$ into W and $(\alpha, u) \mapsto \alpha u$ from $\mathbb{R} \times W$ into W, and let $T \colon V \to W$. If T is bijective and it preserves operations (that is, $T(x + y) = Tx + Ty$ and $T(\alpha x) = \alpha Tx$ for all x, $y \in V$ and $\alpha \in \mathbb{R}$), then W is a vector space that is isomorphic to V. *Hint.* Verify the eight defining axioms for a vector space. The first axiom is associativity of addition. Let u, v, $w \in W$. Write $(u + v) + w$ as $\left(T(T^{-1}u) + T(T^{-1}v)\right) + T(T^{-1}w)$ and use the hypothesis that T is operation preserving.

21.2.13. Problem. Let V be a vector space, let W be a set, and let $T \colon V \to W$ be a bijection. Explain carefully how W can be made into a vector space isomorphic to V. *Hint.* Use Problem 21.2.12.

21.3. Matrices

The purpose of this section and the next two is almost entirely computational. Many (but by no means all!) of the linear transformations we will consider in the rest of the book are maps between various Euclidean spaces; that is, between $\mathbb{R}^n$ and $\mathbb{R}^m$ where $m, n \in \mathbb{N}$. Such transformations may be represented by matrices. This is of great convenience in dealing with specific examples because matrix computations are so very simple. We begin by reviewing a few elementary facts about matrices and matrix operations.

For each $n \in \mathbb{N}$, let $\mathbb{N}_n$ be $\{1, \ldots, n\}$. An $m \times n$ (read "m by n") MATRIX is a function whose domain is $\mathbb{N}_m \times \mathbb{N}_n$. We deal here only with matrices of real numbers; that is, with real valued functions on $\mathbb{N}_m \times \mathbb{N}_n$. If $a \colon \mathbb{N}_m \times \mathbb{N}_n \to \mathbb{R}$ is an $m \times n$ matrix, its value at $(i, j) \in \mathbb{N}_m \times \mathbb{N}_n$ will be denoted by a^i_j. (Occasionally we use the notation a_{ij} instead.) The matrix a itself may be denoted by $\left[a^i_j\right]_{i=1\,j=1}^{m\quad n}$, by $\left[a^i_j\right]$, or by a rectangular array whose entry in row i and column j is a^i_j.

$$\begin{bmatrix} a^1_1 & a^1_2 & \ldots & a^1_n \\ a^2_1 & a^2_2 & \ldots & a^2_n \\ \vdots & \vdots & \ddots & \vdots \\ a^m_1 & a^m_2 & \ldots & a^m_n \end{bmatrix}$$

In light of this notation it is reasonable to refer to the index i in the expression a^i_j as the ROW INDEX and to call j the COLUMN INDEX. (If you are accustomed to thinking of a matrix as *being* a rectangular array, no harm will result. The reason for defining a matrix as a function is to make good on the boast made in Appendix B that everything in what follows can be defined ultimately in terms of sets.) We denote the family of all $m \times n$ matrices of real numbers by $\mathfrak{M}_{m \times n}$. For families of square matrices we shorten $\mathfrak{M}_{n \times n}$ to $\mathfrak{M}_n$.

Two $m \times n$ matrices a and b may be added. Addition is done pointwise. The sum $a + b$ is the $m \times n$ matrix whose value at (i, j) is $a^i_j + b^i_j$, That is,

$$(a + b)^i_j = a^i_j + b^i_j$$

for $1 \le i \le m$ and $1 \le j \le n$. Scalar multiplication is also defined pointwise. If a is an $m \times n$ matrix and $\alpha \in \mathbb{R}$, then αa is the $m \times n$ matrix whose value at (i, j) is αa^i_j. That is,

$$(\alpha a)^i_j = \alpha a^i_j$$

for $1 \le i \le m$ and $1 \le j \le n$. We may also subtract matrices. By $-b$ we mean $(-1)b$, and by $a - b$ we mean $a + (-b)$.

21.3.1. Exercise. Let

$$\begin{bmatrix} 4 & 2 & 0 & -1 \\ -1 & -3 & 1 & 5 \end{bmatrix} \quad \text{and} \quad \begin{bmatrix} 1 & -5 & 3 & -1 \\ 3 & 1 & 0 & -1 \end{bmatrix}.$$

Find $a + b$, $3a$, and $a - 2b$.

If a is an $m \times n$ matrix and b is an $n \times p$ matrix, the product of a and b is the $m \times p$ matrix whose value at (i, j) is $\sum_{k=1}^{n} a_k^i b_j^k$. That is,

$$(ab)_j^i = \sum_{k=1}^{n} a_k^i b_j^k$$

for $1 \leq i \leq m$ and $1 \leq j \leq p$. Notice that, in order for the product ab to be defined, the number of columns of a must be the same as the number of rows of b. Here is a slightly different way of thinking of the product of a and b. Define the INNER PRODUCT (or DOT PRODUCT) of two n-tuples $(x_1, x_2, \ldots, x_n)$ and $(y_1, y_2, \ldots, y_n)$ to be $\sum_{k=1}^{n} x_k y_k$. Regard the rows of the matrix a as n-tuples (read from left to right) and the columns of b as n-tuples (read from top to bottom). Then the entry in the ith row and jth column of the product ab is the dot product of the ith row of a and the jth column of b.

21.3.2. Example. Matrix multiplication is not commutative. If a is a 2×3 matrix and b is a 3×4 matrix, then ab is defined but ba is not. Even in situations where both products ab and ba are defined, they need not be equal. For example, if

$$a = \begin{bmatrix} 1 & 2 \\ 1 & 0 \end{bmatrix} \text{ and } b = \begin{bmatrix} -1 & 1 \\ 2 & 3 \end{bmatrix}, \text{ then } ab = \begin{bmatrix} 3 & 7 \\ -1 & 1 \end{bmatrix}, \text{ whereas } ba = \begin{bmatrix} 0 & -2 \\ 5 & 4 \end{bmatrix}.$$

21.3.3. Exercise. Let $a = \begin{bmatrix} 2 & 3 & -1 \\ 0 & 1 & 4 \end{bmatrix}$ and $b = \begin{bmatrix} 1 & 0 \\ 2 & -1 \\ 1 & -2 \end{bmatrix}$. Find ab.

21.3.4. Problem. Let $a = \begin{bmatrix} 4 & 3 & 1 & 2 \\ 0 & -1 & -1 & 1 \\ 2 & 0 & 1 & 3 \end{bmatrix}$ and $b = \begin{bmatrix} 2 & -1 \\ 0 & 1 \\ 1 & 0 \\ -3 & 2 \end{bmatrix}$.

(a) Find the product ab (if it exists).

(b) Find the product ba (if it exists).

21.3.5. Definition. Let a be an $m \times n$ matrix. The TRANSPOSE of a, denoted by a^t, is the $n \times m$ matrix obtained by interchanging the rows and columns of a. That is, if $b = a^t$, then $b_j^i = a_i^j$ for $1 \leq i \leq n$ and $1 \leq j \leq m$.

21.3.6. Example. Let $a = \begin{bmatrix} 1 & 2 & 0 & -4 \\ 3 & 0 & -1 & 5 \end{bmatrix}$. Then $a^t = \begin{bmatrix} 1 & 3 \\ 2 & 0 \\ 0 & -1 \\ -4 & 5 \end{bmatrix}$.

For material in what follows the most important role played by matrices will be as (representations of) linear transformations on finite-dimensional vector spaces. Here is how it works.

21.3.7. Definition. We define the *action* of a matrix on a vector. If $a \in \mathfrak{M}_{m \times n}$ and $x \in \mathbb{R}^n$, then ax, the RESULT OF a ACTING ON x, is defined to be the vector in $\mathbb{R}^m$ whose jth coordinate is $\sum_{k=1}^{n} a_k^j x_k$ (this is just the dot product of the jth row of a with x). That is,

$$(ax)_j := \sum_{k=1}^{n} a_k^j x_k$$

for $1 \leq j \leq m$. Here is another way of saying the same thing. Regard x as an $n \times 1$ matrix

$$\begin{bmatrix} x_1 \\ x_2 \\ \vdots \\ x_n \end{bmatrix}$$

(sometimes called a COLUMN VECTOR). Now multiply the $m \times n$ matrix a by the $n \times 1$ matrix x. The result will be an $m \times 1$ matrix (another column vector), say

$$\begin{bmatrix} y_1 \\ y_2 \\ \vdots \\ y_m \end{bmatrix} .$$

Then ax is the m-tuple $(y_1, \ldots, y_m)$. Thus a may be thought of as a mapping from $\mathbb{R}^n$ into $\mathbb{R}^m$.

21.3.8. Exercise. Let $a = \begin{bmatrix} 3 & 0 & -1 & -4 \\ 2 & 1 & -1 & -2 \\ 1 & -3 & 0 & 2 \end{bmatrix}$ and $x = (2, 1, -1, 1)$. Find ax.

21.3.9. Problem. Let $a = \begin{bmatrix} 2 & 0 \\ 1 & -3 \\ 5 & 1 \end{bmatrix}$ and $x = (1, -2)$. Find ax.

From the definition of the action of a matrix on a vector, we derive several formulas that will be useful going forward. Each is a simple computation.

21.3.10. Proposition. *Let* a, $b \in \mathfrak{M}_{m \times n}$, $c \in \mathfrak{M}_{n \times p}$, x, $y \in \mathbb{R}^n$, $z \in \mathbb{R}^p$, *and* $\alpha \in \mathbb{R}$. *Then*

(a) $a(x + y) = ax + ay$,

(b) $a(\alpha x) = \alpha(ax)$,

(c) $(a + b)x = ax + bx$,

(d) $(\alpha a)x = \alpha(ax)$,

(e) $(ac)z = a(cz)$.

Proof. Part (a) is an exercise. Parts (b)–(e) are problems.

Next we show that a sufficient (and obviously necessary) condition for two $m \times n$ matrices to be equal is that they have the same action on the standard basis vectors in $\mathbb{R}^n$.

21.3.11. Proposition. *Let a and b be $m \times n$ matrices and, as usual, let $e^1, \ldots, e^n$ be the standard basis vectors in $\mathbb{R}^n$. If $ae^k = be^k$ for $1 \leq k \leq n$, then $a = b$.*

Proof. Problem. *Hint.* Compute $(ae^k)_j$ and $(be^k)_j$ for $1 \leq j \leq m$ and $1 \leq k \leq n$. Remember that $(e^k)_l = 0$ if $k \neq l$ and that $(e^k)_k = 1$.

Remark. The Definition 21.3.7 of the action of a matrix on a vector technically requires us to think of vectors as "column vectors". It is probably more likely that most of us think of vectors in $\mathbb{R}^n$ as "row vectors", that is, as n-tuples or as $1 \times n$ matrices. Then for the matrix multiplication ax to make sense and for the result to again be a "row vector", we really should write

$$\left(a(x^t)\right)^t$$

for the action of the matrix $a \in \mathfrak{M}_{m \times n}$ on the vector $x \in \mathbb{R}^n$. We will not do this. We will regard vectors as "row vectors" or "column vectors" as convenience dictates.

21.3.12. Definition. In our later work we will have occasion to consider the action of a SQUARE MATRIX (one with the same number of rows as columns) on a pair of vectors. Let $a \in \mathfrak{M}_n$ and $x, y \in \mathbb{R}^n$. We denote by xay the number $\sum_{j,k=1}^n a_k^j x_j y_k$.

Since

$$\sum_{j,k=1}^n a_k^j x_j y_k = \sum_{j=1}^n x_j \sum_{k=1}^n a_k^j y_k$$

and since $\sum_{k=1}^n a_k^j y_k$ is just $(ay)_j$, we may write

$$xay = \sum_{j=1}^n x_j (ay)_j.$$

In other words xay is just the dot product of the vectors x and ay. If we identify n-tuples (row vectors) with $1 \times n$ matrices, then xay is the product of the three matrices x, a, and y^t. That is,

$$xay = \begin{bmatrix} x_1 \ldots x_n \end{bmatrix} \begin{bmatrix} a_1^1 & \cdots & a_n^1 \\ \vdots & \ddots & \vdots \\ a_1^n & \cdots & a_n^n \end{bmatrix} \begin{bmatrix} y_1 \\ \vdots \\ y_n \end{bmatrix}.$$

21.3.13. Exercise. Let $a = \begin{bmatrix} 1 & 3 & -1 \\ 0 & 2 & 4 \\ 1 & -1 & 1 \end{bmatrix}$, $x = (1, -2, 0)$, and $y = (3, 0, 1)$. Find the action of a on the pair of vectors x and y; that is, find xay.

21.3.14. Problem. Let $a = \begin{bmatrix} 1 & 2 & 0 & -1 \\ 3 & -3 & 1 & 0 \\ 2 & 0 & 1 & -4 \\ -1 & 1 & -1 & 1 \end{bmatrix}$, $x = (1, -1, 0, 2)$, and $y = (1, 0, 3, 1)$. Find xay.

21.3.15. Definition. The MAIN (or PRINCIPAL) DIAGONAL of a square matrix is the diagonal running from the upper left corner to the lower right corner. That is, it consists of all the elements of the form a_k^k. If each entry on the main diagonal of an $n \times n$ matrix is 1 and all its other entries are 0, then the matrix is the $n \times n$ IDENTITY MATRIX. This matrix is denoted by I_n (or just by I if no confusion will result). If c is a real number, it is conventional to denote the matrix cI_n by c. It is clear that $a I_n = I_n a = a$ for every $n \times n$ matrix a. The $m \times n$ ZERO MATRIX is the $m \times n$ matrix all of whose entries are 0. It is denoted by $\mathbf{0}_{m \times n}$ or just by $\mathbf{0}$. Certainly, $\mathbf{0} + a = a + \mathbf{0} = a$ for every $m \times n$ matrix a.

21.3.16. Definition. A square matrix a in $\mathfrak{M}_{n \times n}$ is invertible if there exists an $n \times n$ matrix a^{-1} such that

$$aa^{-1} = a^{-1}a = I_n.$$

The matrix a^{-1} is the INVERSE of a.

21.3.17. Proposition. *An $n \times n$-matrix has at most one inverse.*

Proof. Exercise.

21.3.18. Exercise. Show that the matrix $b = \begin{bmatrix} -1/2 & 1/2 & 3/2 \\ 1/4 & 1/4 & -1/4 \\ 3/4 & -1/4 & -3/4 \end{bmatrix}$ is the inverse of the matrix $a = \begin{bmatrix} 1 & 0 & 2 \\ 0 & 3 & -1 \\ 1 & -1 & 1 \end{bmatrix}$.

21.3.19. Problem. Show that the matrix $a = \begin{bmatrix} 1 & 3 & -1 \\ 0 & 2 & 1 \\ 1 & -2 & 1 \end{bmatrix}$ satisfies the equation

$$a^3 - 4a^2 + 8a - 9 = \mathbf{0}.$$

Use this fact to find the inverse of a.

21.4. Determinants

A careful development of the properties of the determinant function on $n \times n$ matrices is not a central concern of this course. In this section we record without proof some of its elementary properties. (Proofs of these facts can be found in almost any linear algebra text. Two elegant (if not entirely elementary) presentations can be found in [**Hal58**] and [**HK71**].)

21.4.1. Fact. Let $n \in \mathbb{N}$. There is exactly one function

$$\det\colon \mathfrak{M}_{n \times n} \to \mathbb{R}\colon a \mapsto \det a$$

that satisfies the following.

(a) $\det I_n = 1$.

(b) If $a \in \mathfrak{M}_{n \times n}$ and a' is the matrix obtained by interchanging two rows of a, then $\det a' = -\det a$.

(c) If $a \in \mathfrak{M}_{n \times n}$, $c \in \mathbb{R}$, and a' is the matrix obtained by multiplying each element in one row of a by c, then $\det a' = c \det a$.

(d) If $a \in \mathfrak{M}_{n \times n}$, $c \in \mathbb{R}$, and a' is the matrix obtained from a by multiplying one row of a by c and adding it to another row of a (that is, choose $i, j \in \mathbb{N}_n$ with $i \neq j$ and replace a_k^j by $a_k^j + c a_k^i$ for each k in $\mathbb{N}_n$), then $\det a' = \det a$.

21.4.2. Definition. The unique function $\det\colon \mathfrak{M}_{n \times n} \to \mathbb{R}$ described above is the $n \times n$ DETERMINANT FUNCTION.

21.4.3. Fact. If $a \in \mathbb{R}$ ($= \mathfrak{M}_{1 \times 1}$), then $\det a = a$; if $a \in \mathfrak{M}_{2 \times 2}$, then $\det a = a_1^1 a_2^2 - a_2^1 a_1^2$.

21.4.4. Fact. If $a, b \in \mathfrak{M}_{n \times n}$, then $\det(ab) = (\det a)(\det b)$.

21.4.5. Fact. If $a \in \mathfrak{M}_{n \times n}$, then $\det a^t = \det a$. (An obvious corollary of this is that in conditions (b), (c), and (d) of Fact 21.4.1, the word "columns" may be substituted for the word "rows".)

21.4.6. Definition. Let a be an $n \times n$ matrix. The MINOR of the element a_k^j, denoted by M_k^j, is the determinant of the $(n-1) \times (n-1)$ matrix that results from the deletion of the jth row and kth column of a. The COFACTOR of the element a_k^j, denoted by C_k^j, is defined by

$$C_k^j := (-1)^{j+k} M_k^j.$$

21.4.7. Fact. If $a \in \mathfrak{M}_{n \times n}$ and $1 \leq j \leq n$, then

$$\det a = \sum_{k=1}^{n} a_k^j C_k^j.$$

This is the (LAPLACE) EXPANSION of the determinant along the jth row.

In light of Fact 21.4.5 it is clear that expansion along columns works as well as expansion along rows. That is,

$$\det a = \sum_{j=1}^{n} a_k^j C_k^j$$

for any k between 1 and n. This is the (LAPLACE) EXPANSION of the determinant along the kth column.

21.4.8. Fact. An $n \times n$ matrix a is invertible if and only if $\det a \neq 0$. If a is invertible, then

$$a^{-1} = (\det a)^{-1} C^t,$$

where $C = \left[C_k^j \right]$ is the matrix of cofactors of elements of a.

21.4.9. Exercise. Let $a = \begin{bmatrix} 1 & 0 & 2 \\ 0 & 3 & -1 \\ 1 & -1 & 1 \end{bmatrix}$. Use the preceding facts to show that a is invertible and to compute the inverse of a.

21.4.10. Problem. Let a be the matrix given in Problem 21.3.19. Use the facts stated in section 21.4 to show that a is invertible and to compute a^{-1}.

21.5. Matrix representations of linear transformations

We are now in a position to represent members of $\mathfrak{L}(\mathbb{R}^n, \mathbb{R}^m)$ by means of matrices. This will simplify computations involving such linear transformations.

21.5.1. Definition. If $T \in \mathfrak{L}(\mathbb{R}^n, \mathbb{R}^m)$, we define $[T]$ to be the $m \times n$ matrix whose entry in the jth row and kth column is $(Te^k)_j$, the jth component of the vector Te^k in $\mathbb{R}^m$. That is, if $a = [T]$, then $a_k^j = (Te^k)_j$. The matrix $[T]$ is the MATRIX REPRESENTATION of T (with respect to the standard bases in $\mathbb{R}^n$ and $\mathbb{R}^m$).

21.5.2. Example. Let $T \colon \mathbb{R}^4 \to \mathbb{R}^3 \colon (w, x, y, z) \mapsto (w + 2x + 3y, 5w + 6x + 7y + 8z, -2x - 3y - 4z)$. Then T is linear and

$$\begin{aligned} Te^1 &= T(1,0,0,0) = (1,5,0), \\ Te^2 &= T(0,1,0,0) = (2,6,-2), \\ Te^3 &= T(0,0,1,0) = (3,7,-3), \\ Te^4 &= T(0,0,0,1) = (0,8,-4). \end{aligned}$$

Having computed $Te^1, \ldots, Te^4$, we use these as the successive columns of $[T]$. Thus

$$T = \begin{bmatrix} 1 & 2 & 3 & 0 \\ 5 & 6 & 7 & 8 \\ 0 & -2 & -3 & -4 \end{bmatrix}.$$

21.5.3. Example. If $I \colon \mathbb{R}^n \to \mathbb{R}^n$ is the identity map on $\mathbb{R}^n$, then its matrix representation $[I]$ is just the $n \times n$ identity matrix I_n.

21.5.4. Exercise. Let $T \colon \mathbb{R}^2 \to \mathbb{R}^4 \colon (x, y) \mapsto (x - 3y, 7y, 2x + y, -4x + 5y)$. Find $[T]$.

The point of the representation just defined is that if we compute the action of the matrix $[T]$ on a vector x (as in Definition 21.3.7), what we get is the value of T at x. Moreover, this representation is unique; that is, two distinct matrices cannot represent the same linear map.

21.5.5. Proposition. *If $T \in \mathfrak{L}(\mathbb{R}^n, \mathbb{R}^m)$, then for all x in $\mathbb{R}^n$*

$$Tx = [T]x.$$

Furthermore, if a is any $m \times n$ matrix that satisfies

$$Tx = ax \qquad \text{for all } x \in \mathbb{R}^n,$$

then $a = [T]$.

Proof. Exercise. *Hint.* For simplicity of notation let $b = [T]$. The map $S \colon \mathbb{R}^n \to \mathbb{R}^m \colon x \mapsto bx$ is linear. Why? To show that $Sx = Tx$ for all x in $\mathbb{R}^n$, it suffices to show that $(Se^k)_j = (Te^k)_j$ for $1 \le k \le n$ and $1 \le j \le m$. Why?

21.5.6. Proposition. *Let m, $n \in \mathbb{N}$. The map $T \mapsto [T]$ from $\mathfrak{L}(\mathbb{R}^n, \mathbb{R}^m)$ into $\mathfrak{M}_{m \times n}$ is a bijection.*

Proof. Exercise.

21.5.7. Proposition. *Let m, $n \in \mathbb{N}$, let S, $T \in \mathfrak{L}(\mathbb{R}^n, \mathbb{R}^m)$, and let $\alpha \in \mathbb{R}$. Then*

(a) $[S + T] = [S] + [T]$, *and*

(b) $[\alpha T] = \alpha[T]$.

Proof. Exercise. *Hint.* For (a) use Propositions 21.3.11, 21.5.5, and 21.3.10(c).

21.5.8. Theorem. *Under the operations of addition and scalar multiplication (defined in section 21.3), $\mathfrak{M}_{m \times n}$ is a vector space and the map $T \mapsto [T]$, which takes a linear transformation to its matrix representation, is an isomorphism between $\mathfrak{L}(\mathbb{R}^n, \mathbb{R}^m)$ and $\mathfrak{M}_{m \times n}$.*

Proof. Problem. *Hint.* Use Problem 21.2.12.

21.5.9. Problem. Define $T \colon \mathbb{R}^3 \to \mathbb{R}^4$ by

$$T(x, y, z) = (x - 2y, x + y - 3z, y + 4z, 3x - 2y + z).$$

Find $[T]$.

21.5.10. Problem. Define $T \colon \mathbb{R}^4 \to \mathbb{R}^3$ by

$$T(w, x, y, z) = (w - 3x + z, 2w + x + y - 4z, w + y + z).$$

(a) Find $[T]$.

(b) Use Proposition 21.5.5 to calculate $T(4, 0, -3, 1)$.

21.5.11. Problem. Let $f \colon \mathbb{R}^2 \to \mathbb{R}^4$ be defined by

$$f(x) = (x_1 x_2, (x_1)^2 - 4(x_2)^2, (x_1)^3, x_1 \sin(\pi x_2))$$

for all $x = (x_1, x_2)$ in $\mathbb{R}^2$, and let $T \colon \mathbb{R}^2 \to \mathbb{R}^4$ be the linear transformation whose matrix representation is

$$\begin{bmatrix} 4 & 0 \\ 2 & -1 \\ 5 & -8 \\ -1 & 2 \end{bmatrix}.$$

Find $f(a + h) - f(a) - Th$ when $a = (-2, 1/2)$ and $h = (-1, 1)$.

21.5.12. Proposition. *If $S \in \mathfrak{L}(\mathbb{R}^p, \mathbb{R}^n)$ and $T \in \mathfrak{L}(\mathbb{R}^n, \mathbb{R}^m)$, then*

$$[TS] = [T][S].$$

Proof. Problem. *Hint.* Why does it suffice to show that $[TS]x = ([T][S])x$ for all x in $\mathbb{R}^p$? Use Propositions 21.5.5 and 21.3.10(e).

21.5.13. Problem. Let

$$T \colon \mathbb{R}^3 \to \mathbb{R}^4 \colon (x, y, z) \mapsto (2x + y, x - z, y + z, 3x)$$

and

$$S \colon \mathbb{R}^4 \to \mathbb{R}^3 \colon (w, x, y, z) \mapsto (x - y, y + z, z - w).$$

(a) Use Proposition 21.5.12 to find $[TS]$.

(b) Use Proposition 21.5.12 to find $[ST]$.

21.5.14. Problem. Show that matrix multiplication is associative; that is, show that if $a \in \mathfrak{M}_{m \times n}$, $b \in \mathfrak{M}_{n \times p}$, and $c \in \mathfrak{M}_{p \times r}$, then $(ab)c = a(bc)$. *Hint.* Don't make a complicated and messy computation of this by trying to prove it directly. Use Propositions L.2.3, 21.5.6, and 21.5.12.

21.5.15. Problem. Show that $\mathfrak{M}_{n \times n}$ is a unital algebra. *Hint.* Use Definition 21.2.6. Notice that Problem 21.5.14 establishes condition (c) of Definition 21.2.6. Verify the other conditions in a similar fashion.

21.5.16. Proposition. *A linear map $T \in \mathfrak{L}(\mathbb{R}^n, \mathbb{R}^n)$ is invertible if and only if $\det[T] \neq 0$. If T is invertible, then $[T^{-1}] = [T]^{-1}$.*

Proof. Problem. *Hint.* Show that T is invertible if and only if its matrix representation is. Then use Fact 21.4.8.

21.5.17. Problem. Let $T \colon \mathbb{R}^3 \to \mathbb{R}^3 \colon (x, y, z) \mapsto (x + 2z, y - z, x + y)$.

(a) Compute $[T]$ by calculating T and then writing down its matrix representation.

(b) Use Proposition 21.5.16 to find $[T]$.

21.5.18. Problem. Let $\mathcal{P}_4$ be the family of all polynomial functions on $\mathbb{R}$ with degree (strictly) less than 4.

(a) Show that (under the usual pointwise operations) $\mathcal{P}_4$ is a vector space that is isomorphic to $\mathbb{R}^4$. *Hint.* Problem 21.2.12.

(b) Let $D \colon \mathcal{P}_4 \to \mathcal{P}_4 \colon f \mapsto f'$ (where f' is the derivative of f). Using part (a) to identify the spaces $\mathcal{P}_4$ and $\mathbb{R}^4$, find a matrix representation for the (obviously linear) differentiation operator D.

(c) Use your answer to part (b) to differentiate the polynomial $7x^3 - 4x^2 + 5x - 81$.

21.5.19. Problem. Let $\mathcal{P}_4$ be as in Problem 21.5.18. Consider the map

$$K \colon \mathcal{P}_4 \to \mathbb{R} \colon f \mapsto \int_0^1 f(x) \, dx.$$

(a) Show that K is linear.

(b) Find a way to represent K as a matrix. *Hint.* Use Problem 21.5.18(a).

(c) Use your answer to part (b) to integrate the polynomial $8x^3 - 5x^2 - 4x + 6$ over the interval $[0, 1]$.

(d) Let D be as in Problem 21.5.18. Find $[KD]$ by two different techniques.

21.5.20. Problem. Let $T \in \mathfrak{L}(\mathbb{R}^4, \mathbb{R}^4)$ satisfy

$$Te^1 = (1, 2, 0, -1),$$
$$Te^2 = (1, 0, -3, 2),$$
$$Te^3 = (1, -1, -1, 1),$$
$$Te^4 = (0, 2, -1, 0).$$

Also let $x = (1, -2, 3, -1)$ and $y = (0, 1, 2, 1)$. Find $x[T]y$.

Chapter 22

Norms

22.1. Norms on linear spaces

The last two chapters have been pure algebra. In order to deal with topics in analysis (e.g., differentiation, integration, infinite series), we need also a notion of convergence; that is, we need topology as well as algebra. As in earlier chapters we consider only those topologies generated by metrics, in fact only those that arise from norms on vector spaces. Norms, which we introduce in this chapter, are very natural objects; in many concrete situations they abound. Furthermore, they possess one extremely pleasant property: a norm on a vector space generates a metric on the space, and this metric is compatible with the algebraic structure in the sense that it makes the vector space operations of addition and scalar multiplication continuous. Just as metric is a generalization of ordinary Euclidean distance, the concept of norm generalizes on vector spaces the idea of length.

22.1.1. Definition. Let V be a vector space. A function $\| \quad \| : V \to \mathbb{R} : x \mapsto \|x\|$ is a NORM on V if

(a) $\|x + y\| \leq \|x\| + \|y\| \quad$ for all $x, y \in V$,

(b) $\|\alpha x\| = |\alpha| \, \|x\| \quad$ for all $x \in V$ and $\alpha \in \mathbb{R}$, and

(c) if $\|x\| = 0$, then $x = \mathbf{0}$.

The expression $\|x\|$ may be read as "the NORM of x" or "the LENGTH of x". A vector space on which a norm has been defined is a NORMED LINEAR SPACE (or NORMED VECTOR SPACE). A vector in a normed linear space that has norm 1 is a UNIT VECTOR.

22.1.2. Example. The absolute value function is a norm on $\mathbb{R}$.

22.1.3. Example. For $x = (x_1, \ldots, x_n) \in \mathbb{R}^n$, let $\|x\| = \left(\sum_{k=1}^{n} x_k^{\,2} \right)^{1/2}$. The only non-obvious part of the proof that this defines a norm on $\mathbb{R}^n$ is the verification of the TRIANGLE INEQUALITY (that is, condition (a) in the preceding definition). But we have already done this: it is just Minkowski's inequality (Proposition 9.2.7). This is the USUAL NORM (or EUCLIDEAN NORM) on $\mathbb{R}^n$; unless the contrary is explicitly stated, $\mathbb{R}^n$ when regarded as a normed linear space will always be assumed to possess this norm.

22.1.4. Example. For $x = (x_1, \dots, x_n) \in \mathbb{R}^n$, let $\|x\|_1 = \sum_{k=1}^{n} |x_k|$. The function $x \mapsto \|x\|_1$ is easily seen to be a norm on $\mathbb{R}^n$. It is sometimes called the 1-NORM on $\mathbb{R}^n$.

22.1.5. Example. For $x = (x_1, \dots, x_n) \in \mathbb{R}^n$, let $\|x\|_u = \max\{|x_k| : 1 \le k \le n\}$. Again it is easy to see that this defines a norm on $\mathbb{R}^n$; it is the UNIFORM NORM on $\mathbb{R}^n$.

22.1.6. Exercise. Let

$$f \colon \mathbb{R}^3 \to \mathbb{R}^4 \colon (x, y, z) \mapsto (xz, x^2 + 3y, -x + y^2 - 3z, xyz - \sqrt{2}\,x).$$

Find $\|f(a + \lambda h)\|$ when $a = (4, 2, -4)$, $h = (2, 4, -4)$, and $\lambda = -1/2$.

22.1.7. Exercise. Let $f \colon \mathbb{R}^3 \to \mathbb{R}^2 \colon x \mapsto (3x_1{}^2, x_1 x_2 - x_3)$, and let $m = \begin{bmatrix} 6 & 0 & 0 \\ 0 & 1 & -1 \end{bmatrix}$.

Find $\|f(a + h) - f(a) - mh\|$ when $a = (1, 0, -2)$ and h is an arbitrary vector in $\mathbb{R}^3$.

22.1.8. Problem. Let $f \colon \mathbb{R}^3 \to \mathbb{R}^3$ be defined by

$$f(x, y, z) = (xy^3 + yz^2, x\sin(3\pi y), 2z).$$

Find $\|f(a)\|$ when $a = (16, 1/2, 2)$.

22.1.9. Example. Let S be a nonempty set. For f in $\mathcal{B}(S, \mathbb{R})$ let

$$\|f\|_u := \sup\{|f(x)| \colon x \in S\}.$$

This is the UNIFORM NORM on $\mathcal{B}(S, \mathbb{R})$. Notice that Example 22.1.5 is a special case of this one. (An n-tuple may be regarded as a function on the set $\{1, \dots, n\}$; thus, $\mathbb{R}^n = \mathcal{B}(S, \mathbb{R})$ where $S = \{1, \dots, n\}$.)

22.1.10. Exercise. Define f, $g \colon [0, 2\pi] \to \mathbb{R}$ by $f(x) = \sin x$ and $g(x) = \cos x$. Find $\|f + g\|_u$.

22.1.11. Problem. Let $f(x) = x + x^2 - x^3$ for $0 \le x \le 3$. Find $\|f\|_u$.

The following proposition lists some almost obvious properties of norms.

22.1.12. Proposition. *If V is a normed linear space, then*

(a) $\|\mathbf{0}\| = 0$,

(b) $\|-x\| = \|x\|$ *for all $x \in V$, and*

(c) $\|x\| \ge 0$ *for all $x \in V$.*

Proof. Exercise. *Hint.* For part (a) use Exercise 20.1.5; for part (b) use Problem 20.1.7; and for (c) use (a), (b), and the fact that $x + (-x) = \mathbf{0}$.

22.2. Norms induce metrics

We now introduce a crucial fact: *every normed linear space is a metric space.* That is, the norm on a normed linear space induces a metric d defined by $d(x, y) = \|x-y\|$. The distance between two vectors is the length of their difference.

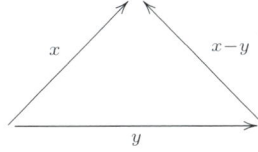

If no other metric is specified, we always regard a normed linear space as a metric space under this induced metric. Thus the concepts of compactness, open sets, continuity, completeness, and so on make sense on any normed linear space.

22.2.1. Proposition. *Let V be a normed linear space. Define $d\colon V \times V \to \mathbb{R}$ by $d(x, y) = \|x - y\|$. Then d is a metric on V.*

Proof. Problem.

The existence of a metric on a normed linear space V makes it possible to speak of neighborhoods of points in V. These neighborhoods satisfy some simple algebraic properties.

22.2.2. Proposition. *Let V be a normed linear space, let $x \in V$, and let $r, s > 0$. Then*

(a) $B_r(0) = -B_r(0)$,

(b) $B_{rs}(0) = r\,B_s(0)$,

(c) $x + B_r(0) = B_r(x)$, *and*

(d) $B_r(0) + B_r(0) = 2B_r(0)$.

Proof. Part (a) is an exercise. Parts (b), (c), and (d) are problems. *Hint.* For (d) divide the proof into two parts: $B_r(0) + B_r(0) \subseteq 2B_r(0)$ and the reverse inclusion. For the first, suppose x belongs to $B_r(0) + B_r(0)$. Then there exist $u,\ v \in B_r(0)$ such that $x = u + v$. Show that $x = 2w$ for some w in $B_r(0)$. You may wish to use Problem 22.2.5.

22.2.3. Proposition. *If V is a normed linear space, then the following hold.*

(a) $\big|\, \|x\| - \|y\| \,\big| \leq \|x - y\|$ *for all $x,\ y \in V$.*

(b) *The norm $x \mapsto \|x\|$ is a continuous function on V.*

(c) *If $x_n \to a$ in V, then $\|x_n\| \to \|a\|$.*

(d) $x_n \to 0$ *in V if and only if $\|x_n\| \to 0$ in $\mathbb{R}$.*

Proof. Problem.

22.2.4. Problem. Give an example to show that the converse of part (c) of Proposition 22.2.3 does not hold.

22.2.5. Problem. Prove that in a normed linear space every open ball is a convex set. And so is every closed ball.

22.3. Products

In this and the next three sections we substantially increase our store of examples of normed linear spaces by creating new spaces from old ones. In particular we will show that each of the following can be made into a normed linear space:

 (i) a vector subspace of a normed linear space,

 (ii) the product of two normed linear spaces,

 (iii) the set of bounded functions from a nonempty set into a normed linear space, and

 (iv) the set of continuous linear maps between two normed linear spaces.

It is obvious that (i) is a normed linear space: if V is a normed linear space with norm $\| \ \|$ and W is a vector subspace of V, then the restriction of $\| \ \|$ to W is a norm on W. Now consider (ii). Given normed linear spaces V and W, we wish to make the product vector space (see Example 20.1.20) into a normed linear space. As a preliminary we discuss *equivalent norms*.

22.3.1. Definition. Two norms on a vector space are EQUIVALENT if they induce equivalent metrics. If two norms on a vector space V are equivalent, then, since they induce equivalent metrics, they induce identical topologies on V. Thus, properties such as continuity, compactness, and connectedness are unaltered when norms are replaced by equivalent ones (see Proposition 11.2.3 and the discussion preceding it). In the next proposition we give a very simple necessary and sufficient condition for two norms to be equivalent.

22.3.2. Proposition. *Two norms $\| \ \|_1$ and $\| \ \|_2$ on a vector space V are equivalent if and only if there exist numbers α, $\beta > 0$ such that*

$$\|x\|_1 \leq \alpha \|x\|_2 \qquad and \qquad \|x\|_2 \leq \beta \|x\|_1$$

for all $x \in V$.

Proof. Exercise.

If V and W are normed linear spaces with norms $\| \ \|_V$ and $\| \ \|_W$, respectively, how do we provide the product vector space $V \times W$ (see Example 20.1.20) with a norm? There are at least three more or less obvious candidates: for $v \in V$ and $w \in W$, let

$$\|(v,w)\| := \left(\|v\|_V^{\ 2} + \|w\|_W^{\ 2} \right)^{1/2},$$
$$\|(v,w)\|_1 := \|v\|_V + \|w\|_W, \qquad \text{and}$$
$$\|(v,w)\|_u := \max\{\|v\|_V, \|w\|_W\}.$$

First of all, are these really norms on $V \times W$? Routine computations show that the answer is "yes". By way of illustration we write out the three verifications required for the first of the candidate norms. If v, $x \in V$, if w, $y \in W$, and if $\alpha \in \mathbb{R}$, then

$$
\begin{aligned}
\text{(a)} \quad \|(v, w) + (x, y)\| &= \|(v + x, w + y)\| \\
&= \left(\|v + x\|_V^2 + \|w + y\|_W^2 \right)^{1/2} \\
&\leq \left((\|v\|_V + \|x\|_V)^2 + (\|w\|_W + \|y\|_W)^2 \right)^{1/2} \\
&\leq \left(\|v\|_V^2 + \|w\|_W^2 \right)^{1/2} + \left(\|x\|_V^2 + \|y\|_W^2 \right)^{1/2} \\
&= \|(v, w)\| + \|(x, y)\|.
\end{aligned}
$$

The last inequality in this computation is, of course, Minkowski's inequality (Proposition 9.2.7).

$$
\begin{aligned}
\text{(b)} \quad \|\alpha(v, w)\| &= \|(\alpha v, \alpha w)\| \\
&= \left(\|\alpha v\|_V^2 + \|\alpha w\|_W^2 \right)^{1/2} \\
&= \left((|\alpha| \|v\|_V)^2 + (|\alpha| \|w\|_W)^2 \right)^{1/2} \\
&= |\alpha| \left(\|v\|_V^2 + \|w\|_W^2 \right)^{1/2} \\
&= |\alpha| \|(v, w)\|.
\end{aligned}
$$

(c) If $\|(v, w)\| = 0$, then $\|v\|_V^2 + \|w\|_W^2 = 0$. This implies that $\|v\|_V$ and $\|w\|_W$ are both zero. Thus v is the zero vector in V and w is the zero vector in W; so $(v, w) = (0, 0)$, the zero vector in $V \times W$.

Now which of these norms should we choose to be the product norm on $V \times W$? The next proposition shows that at least as far as topological considerations (continuity, compactness, connectedness, etc.) are concerned, it really doesn't matter.

22.3.3. Proposition. *The three norms on $V \times W$ defined above are equivalent.*

Proof. Notice that the norms $\| \ \|$, $\| \ \|_1$, and $\| \ \|_u$ defined above induce the metrics d, d_1, and d_u, respectively, defined in Chapter 9. In Proposition 9.3.2 we proved that these three metrics are equivalent. Thus the norms that induce them are equivalent (see Definition 22.3.1). $\qquad \square$

22.3.4. Definition. Since d_1 was chosen (in Definition 12.3.3) as our "official" product metric, we choose $\| \ \|_1$, which induces d_1, as the PRODUCT NORM on $V \times W$. In Proposition 22.3.8 you are asked to show that with this definition of the product norm, the operation of addition on a normed linear space is continuous. In the next proposition we verify that scalar multiplication (regarded as a map from $\mathbb{R} \times V$ into V) is also continuous.

22.3.5. Proposition. *If V is a normed linear space, then the mapping $(\beta, x) \mapsto \beta x$ from $\mathbb{R} \times V$ into V is continuous.*

Proof. Exercise. *Hint.* To show that a map $f \colon U \to W$ between two normed linear spaces is continuous at a point a in U, it must be shown that for every $\epsilon > 0$ there exists $\delta > 0$ such that $\|u - a\|_U < \delta$ implies $\|f(u) - f(a)\|_W < \epsilon$.

22.3.6. Corollary. *Let* (β_n) *be a sequence of real numbers, and let* (x_n) *be a sequence of vectors in a normed linear space* V. *If* $\beta_n \to \alpha$ *in* $\mathbb{R}$ *and* $x_n \to a$ *in* V, *then* $\beta_n x_n \to \alpha a$ *in* V.

Proof. Exercise.

22.3.7. Corollary. *If* V *is a normed linear space and* α *is a nonzero scalar, the map*
$$M_\alpha : V \to V : x \mapsto \alpha x$$
is a homeomorphism.

Proof. Problem.

22.3.8. Proposition. *Let* V *be a normed linear space. The operation of addition*
$$A : V \times V \to V : (x, y) \mapsto x + y$$
is continuous.

Proof. Problem.

22.3.9. Problem. Let V be a normed linear space. Prove the following.

(a) If $x_n \to a$ and $y_n \to b$ in V, then $x_n + y_n \to a + b$.

(b) If S is a vector subspace of V, then so is $\overline{S}$.

22.3.10. Problem. If K and L are compact subsets of a normed linear space, then the set
$$K + L := \{k + l : k \in K \text{ and } l \in L\}$$
is compact. *Hint.* Let A be as in Proposition 22.3.8. What is $A^\to(K \times L)$?

22.3.11. Problem. Let B and C be subsets of a normed linear space, and let $\alpha \in \mathbb{R}$. Prove the following.

(a) $\overline{\alpha B} = \alpha \overline{B}$.

(b) $\overline{B} + \overline{C} \subseteq \overline{B + C}$.

(c) $B + C$ need not be closed even if B and C are; thus, equality need not hold in (b). *Hint.* In $\mathbb{R}^2$ try part of the curve $y = 1/x$ and the negative x-axis.

22.3.12. Problem. Show that a linear bijection $f : V \to W$ between normed linear spaces is an isometry if and only if it is norm preserving (that is, if and only if $\|f(x)\|_W = \|x\|_V$ for all $x \in V$).

22.3.13. Definition. Let a be a vector in a vector space V. The map
$$T_a : V \to V : x \mapsto x + a$$
is called TRANSLATION by a.

22.3.14. Problem. Show that every translation map on a normed linear space is an isometry and, therefore, a homeomorphism.

22.3.15. Problem. Let U be a nonempty open set in a normed linear space. Then $U - U$ contains a neighborhood of 0. (By $U - U$ we mean $\{u - v : u, v \in U\}$.) *Hint.* Consider the union of all sets of the form $\left(T_{-v}\right)^\to(U)$ where $v \in U$. (As in Problem 22.3.14 the function T_{-v} is a translation map.)

22.3.16. Problem. Show that if B is a closed subset of a normed linear space V and C is a compact subset of V, then $B + C$ is closed. (Recall that part (c) of Problem 22.3.11 showed that this conclusion cannot be reached by assuming only that B and C are closed.) *Hint.* Use the sequential characterization of "closed" given in Proposition 12.2.2. Let (a_n) be a sequence in $B + C$ that converges to a point in V. Write $a_n = b_n + c_n$ where $b_n \in B$ and $c_n \in C$. Why does (c_n) have a subsequence (c_{n_k}) that converges to a point in C? Does (b_{n_k}) converge?

22.3.17. Problem. Let V and W be normed linear spaces, and let $A \subseteq V$, $a \in A'$, $\alpha \in \mathbb{R}$, and $f, g \colon A \to W$. Prove the following.

(a) If the limits of f and g exist as x approaches a, then so does the limit of $f + g$ and
$$\lim_{x \to a} (f + g)(x) = \lim_{x \to a} f(x) + \lim_{x \to a} g(x).$$

(b) If the limit of f exists as x approaches a, then so does the limit of αf and
$$\lim_{x \to a} (\alpha f)(x) = \alpha \lim_{x \to a} f(x).$$

22.3.18. Problem. Let V and W be normed linear spaces, and let $A \subseteq V$, $a \in A'$, and $f \colon A \to W$. Show that

(a) $\lim_{x \to a} f(x) = 0$ if and only if $\lim_{x \to a} \|f(x)\| = 0$; and

(b) $\lim_{h \to 0} f(a + h) = \lim_{x \to a} f(x)$.

Hint. These require only the most trivial modifications of the solutions to Problem 14.3.11 and Proposition 14.3.5.

22.3.19. Problem. Let V and W be normed linear spaces, and let $A \subseteq V$ and $f \colon A \to W$. Suppose that a is an accumulation point of A and that $l = \lim_{x \to a} f(x)$ exists in W.

(a) Show that if the norm on V is replaced by an equivalent one, then a is still an accumulation point of A.

(b) Show that if both the norm on V and the one on W are replaced by equivalent ones, then it is still true that $f(x) \to l$ as $x \to a$.

22.3.20. Problem. Let $f \colon U \times V \to W$ where U, V, and W are normed linear spaces. If the limit
$$l := \lim_{(x,y) \to (a,b)} f(x, y)$$
exists and if $\lim_{x \to a} f(x, y)$ and $\lim_{y \to b} f(x, y)$ exist for all $y \in V$ and $x \in U$, respectively, then the iterated limits
$$\lim_{x \to a} \left(\lim_{y \to b} f(x, y) \right) \quad \text{and} \quad \lim_{y \to b} \left(\lim_{x \to a} f(x, y) \right)$$
exist and are equal to l.

22.3.21. Problem. All norms on $\mathbb{R}^n$ are equivalent. *Hint.* It is enough to show that an arbitrary norm $\| \ \|$ on R^n is equivalent to $\| \ \|_1$, where $\|x\|_1 = \sum_{k=1}^n |x_k|$. Use Proposition 22.3.2. To find $\alpha > 0$ such that $\|x\| \leq \alpha \|x\|_1$ for all x, write $x = \sum_{k=1}^n x_k e^k$, where $e^1, \dots, e^n$ are the standard basis vectors on $\mathbb{R}^n$. To find $\beta > 0$ such that $\|x\|_1 \leq \beta \|x\|$, let $\mathbb{R}_1^n$ be the normed linear space $\mathbb{R}^n$ under the norm $\| \ \|_1$. Show that the function $x \mapsto \|\|x\|\|$ from $\mathbb{R}^n_1$ into $\mathbb{R}$ is continuous. Show that the unit sphere $S = \{x \in \mathbb{R}^n : \|x\|_1 = 1\}$ is compact in $\mathbb{R}^n_1$.

22.4. The space $\mathcal{B}(S, V)$

Throughout this section S will be a nonempty set and V will be a normed linear space. In the first section of this chapter we listed $\mathcal{B}(S, \mathbb{R})$ as an example of a normed linear space. Here we do little more than observe that the fundamental facts presented in Chapter 13, concerning pointwise and uniform convergence in the space $\mathcal{B}(S, \mathbb{R})$, all remain true when the set $\mathbb{R}$ is replaced by an arbitrary normed linear space. It is very easy to generalize these results: replace absolute values by norms.

22.4.1. Definition. Let S be a set , and let V be a normed linear space. A function $f \colon S \to V$ is BOUNDED if there exists a number $M > 0$ such that

$$\|f(x)\| \leq M$$

for all x in S. We denote by $\mathcal{B}(S, V)$ the family of all bounded V valued functions on S.

22.4.2. Exercise. Under the usual pointwise operations, $\mathcal{B}(S, V)$ is a vector space.

22.4.3. Definition. Let S be a set, and let V be a normed linear space. For every f in $\mathcal{B}(S, V)$ define

$$\|f\|_u := \sup\{\|f(x)\| : x \in S\} \, .$$

The function $f \mapsto \|f\|_u$ is called the UNIFORM NORM on $\mathcal{B}(S, V)$. This is the usual norm on $\mathcal{B}(S, V)$.

In the following problem you are asked to show that this function really is a norm. The metric d_u induced by the uniform norm $\| \ \|_u$ on $\mathcal{B}(S, V)$ is the UNIFORM METRIC on $\mathcal{B}(S, V)$.

22.4.4. Problem. Show that the uniform norm as in Definition 22.4.3 is in fact a norm on $\mathcal{B}(S, V)$.

22.4.5. Definition. Let (f_n) be a sequence of functions in $\mathcal{F}(S, V)$. If there is a function g in $\mathcal{F}(S, V)$ such that

$$\sup\{\|f_n(x) - g(x)\| : x \in S\} \to 0 \quad \text{as } n \to \infty,$$

then we say that the sequence (f_n) CONVERGES UNIFORMLY to g and write $f_n \to g$ (unif). The function g is the UNIFORM LIMIT of the sequence (f_n). Notice that if g and all the f_n's belong to $\mathcal{B}(S, V)$, then uniform convergence of (f_n) to g is just convergence of (f_n) to g with respect to the uniform metric. Notice also that the preceding repeats verbatim Definition 13.1.12, except that $\mathbb{R}$ has been replaced by V and absolute values by norms. We may similarly generalize Definition 13.2.1.

22.4.6. Definition. Let (f_n) be a sequence in $\mathcal{F}(S, V)$. If there is a function g such that

$$f_n(x) \to g(x) \qquad \text{for all } x \in S,$$

then (fn) CONVERGES POINTWISE to g. In this case we write

$$f_n \to g \text{ (ptws)}.$$

The function g is the POINTWISE LIMIT of the f_n's. Problem 22.4.7 repeats Proposition 13.2.2—uniform convergence implies pointwise convergence—except that it now holds for V-valued functions (not just real valued ones). Problem 22.4.8 generalizes Proposition 13.2.4(a).

22.4.7. Problem. If a sequence (f_n) in $\mathcal{F}(S,V)$ converges uniformly to a function g in $\mathcal{F}(S,V)$, then $f_n \to g$ (ptws).

22.4.8. Problem. Let (f_n) be a sequence in $\mathcal{B}(S,V)$, and let g be a member of $\mathcal{F}(S,V)$. If $f_n \to g$ (unif), then g is bounded.

22.4.9. Problem. Define

$$f(t) = \begin{cases} (t,0) & \text{if } 0 \le t \le 1, \\ (1, t-1) & \text{if } 1 < t \le 2, \\ (3-t, 1) & \text{if } 2 < t \le 3, \\ (0, 4-t) & \text{if } 3 < t \le 4. \end{cases}$$

Regarding f as a member of the space $\mathcal{B}([0,4], \mathbb{R}^2)$, find $\|f\|_u$.

22.4.10. Example. If M is a compact metric space, then the family $\mathcal{C}(M,V)$ of all continuous V-valued functions on M is a normed linear space.

Proof. Problem.

22.4.11. Problem. Let M be a compact metric space. Show that the family $\mathcal{C}(M, \mathbb{R})$ of all continuous real valued functions on M is a unital algebra and that $\|fg\|_u \le \|f\|_u \|g\|_u$. Show also that if A is a subalgebra of $\mathcal{C}(M, \mathbb{R})$, then so is $\overline{A}$.

22.4.12. Proposition. *If (f_n) is a sequence of continuous V-valued functions on a metric space M and if this sequence converges uniformly to a V-valued function g on M, then g is continuous.*

Proof. Problem. *Hint.* Modify the proof of Proposition 14.2.15.

Chapter 23

Continuity and linearity

23.1. Bounded linear transformations

Normed linear spaces have both algebraic and topological structure. It is therefore natural to be interested in those functions between normed linear spaces that preserve both types of structure, that is, that are both linear and continuous. In this section we study such functions.

23.1.1. Definition. A linear transformation $T: V \to W$ between normed linear spaces is BOUNDED if there exists a number $M > 0$ such that

$$\|Tx\| \leq M\|x\|$$

for all $x \in V$. The family of all bounded linear transformations from V into W is denoted by $\mathfrak{B}(V, W)$.

CAUTION. There is a possibility of confusion. Here we have defined bounded linear transformations; in section 22.4 we gave a quite different definition for "bounded" as it applies to arbitrary vector valued functions; and certainly linear transformations are such functions. The likelihood of becoming confused by these two different notions of boundedness is very small once one has made the following observation: Except for the zero function, it is impossible for a linear transformation to be bounded in the sense of section 22.4. (*Proof.* Let $T: V \to W$ be a nonzero linear transformation between normed linear spaces. Choose a in V so that $Ta \neq 0$. Then $\|Ta\| > 0$, so that by the Archimedean principle (Proposition J.4.1) the number $\|T(na)\| = n\|Ta\|$ can be made as large as desired by choosing n sufficiently large. Thus there is certainly no $M > 0$ such that $\|Tx\| \leq M$ for all x.) Since nonzero linear transformations cannot be bounded in the sense of section 22.4, an assertion that a linear map is bounded should always be interpreted in the sense of boundedness introduced in this section.

A function $f: S \to W$, which maps a set S into a normed linear space W, is bounded if and only if it maps *every* subset of S into a bounded subset of W.

However, a *linear* map $T\colon V \to W$ from one normed linear space into another is bounded if and only if it maps every bounded subset of V into a bounded subset of W.

23.1.2. Proposition. *A linear transformation $T\colon V \to W$ between normed linear spaces is bounded if and only if $T^{\to}(A)$ is a bounded subset of W whenever A is a bounded subset of V.*

Proof. Problem.

23.1.3. Example. Let $T\colon \mathbb{R}^2 \to \mathbb{R}^3\colon (x, y) \mapsto (3x + y, x - 3y, 4y)$. It is easily seen that T is linear. For all (x, y) in $\mathbb{R}^2$

$$
\begin{aligned}
\|T(x, y)\| &= \|(3x + y, x - 3y, 4y)\| \\
&= \left((3x + y)^2 + (x - 3y)^2 + (4y)^2\right)^{\frac{1}{2}} \\
&= (10x^2 + 26y^2)^{\frac{1}{2}} \\
&\le \sqrt{26}(x^2 + y^2)^{\frac{1}{2}} \\
&= \sqrt{26}\|(x, y)\|\,.
\end{aligned}
$$

So the linear transformation T is bounded.

Why is boundedness of linear transformations an important concept? Because it turns out to be equivalent to continuity and is usually easier to establish.

23.1.4. Proposition. *Let $T\colon V \to W$ be a linear transformation between normed linear spaces. The following are equivalent.*

(a) *T is continuous.*

(b) *There is at least one point at which T is continuous.*

(c) *T is continuous at $\mathbf{0}$.*

(d) *T is bounded.*

Proof. Exercise. *Hint.* To prove that (c) implies (d), argue by contradiction. Show that for each n there exists x_n in V such that $\|Tx_n\| > n\|x_n\|$. Let $y_n = (n\|x_n\|)^{-1} x_n$. Give one argument to show that the sequence (Ty_n) converges to zero as $n \to \infty$. Give another to show that it does not.

23.1.5. Definition. Let $T \in \mathfrak{B}(V, W)$, where V and W are normed linear spaces. Define

$$
\|T\| := \inf\{M > 0\colon \|Tx\| \le M\|x\| \quad \text{for all } x \in V\}.
$$

This number is called the NORM of T. We show in Proposition 23.1.14 that the map $T \mapsto \|T\|$ really is a norm on $\mathfrak{B}(V, W)$.

There are at least four ways to compute the norm of a linear transformation T. Use the definition or any of the three formulas given in the next lemma.

23.1.6. Lemma. *Let T be a bounded linear map between nonzero normed linear spaces. Then*

$$\|T\| = \sup\{\|x\|^{-1}\|Tx\| : x \neq \mathbf{0}\}$$
$$= \sup\{\|Tu\| : \|u\| = 1\}$$
$$= \sup\{\|Tu\| : \|u\| \leq 1\}.$$

Proof. Exercise. *Hint.* To obtain the first equality, use the fact that if a subset A of $\mathbb{R}$ is bounded above, then $\sup A = \inf\{M : M$ is an upper bound for $A\}$.

23.1.7. Corollary. *If $T \in \mathfrak{B}(V,W)$ where V and W are normed linear spaces, then*

$$\|Tx\| \leq \|T\|\,\|x\|$$

for all x in V.

Proof. By the preceding lemma $\|T\| \geq \|x\|^{-1}\|Tx\|$ for all $x \neq 0$. Thus $\|Tx\| \leq \|T\|\,\|x\|$ for all x. $\qquad\square$

The following example shows how to use Lemma 23.1.6 (in conjunction with the definition) to compute the norm of a linear transformation.

23.1.8. Example. Let T be the linear map defined in Example 23.1.3. We have already seen that $\|T(x,y)\| \leq \sqrt{26}\|(x,y)\|$ for all (x,y) in $\mathbb{R}^2$. Since $\|T\|$ is defined to be the infimum of the set of all numbers M such that $\|T(x,y)\| \leq M\|(x,y)\|$ for all $(x,y) \in \mathbb{R}^2$, and since $\sqrt{26}$ is such a number, we know that

$$(25) \qquad\qquad \|T\| \leq \sqrt{26}.$$

On the other hand Lemma 23.1.6 tells us that $\|T\|$ is the supremum of the set of all numbers $\|Tu\|$ where u is a unit vector. Since $(0,1)$ is a unit vector and $\|T(0,1)\| = \|(1,-3,4)\| = \sqrt{26}$, we conclude that

$$(26) \qquad\qquad \|T\| \geq \sqrt{26}.$$

Conditions (25) and (26) imply that $\|T\| = \sqrt{26}$.

23.1.9. Problem. Let $T\colon \mathbb{R}^2 \to \mathbb{R}^3$ be the linear transformation defined by $T(x,y) = (3x, x+2y, x-2y)$. Find $\|T\|$.

23.1.10. Problem. Let $T\colon \mathbb{R}^3 \to \mathbb{R}^4\colon x \mapsto (x_1-4x_2, 2x_1+3x_3, x_1+4x_2, x_1-6x_3)$. Find $\|T\|$.

23.1.11. Exercise. Find the norm of each of the following.

(a) The identity map on a normed linear space.

(b) The zero map in $\mathfrak{B}(V,W)$.

(c) A coordinate projection $\pi_k : V_1 \times V_2 \to V_k$ ($k = 1,2$), where V_1 and V_2 are nontrivial normed linear spaces, that is, normed linear spaces that contain vectors other than the zero vector.

23.1.12. Exercise. Let $\mathcal{C} = \mathcal{C}([a, b], \mathbb{R})$. Define $J \colon \mathcal{C} \to \mathbb{R}$ by

$$Jf = \int_a^b f(x)\, dx\,.$$

Show that $J \in \mathfrak{B}(\mathcal{C}, \mathbb{R})$ and find $\|J\|$.

23.1.13. Problem. Let $\mathcal{C}^1$ and $\mathcal{C}$ be as in Problem 21.1.23. Let D be the differentiation operator

$$D \colon \mathcal{C}^1 \to \mathcal{C} \colon f \mapsto f',$$

where f' is the derivative of f. Let both $\mathcal{C}^1$ and $\mathcal{C}$ have the uniform norm. Is the linear transformation D bounded? *Hint.* Let $[a, b] = [0, 1]$ and consider the functions $f_n(x) = x^n$ for $n \in \mathbb{N}$ and $0 \le x \le 1$.

Next we show that the set $\mathfrak{B}(V, W)$ of all bounded linear transformations between two normed linear spaces is itself a normed linear space.

23.1.14. Proposition. *If V and W are normed linear spaces, then, under pointwise operations, $\mathfrak{B}(V, W)$ is a vector space and the map $T \mapsto \|T\|$ from $\mathfrak{B}(V, W)$ into $\mathbb{R}$ defined above is a norm on $\mathfrak{B}(V, W)$.*

Proof. Exercise.

One obvious fact that we state for future reference is that the composite of bounded linear transformations is bounded and linear.

23.1.15. Proposition. *If $S \in \mathfrak{B}(U, V)$ and $T \in \mathfrak{B}(V, W)$, then $TS \in \mathfrak{B}(U, W)$ and $\|TS\| \le \|T\|\,\|S\|$*

Proof. Exercise.

In Propositions 21.1.16 and 21.1.17 we saw that the kernel and range of a linear map T are vector subspaces of the domain and codomain of T, respectively. It is interesting to note that the kernel is always a closed subspace while the range need not be.

23.1.16. Proposition. *If V and W are normed linear spaces and $T \in \mathfrak{B}(V, W)$, then $\ker T$ is a closed linear subspace of V.*

Proof. Problem.

23.1.17. Example. Let c_0 be the vector space of all sequences x of real numbers (with pointwise operations) that converge to zero. Give c_0 the uniform norm (see Example 22.1.9) so that for every $x \in c_0$

$$\|x\|_u = \sup\{x_k : k \in \mathbb{N}\}\,.$$

The family l of sequences of real numbers that have only finitely many nonzero coordinates is a vector subspace of c_0, but it is not a closed subspace. Thus the range of the inclusion map of l into c_0 does not have closed range.

Proof. Problem.

In general the calculus of infinite-dimensional spaces is no more complicated than calculus on $\mathbb{R}^n$. One respect in which the Euclidean spaces $\mathbb{R}^n$ turn out to be simpler, however, is the fact that every linear map from $\mathbb{R}^n$ into $\mathbb{R}^m$ is automatically continuous. Between finite-dimensional spaces there are no discontinuous linear maps. And this is true regardless of the particular norms placed on these spaces.

23.1.18. Proposition. *Let $\mathbb{R}^n$ and $\mathbb{R}^m$ have any norms whatever. If $T\colon \mathbb{R}^n \to \mathbb{R}^m$ is linear, then it is continuous.*

Proof. Problem. *Hint.* Let $\left[t_k^j\right] = [T]$ be the matrix representation of T, and let

$$M = \max\{|t_k^j| \colon 1 \le j \le m \text{ and } 1 \le k \le n\}.$$

Let $\mathbb{R}_u^m$ be $\mathbb{R}^m$ provided with the uniform norm

$$\|x\|_u := \max\{|x_k| \colon 1 \le k \le m\},$$

and let $\mathbb{R}_1^n$ be $\mathbb{R}^n$ equipped with the norm

$$\|x\|_1 := \sum_{k=1}^n |x_k|.$$

Show that T regarded as a map from $\mathbb{R}_1^n$ to $\mathbb{R}_u^m$ is bounded (with $\|T\| \le M$). Then use Problem 22.3.21.

23.1.19. Problem. Let V and W be normed linear spaces, and let $x \in V$. Define a map E_x (called EVALUATION AT x) by

$$E_x \colon \mathfrak{B}(V, W) \to W \colon T \mapsto Tx.$$

Show that $E_x \in \mathfrak{B}\big(\mathfrak{B}(V, W), W\big)$ and that $\|E_x\| \le \|x\|$.

23.1.20. Problem. What changes in the preceding problem if we let M be a compact metric space and W be a nonzero normed linear space and consider the evaluation map $E_x \colon \mathcal{C}(M, W) \to W \colon f \mapsto f(x)$?

23.1.21. Problem. Let S be a nonempty set, and let $T \colon V \to W$ be a bounded linear transformation between normed linear spaces. Define a function C_T on the normed linear space $\mathcal{B}(S, V)$ by

$$C_T(f) := T \circ f$$

for all f in $\mathcal{B}(S, V)$.

(a) Show that $C_T(f)$ belongs to $\mathcal{B}(S, W)$ whenever f is a member of $\mathcal{B}(S, V)$.

(b) Show that the map $C_T \colon \mathcal{B}(S, V) \to \mathcal{B}(S, W)$ is linear and continuous.

(c) Find $\|C_T\|$.

(d) Show that if $f_n \to g$ (unif) in $\mathcal{B}(S, V)$, then $T \circ f_n \to T \circ g$ (unif) in $\mathcal{B}(S, W)$.

(e) Show that C_T is injective if and only if T is.

23.1.22. Problem. Let M and N be compact metric spaces, and let $\phi \colon M \to N$ be continuous. Define T_ϕ on $\mathcal{C}(N, \mathbb{R})$ by

$$T_\phi(g) := g \circ \phi$$

for all g in $\mathcal{C}(N, \mathbb{R})$.

(a) T_ϕ maps $\mathcal{C}(N, \mathbb{R})$ into $\mathcal{C}(M, \mathbb{R})$.

(b) T_ϕ is a bounded linear transformation.

(c) $\|T_\phi\| = 1$.

(d) If ϕ is surjective, then T_ϕ is injective.

(e) If T_ϕ is injective, then ϕ is surjective. *Hint.* Suppose ϕ is not surjective. Choose y in $N \setminus \operatorname{ran} \phi$. Show that problem 14.1.31 can be applied to the sets $\{y\}$ and $\operatorname{ran} \phi$.

(f) If T_ϕ is surjective, then ϕ is injective. *Hint.* Here again Proposition 14.1.31 is useful. (It is also true that if ϕ is injective, then T_ϕ is surjective. But more machinery is needed before we can prove this.)

23.1.23. Problem. Let (S_k) be a sequence in $\mathfrak{B}(V, W)$, and let U be a member of $\mathfrak{B}(W, X)$ where V, W, and X are normed linear spaces. If $S_k \to T$ in $\mathfrak{B}(V, W)$, then $US_k \to UT$. Also, state and prove a similar result whose conclusion is, "then $S_k U \to TU$".

23.1.24. Definition. A family $\mathfrak{T}$ of linear maps from a vector space into itself is a COMMUTING FAMILY if $ST = TS$ for all S, $T \in \mathfrak{T}$.

23.1.25. Problem (Markov–Kakutani fixed point theorem). Prove: If $\mathfrak{T}$ is a commuting family of bounded linear maps from a normed linear space V into itself and K is a nonempty convex compact subset of V that is mapped into itself by every member of $\mathfrak{T}$, then there is at least one point in K that is fixed under every member of $\mathfrak{T}$. *Hint.* For every T in $\mathfrak{T}$ and n in $\mathbb{N}$ define

$$T_n = n^{-1} \sum_{j=0}^{n-1} T^j,$$

where $T^0 := I$. Let $\mathfrak{U} = \{T_n \colon T \in \mathfrak{T} \text{ and } n \in \mathbb{N}\}$. Show that $\mathfrak{U}$ is a commuting family of bounded linear maps on V, each of which maps K into itself, and that if $U_1, \dots, U_n \in \mathfrak{U}$, then

$$(U_1 \dots U_n)^{\rightarrow}(K) \subseteq \bigcap_{j=1}^{n} U_j^{\rightarrow}(K).$$

Let $\mathfrak{C} = \{U^{\rightarrow}(K) \colon U \in \mathfrak{U}\}$, and use Problem 15.3.2 to show that $\bigcap \mathfrak{C} \neq \emptyset$.

Finally, show that every element of $\bigcap \mathfrak{C}$ is fixed under each T in $\mathfrak{T}$; that is, if $a \in \bigcap \mathfrak{C}$ and $T \in \mathfrak{T}$, then $Ta = a$. To this end argue that for every $n \in \mathbb{N}$ there exists $c_n \in K$ such that $a = T_n c_n$ and therefore $Ta - a$ belongs to $n^{-1}(K - K)$ for each n. Use Problems 22.3.10 and 15.1.5 to show that every neighborhood of 0 contains, for sufficiently large n, sets of the form $n^{-1}(K - K)$. What do these last two observations say about $Ta - a$?

23.2. The Stone–Weierstrass theorem

In Example 15.3.5 we found that the square root function can be uniformly approximated on $[0, 1]$ by polynomials. In this section we prove the remarkable *Weierstrass approximation theorem*, which says that *every* continuous real valued function can be uniformly approximated on compact intervals by polynomials. We will in fact prove an even stronger result due to M. H. Stone that generalizes the Weierstrass theorem to arbitrary compact metric spaces.

23.2.1. Proposition. *Let A be a subalgebra of $\mathcal{C}(M,\mathbb{R})$, where M is a compact metric space. If $f \in A$, then $|f| \in \overline{A}$.*

Proof. Exercise. *Hint.* Let (p_n) be a sequence of polynomials converging uniformly on $[0,1]$ to the square root function. (See Example 15.3.5.) What can you say about the sequence $(p_n \circ g^2)$ where $g = f/\|f\|$?

23.2.2. Corollary. *If A is a subalgebra of $\mathcal{C}(M,\mathbb{R})$ where M is a compact metric space and if $f, g \in A$, then $f \vee g$ and $f \wedge g$ belong to $\overline{A}$.*

Proof. As in the solution to Problem 14.2.10, write $f \vee g = \frac{1}{2}(f + g + |f - g|)$ and $f \wedge g = \frac{1}{2}(f + g - |f - g|)$; then apply the preceding proposition. $\square$

23.2.3. Definition. A family $\mathcal{F}$ of real valued functions defined on a set S is a SEPARATING family if corresponding to every pair of distinct points x and y and S there is a function f in $\mathcal{F}$ such that $f(x) \neq f(y)$. In this circumstance we may also say that the family $\mathcal{F}$ SEPARATES POINTS of S.

23.2.4. Proposition. *Let A be a separating unital subalgebra of $\mathcal{C}(M,\mathbb{R})$ where M is a compact metric space. If a and b are distinct points in M and $\alpha, \beta \in \mathbb{R}$, then there exists a function $f \in A$ such that $f(a) = \alpha$ and $f(b) = \beta$.*

Proof. Problem. *Hint.* Let g be any member of A such that $g(a) \neq g(b)$. Notice that if k is a constant, then the function $f\colon x \mapsto \alpha + k(g(x) - g(a))$ satisfies $f(a) = \alpha$. Choose k so that $f(b) = \beta$.

Suppose that M is a compact metric space. The *Stone–Weierstrass theorem* says that any separating unital subalgebra of the algebra of continuous real valued functions on M is dense. That is, if A is such a subalgebra, then we can approximate each function $f \in \mathcal{C}(M,\mathbb{R})$ arbitrarily closely by members of A. The proof falls rather naturally into two steps. First (in Lemma 23.2.5) we find a function in $\overline{A}$ that does not exceed f by much; precisely, given $a \in M$ and $\epsilon > 0$, we find a function g in $\overline{A}$ that agrees with f at a and satisfies $g(x) < f(x) + \epsilon$ elsewhere. Then (in Theorem 23.2.6) given $\epsilon > 0$, we find $h \in \overline{A}$ such that $f(x) - \epsilon < h(x) < f(x) + \epsilon$.

23.2.5. Lemma. *Let A be a unital separating subalgebra of $\mathcal{C}(M,\mathbb{R})$ where M is a compact metric space. For every $f \in \mathcal{C}(M,\mathbb{R})$, every $a \in M$, and every $\epsilon > 0$ there exists a function $g \in \overline{A}$ such that $g(a) = f(a)$ and $g(x) < f(x) + \epsilon$ for all $x \in M$.*

Proof. Exercise. *Hint.* For each $y \in M$ find a function ϕ_y that agrees with f at a and at y. Then $\phi_y(x) < f(x) + \epsilon$ for all x in some neighborhood U_y of y. Find finitely many of these neighborhoods $U_{y_1}, \ldots, U_{y_n}$ that cover M. Let $g = \phi_{y_1} \wedge \cdots \wedge \phi_{y_n}$.

23.2.6. Theorem (Stone–Weierstrass theorem). *Let A be a unital separating subalgebra of $\mathcal{C}(M,\mathbb{R})$ where M is a compact metric space. Then A is dense in $\mathcal{C}(M,\mathbb{R})$.*

Proof. All we need to show is that $\mathcal{C}(M,\mathbb{R}) \subseteq \overline{A}$. So we choose $f \in \mathcal{C}(M,\mathbb{R})$ and try to show that $f \in \overline{A}$. It will be enough to show that for every $\epsilon > 0$, we can find a function $h \in \overline{A}$ such that $\|f - h\|_u < \epsilon$. (*Reason.* Then f belongs to $\overline{\overline{A}}$ (by Proposition 11.1.22) and therefore to $\overline{A}$ (by Proposition 10.3.2(b)).)

Let $\epsilon > 0$. For each $x \in M$, we may, according to Lemma 23.2.5, choose a function $g_x \in \overline{A}$ such that $g_x(x) = f(x)$ and

$$(27) \qquad\qquad\qquad g_x(y) < f(y) + \epsilon$$

for every $y \in M$. Since both f and g_x are continuous and they agree at x, there exists an open set U_x containing x such that

$$(28) \qquad\qquad\qquad f(y) < g_x(y) + \epsilon$$

for every $y \in U_x$. (Why?)

Since the family $\{U_x : x \in M\}$ covers M and M is compact, there exist points $x_1, \ldots, x_n$ in M such that $\{U_{x_1}, \ldots, U_{x_n}\}$ covers M. Let $h = g_{x_1} \vee \cdots \vee g_{x_n}$. We know from Definition 22.4.11 that $\overline{A}$ is a subalgebra of $\mathcal{C}(M, \mathbb{R})$. So according to Corollary 23.2.2, $h \in \overline{\overline{A}} = \overline{A}$.

By inequality (27)

$$g_{x_k}(y) < f(y) + \epsilon$$

holds for all $y \in M$ and $1 \le k \le n$. Thus,

$$(29) \qquad\qquad\qquad h(y) < f(y) + \epsilon$$

for all $y \in M$. Each $y \in M$ belongs to at least one of the open sets U_{x_k}. Thus, by (28)

$$(30) \qquad\qquad\qquad f(y) < g_{x_k}(y) + \epsilon < h(y) + \epsilon$$

for every $y \in M$. Together (29) and (30) show that

$$-\epsilon < f(y) - h(y) < \epsilon$$

for all $y \in M$. That is, $\|f - h\|_u < \epsilon$. $\square$

23.2.7. Problem. Give the missing reason for inequality (28) in the preceding proof.

23.2.8. Theorem (Weierstrass approximation theorem). *Every continuous real valued function on $[a, b]$ can be uniformly approximated by polynomials.*

Proof. Problem.

23.2.9. Problem. Let $\mathcal{G}$ be the set of all functions f in $\mathcal{C}\big([0,1]\big)$ such that f is differentiable on $(0,1)$ and $f'(\frac{1}{2}) = 0$. Show that $\mathcal{G}$ is dense in $\mathcal{C}\big([0,1]\big)$.

23.2.10. Proposition. *If M is a closed and bounded subset of $\mathbb{R}$, then the normed linear space $\mathcal{C}(M, \mathbb{R})$ is separable.*

Proof. Problem.

23.3. Banach spaces

23.3.1. Definition. A BANACH SPACE is a normed linear space that is complete with respect to the metric induced by its norm.

23.3.2. Example. In Example 18.2.2 we saw that $\mathbb{R}$ is complete; so it is a Banach space.

23.3.3. Example. In Example 18.2.11 it was shown that the space $\mathbb{R}^n$ is complete with respect to the Euclidean norm. Since all norms on $\mathbb{R}^n$ are equivalent (Problem 22.3.21) and since completeness of a space is not affected by changing to an equivalent metric (Proposition 18.2.10), we conclude that $\mathbb{R}^n$ is a Banach space under all possible norms.

23.3.4. Example. If S is a nonempty set, then (under the uniform norm) $\mathcal{B}(S, \mathbb{R})$ is a Banach space (see Example 18.2.12).

23.3.5. Example. If M is a compact metric space, then (under the uniform norm) $\mathcal{C}(M, \mathbb{R})$ is a Banach space (see Example 18.2.13).

At the beginning of section 22.3 we listed four ways of making new normed linear spaces from old ones. Under what circumstances do these new spaces turn out to be Banach spaces?

 (i) A closed vector subspace of a Banach space is a Banach space (see Proposition 18.2.8).

 (ii) The product of two Banach spaces is a Banach space (see Proposition 18.2.9).

 (iii) If S is a nonempty set and E is a Banach space, then $\mathcal{B}(S, E)$ is a Banach space (see Problem 23.3.7).

 (iv) If V is a normed linear space and F is a Banach space, then $\mathfrak{B}(V, F)$ is a Banach space (see the following proposition).

23.3.6. Proposition. *Let V and W be normed linear spaces. Then $\mathfrak{B}(V, W)$ is complete if W is.*

Proof. Exercise. *Hint.* Show that if (T_n) is a Cauchy sequence in $\mathfrak{B}(V, W)$, then $\lim_{n \to \infty} T_n x$ exists for every x in V. Let $Sx = \lim_{n \to \infty} T_n x$. Show that the map $S \colon x \mapsto Sx$ is linear. If $\epsilon > 0$, then for m and n sufficiently large $\|T_m - T_n\| < \frac{1}{2}\epsilon$. For such m and n show that

$$\|Sx - T_n x\| \leq \|Sx - T_m x\| + \tfrac{1}{2}\epsilon\|x\|,$$

and conclude from this that S is bounded and that $T_n \to S$ in $\mathfrak{B}(V, W)$.

23.3.7. Problem. Let S be a nonempty set, and let V be a normed linear space. If V is complete, so is $\mathcal{B}(S, V)$. *Hint.* Make suitable modifications in the proof of Example 18.2.12.

23.3.8. Problem. Let M be a compact metric space, and let V be a normed linear space. If V is complete, so is $\mathcal{C}(M, V)$. *Hint.* Use Proposition 22.4.12.

23.3.9. Problem. Let m be the set of all bounded sequences of real numbers. (That is, a sequence $(x_1, x_2, \dots)$ of real numbers belongs to m provided that there exists a constant $M > 0$ such that $|x_k| \leq M$ for all $k \in \mathbb{N}$.) If $x = (x_k)$ and $y = (y_k)$ belong to m and α is a scalar, define

$$x + y = (x_1 + y_1, x_2 + y_2, \ldots)$$

and

$$\alpha x = (\alpha x_1, \alpha x_2, \ldots).$$

Also, for each sequence x in m define

$$\|x\|_u = \sup\{|x_k| \colon k \in \mathbb{N}\}.$$

(a) Show that m is a Banach space. *Hint.* The proof should be very short. This is a special case of a previous example.

(b) The space m is not separable. *Hint.* Argue that it suffices to find an uncountable, pairwise disjoint family of open sets in the space. To this end let A be the subset of m comprising all sequences containing only 0's and 1's. Consider the family

$$\mathfrak{U} = \{B_{\frac{1}{2}}(a) \colon a \in A\}.$$

Perhaps the easiest proof of the uncountability of this family uses the fact that real numbers have binary expansions.

(c) The closed unit ball of the space m is closed and bounded but not compact. *Hint.* Use Corollary 16.1.15 and Theorem 16.2.1.

23.4. Dual spaces and adjoints

> In this section U, V, and W are normed linear spaces.

Particularly important among the spaces of bounded linear transformations introduced in section 23.1 are those consisting of maps from a space V into its scalar field $\mathbb{R}$. The space $\mathfrak{B}(V, \mathbb{R})$ is the DUAL SPACE of V and is usually denoted by V^*. Members of V^* are called BOUNDED LINEAR FUNCTIONALS. (When the word "function" is given the suffix "-al", we understand that the function is scalar valued.)

23.4.1. Example. Let $f \colon \mathbb{R}^3 \to \mathbb{R} \colon (x, y, z) \mapsto x + y + z$. Then f is a member of $(\mathbb{R}^3)^*$. (It is obviously linear and is bounded by Proposition 23.1.18.)

23.4.2. Example. The familiar Riemann integral of beginning calculus is a bounded linear functional on the space $\mathcal{C} = \mathcal{C}([a, b], \mathbb{R})$. That is, the functional J defined in Exercise 23.1.12 belongs to $\mathcal{C}^*$.

23.4.3. Example. Let M be a compact metric space, and let $\mathcal{C} = \mathcal{C}(M, \mathbb{R})$. For each x in M the evaluation functional $E_x \colon f \mapsto f(x)$ belongs to $\mathcal{C}^*$. (Take $W = \mathbb{R}$ in Problem 23.1.20.)

23.4.4. Definition. Let $T \in \mathfrak{B}(V, W)$. Define T^* by $T^*(g) = g \circ T$ for every g in W^*. The map T^* is the ADJOINT of T. In the next two propositions and Problem 23.4.7 we state only the most elementary properties of the adjoint map $T \mapsto T^*$. We will see more of it later.

23.4.5. Proposition. *If $T \in \mathfrak{B}(V, W)$, then T^* maps W^* into V^*. Furthermore, if $g \in W^*$, then $\|T^*g\| \leq \|T\| \, \|g\|$.*

Proof. Exercise.

23.4.6. Proposition. *Let $S \in \mathfrak{B}(U, V)$, $T \in \mathfrak{B}(V, W)$, and I_V be the identity map on V. Then*

(a) *T^* belongs to $\mathfrak{B}(W^*, V^*)$ and $\|T^*\| \leq \|T\|$;*

(b) *$(I_V)^*$ is I_{V^*}, the identity map on V^*; and*

(c) *$(TS)^* = S^*T^*$.*

Proof. Problem.

23.4.7. Problem. The adjoint map $T \mapsto T^*$ from $\mathfrak{B}(V, W)$ into $\mathfrak{B}(W^*, V^*)$ is itself a bounded linear transformation, and it has norm not exceeding 1.

Chapter 24

The Cauchy integral

In this chapter we develop a theory of integration for vector valued functions. In one way our integral, the Cauchy integral, will be more general and in another way slightly less general than the classical Riemann integral, which is presented in beginning calculus. The Riemann integral is problematic: the derivation of its properties is considerably more complicated than the corresponding derivation for the Cauchy integral, but very little added generality is obtained as a reward for the extra work. On the other hand the Riemann integral is not nearly general enough for advanced work in analysis; there the full power of the Lebesgue integral is needed. Before starting our discussion of integration, we derive some standard facts concerning uniform continuity.

24.1. Uniform continuity

24.1.1. Definition. A function $f\colon M_1 \to M_2$ between metric spaces is UNIFORMLY CONTINUOUS if for every $\epsilon > 0$ there exists $\delta > 0$ such that $d(f(x), f(y)) < \epsilon$ whenever $d(x, y) < \delta$.

Compare the definitions of "continuity" and "uniform continuity". A function $f\colon M_1 \to M_2$ is continuous if

$$\forall a \in M_1 \, \forall \epsilon > 0 \, \exists \delta > 0 \, \forall x \in M_1 \, d(x,a) < \delta \implies d(f(x), f(a)) < \epsilon \,.$$

We may just as well write this reversing the order of the first two (universal) quantifiers.

$$\forall \epsilon > 0 \, \forall a \in M_1 \, \exists \delta > 0 \, \forall x \in M_1 \, d(x,a) < \delta \implies d(f(x), f(a)) < \epsilon \,.$$

The function f is uniformly continuous if

$$\forall \epsilon > 0 \, \exists \delta > 0 \, \forall a \in M_1 \, \forall x \in M_1 \, d(x,a) < \delta \implies d(f(x), f(a)) < \epsilon \,.$$

Thus the difference between continuity and uniform continuity is the order of two quantifiers. This makes the following result obvious.

24.1.2. Proposition. *Every uniformly continuous function between metric spaces is continuous.*

24.1.3. Example. The function $f \colon \mathbb{R} \to \mathbb{R} \colon x \mapsto 3x - 4$ is uniformly continuous.

Proof. Given $\epsilon > 0$, choose $\delta = \epsilon/3$. If $|x - y| < \delta$, then $|f(x) - f(y)| = |(3x - 4) - (3y - 4)| = 3|x - y| < 3\delta = \epsilon$. $\square$

24.1.4. Example. The function $f \colon [1, \infty) \to \mathbb{R} \colon x \mapsto x^{-1}$ is uniformly continuous.

Proof. Exercise.

24.1.5. Example. The function $g \colon (0, 1] \to \mathbb{R} \colon x \mapsto x^{-1}$ is not uniformly continuous.

Proof. Exercise.

24.1.6. Problem. Let M be an arbitrary positive number. The function

$$f \colon [0, M] \to \mathbb{R} \colon x \mapsto x^2$$

is uniformly continuous. Prove this assertion using only the definition of uniform continuity.

24.1.7. Problem. The function

$$g \colon [0, \infty) \to \mathbb{R} \colon x \mapsto x^2$$

is *not* uniformly continuous.

24.1.8. Proposition. *Norms on vector spaces are always uniformly continuous.*

Proof. Problem.

We have already seen in Proposition 24.1.2 that uniform continuity implies continuity. Example 24.1.5 shows that the converse is not true in general. There are however two special (and important!) cases where the concepts coincide. One is linear maps between normed linear spaces, and the other is functions defined on compact metric spaces.

24.1.9. Proposition. *A linear transformation between normed linear spaces is continuous if and only if it is uniformly continuous.*

Proof. Problem.

Of course, the preceding result does not hold in general metric spaces (where "linearity" makes no sense). The next proposition, for which we give a preparatory lemma, is a metric space result.

24.1.10. Lemma. *Let (x_n) and (y_n) be sequences in a compact metric space. If $d(x_n, y_n) \to 0$ as $n \to \infty$, then there exist convergent subsequences of (x_n) and (y_n) that have the same limit.*

Proof. Exercise.

24.1.11. Proposition. *Let M_1 be a compact metric space, and let M_2 be an arbitrary metric space. Every continuous function $f\colon M_1 \to M_2$ is uniformly continuous.*

Proof. Exercise.

In section 24.3, where we define the Cauchy integral, an important step in the development is the extension of the integral from an exceedingly simple class of functions, the step functions, to a class of functions large enough to contain all the continuous functions. The two basic ingredients of this extension are the density of the step functions in the large class and the uniform continuity of the integral. Theorem 24.1.15 below is the crucial result that allows this extension. First, two preliminary results.

24.1.12. Proposition. *If $f\colon M_1 \to M_2$ is a uniformly continuous map between two metric spaces and (x_n) is a Cauchy sequence in M_1, then $\big(f(x_n)\big)$ is a Cauchy sequence in M_2.*

Proof. Problem.

24.1.13. Problem. Show by example that Proposition 24.1.12 is no longer true if the word "uniformly" is deleted.

24.1.14. Lemma. *Let M_1 and M_2 be metric spaces, let $S \subseteq M_1$, and let $f\colon S \to M_2$ be uniformly continuous. If two sequences (x_n) and (y_n) in S converge to the same limit in M_1 and if the sequence $\big(f(x_n)\big)$ converges, then the sequence $\big(f(y_n)\big)$ converges and $\lim f(x_n) = \lim f(y_n)$.*

Proof. Exercise. *Hint.* Consider the *interlaced* sequence

$$(x_1, y_1, x_2, y_2, x_3, y_3, \dots).$$

We are now in a position to show that a uniformly continuous map f from a subset of a metric space into a complete metric space can be extended in a unique fashion to a continuous function on the closure of the domain of f.

24.1.15. Theorem. *Let M_1 and M_2 be metric spaces, let S be a subset of M_1, and let $f\colon S \to M_2$. If f is uniformly continuous and M_2 is complete, then there exists a unique continuous extension of f to $\overline{S}$. Furthermore, this extension is uniformly continuous.*

Proof. Problem. *Hint.* Define $g\colon \overline{S} \to M_2$ by $g(a) = \lim f(x_n)$ where (x_n) is a sequence in S converging to a. First show that g is well-defined. To this end you must show that

(i) $\lim f(x_n)$ does exist, and

(ii) the value assigned to g at a does not depend on the particular sequence (x_n) chosen. That is, if $x_n \to a$ and $y_n \to a$, then $\lim f(x_n) = \lim f(y_n)$.

Next show that g is an extension of f.

To establish the uniform continuity of g, let a and b be points in $\overline{S}$. If (x_n) is a sequence in S converging to a, then $f(x_n) \to g(a)$. This implies that both $d(x_j, a)$

and $d\big(f(x_j), g(a)\big)$ can be made as small as we please by choosing j sufficiently large. A similar remark holds for a sequence (y_n) in S that converges to b. From this show that x_j is arbitrarily close to y_k (for large j and k), provided we assume that a is sufficiently close to b. Use this in turn to show that $g(a)$ is arbitrarily close to $g(b)$ when a and b are sufficiently close.

The uniqueness argument is very easy.

24.1.16. Proposition. *Let $f\colon M \to N$ be a continuous bijection between metric spaces. If M is complete and f^{-1} is uniformly continuous, then N is complete.*

Proof. Problem.

24.2. The integral of step functions

$\boxed{\textbf{Throughout this section } E \textbf{ is a Banach space.}}$

24.2.1. Definition. An $(n+1)$-tuple $(t_0, t_1, \ldots, t_n)$ of real numbers is a PARTITION of the interval $[a, b]$ in $\mathbb{R}$ provided that

(a) $t_0 = a$,

(b) $t_n = b$, and

(c) $t_{k-1} < t_k$ for $1 \le k \le n$.

If $P = (s_0, \ldots, s_m)$ and $Q = (t_0, \ldots, t_n)$ are partitions of the same interval $[a, b]$ and if
$$\{s_0, \ldots, s_m\} \subseteq \{t_0, \ldots, t_n\},$$
then we say that Q is a REFINEMENT of P and we write $P \preceq Q$.

Let $P = (s_0, \ldots, s_m)$, $Q = (t_0, \ldots, t_n)$, and $R = (u_0, \ldots, u_p)$ be partitions of $[a, b] \subseteq \mathbb{R}$. If
$$\{u_0, \ldots, u_p\} = \{s_0, \ldots, s_m\} \cup \{t_0, \ldots, t_n\},$$
then R is the SMALLEST COMMON REFINEMENT of P and Q and is denoted by $P \vee Q$. It is clear that $P \vee Q$ is the partition with fewest points that is a refinement of both P and Q.

24.2.2. Exercise. Consider partitions
$$P = \Big(0, \frac{1}{4}, \frac{1}{3}, \frac{1}{2}, \frac{2}{3}, \frac{3}{4}, 1\Big) \quad \text{and} \quad Q = \Big(0, \frac{1}{5}, \frac{1}{3}, \frac{2}{3}, \frac{5}{6}, 1\Big)$$
of $[0, 1]$. Find $P \vee Q$.

24.2.3. Definition. Let S be a set, and let A be a subset of S. We define $\chi_A \colon S \to \mathbb{R}$, the CHARACTERISTIC FUNCTION of A, by
$$\chi_A(x) = \begin{cases} 1 & \text{if } x \in A, \\ 0 & \text{if } x \in A^c. \end{cases}$$

If E is a Banach space, then a function $\sigma\colon [a, b] \to E$ is an E-VALUED STEP FUNC-TION on the interval $[a, b]$ if

(a) $\operatorname{ran}\sigma$ is finite and

(b) for every $x \in \operatorname{ran} \sigma$ the set $\sigma^{\leftarrow}(\{x\})$ is the union of finitely many subintervals of $[a, b]$.

We denote by $\mathcal{S}([a, b], E)$ the family of all E-valued step functions defined on $[a, b]$. Notice that $\mathcal{S}([a, b], E)$ is a vector subspace of $\mathcal{B}([a, b], E)$.

It is not difficult to see that $\sigma \colon [a, b] \to E$ is a step function if and only if there exists a partition $(t_0, \ldots, t_n)$ of $[a, b]$ such that σ is constant on each of the open subintervals (t_{k-1}, t_k). If, in addition, we insist that σ be discontinuous at each of the points $t_1, \ldots, t_{n-1}$, then this partition is unique. Thus we speak of the PARTITION ASSOCIATED WITH (or INDUCED BY) a step function σ.

24.2.4. Notation. Let σ be a step function on $[a, b]$, and let $P = (t_0, \ldots, t_n)$ be a partition of $[a, b]$ that is a refinement of the partition associated with σ. We define

$$\sigma_P = (x_1, \ldots, x_n),$$

where x_k is the value of σ on the open interval (t_{k-1}, t_k) for $1 \leq k \leq n$.

24.2.5. Exercise. Define $\sigma \colon [0, 5] \to \mathbb{R}$ by

$$\sigma = \chi_{[1,4]} - \chi_{(2,5]} - \chi_{\{4\}} - 2\chi_{[2,3)} - \chi_{[1,2)} + \chi_{[4,5)} \,.$$

(a) Find the partition P associated with σ.

(b) Find σ_Q, where $Q = (0, 1, 2, 3, 4, 5)$.

24.2.6. Definition. Let σ be an E-valued step function on $[a, b]$, and let $(t_0, \ldots, t_n)$ be the partition associated with σ. For $1 \leq k \leq n$, let $\Delta t_k = t_k - t_{k-1}$, and let x_k be the value of σ on the subinterval (t_{k-1}, t_k). Define

$$\int \sigma := \sum_{k=1}^{n} (\Delta t_k) x_k \,.$$

The vector $\int \sigma$ is the INTEGRAL of σ over $[a, b]$. Other standard notations for $\int \sigma$ are $\int_a^b \sigma$ and $\int_a^b \sigma(t)\, dt$.

24.2.7. Exercise. Find $\int_0^5 \sigma$ where σ is the step function given in Exercise 24.2.5.

24.2.8. Problem. Let $\sigma \colon [0, 10] \to \mathbb{R}$ be defined by

$$\sigma = 2\chi_{[1,5)} - 3\chi_{[2,8)} - 5\chi_{\{6\}} + \chi_{[4,10]} + 4\chi_{[9,10]} \,.$$

(a) Find the partition associated with σ.

(b) If $Q = (0, 1, 2, 3, 4, 5, 6, 7, 8, 9, 10)$, what is σ_Q?

(c) Find $\int \sigma$.

The next lemma is essentially obvious, but it is good practice to write out a proof anyway. It says that, in computing the integral of a step function σ, it doesn't matter whether we work with the partition induced by σ or with a refinement of that partition.

24.2.9. Lemma. *Let σ be an E-valued step function on $[a, b]$. If $Q = (u_0, \ldots, u_m)$ is a refinement of the partition associated with σ and if $\sigma_Q = (y_1, \ldots, y_m)$, then*

$$\int \sigma = \sum_{k=1}^{n} (\Delta u_k) y_k \,.$$

Proof. Exercise.

It follows easily from the preceding lemma that changing the value of a step function at a finite number of points does not affect the value of its integral. Next we show that for a given interval the integral is a bounded linear transformation on the family of all E-valued step functions on the interval.

24.2.10. Proposition. *The map*

$$\int : \mathcal{S}([a,b], E) \to E$$

is bounded and linear with $\|\int\| = b - a$. *Furthermore,*

$$\left\| \int \sigma(t)\, dt \right\| \leq \int \|\sigma(t)\|\, dt$$

for every E-valued step function σ on $[a,b]$.

Proof. Problem. *Hint.* To show that $\int (\sigma + \tau) = \int \sigma + \int \tau$, let P and Q be the partitions associated with σ and τ, respectively. Define the partition $R = (t_0, \ldots, t_n)$ to be $P \vee Q$. Clearly, R is a refinement of the partition associated with $\sigma + \tau$. Suppose $\sigma_R = (x_1, \ldots, x_n)$ and $\tau_R = (y_1, \ldots, y_n)$. Use Lemma 24.2.9 to compute $\int (\sigma + \tau)$. To find $\|\int\|$, use the definition of the norm of a linear map and Lemma 23.1.6.

We now show that in the case of real valued step functions the integral is a positive linear functional; that is, it takes positive functions to positive numbers.

24.2.11. Proposition. *If σ is a real valued step function on $[a,b]$ and if $\sigma(t) \geq 0$ for all t in $[a,b]$, then $\int \sigma \geq 0$.*

Proof. Problem.

24.2.12. Corollary. *If σ and τ are real valued step functions on $[a,b]$ and if $\sigma(t) \leq \tau(t)$ for all t in $[a,b]$, then $\int \sigma \leq \int \tau$.*

Proof. Apply the preceding proposition to $\tau - \sigma$. Then (by Proposition 24.2.10) $\int \tau - \int \sigma = \int (\tau - \sigma) \geq 0$. ☐

Finally we prepare the ground for piecing together and integrating two functions on adjoining intervals.

24.2.13. Proposition. *Let c be an interior point of the interval $[a,b]$. If τ and ρ are E-valued step functions on the intervals $[a,c]$ and $[c,b]$, respectively, define a function $\sigma \colon [a,b] \to E$ by*

$$\sigma(t) = \begin{cases} \tau(t) & \text{if } a \leq t \leq c, \\ \rho(t) & \text{if } c < t \leq b. \end{cases}$$

Then σ is an E-valued step function on $[a,b]$ and

$$\int_a^b \sigma = \int_a^c \tau + \int_c^b \rho.$$

Proof. Exercise.

Notice that if σ is a step function on $[a, b]$, then $\tau := \sigma|_{[a,c]}$ and $\rho := \sigma|_{[c,b]}$ are step functions, and by the preceding proposition

$$\int_a^b \sigma = \int_a^c \tau + \int_c^b \rho.$$

In this context one seldom distinguishes notationally between a function on an interval and the restriction of that function to a subinterval. Thus the preceding equation is usually written

$$\int_a^b \sigma = \int_a^c \sigma + \int_c^b \sigma.$$

24.2.14. Problem. Let $\sigma\colon [a, b] \to E$ be a step function, and let $T\colon E \to F$ be a bounded linear transformation from E into another Banach space F. Then $T \circ \sigma$ is an F-valued step function on $[a, b]$ and

$$\int (T \circ \sigma) = T\left(\int \sigma\right).$$

24.3. The Cauchy integral

We are now ready to extend the integral from the rather limited family of step functions to a class of functions large enough to contain all continuous functions (in fact, all piecewise continuous functions).

Following Dieudonné [**Die62**], we will call members of this larger class *regulated functions*.

> **In this section E is a Banach space, and a and b are real numbers with $a < b$.**

24.3.1. Definition. Recall that the family $\mathcal{S} = \mathcal{S}([a, b], E)$ of E-valued step functions on $[a, b]$ is a subspace of the normed linear space $\mathcal{B} = \mathcal{B}([a, b], E)$ of bounded E-valued functions on $[a, b]$. The closure $\overline{\mathcal{S}}$ of $\mathcal{S}$ in $\mathcal{B}$ is the family of REGULATED FUNCTIONS on $[a, b]$.

It is an interesting fact that the regulated functions on an interval turn out to be exactly those functions that have one-sided limits at every point of the interval. We will not need this fact, but a proof may be found in Dieudonné [**Die62**].

According to Problem 22.3.9(b), the set $\overline{\mathcal{S}}$ is a vector subspace of the Banach space $\mathcal{B}$. Since it is closed in $\mathcal{B}$, it is itself a Banach space. It is on this Banach space that we define the Cauchy integral. The Cauchy integral is not as general as the Riemann integral because the set $\overline{\mathcal{S}}$ is not quite as large as the set of functions on which the Riemann integral is defined. We do not prove this; nor do we prove the fact that when both integrals are defined, they agree. What we are interested in proving is that every continuous function is regulated; that is, every continuous function on $[a, b]$ belongs to $\mathcal{S}$.

24.3.2. Proposition. *Every continuous E-valued function on $[a, b]$ is regulated.*

Proof. *Exercise. Hint.* Use Proposition 24.1.11.

It is not difficult to modify the preceding proof to show that every piecewise continuous function on $[a,b]$ is regulated. (*Definition.* A function $f\colon [a,b] \to E$ is PIECEWISE CONTINUOUS if there exists a partition $(t_0, \ldots, t_n)$ of $[a,b]$ such that f is continuous on each subinterval (t_{k-1}, t_k).)

24.3.3. Corollary. *Every continuous E-valued function on $[a,b]$ is the uniform limit of a sequence of step functions.*

Proof. According to Problem 12.2.5. a function f belongs to the closure of S if and only if there is a sequence of step functions that converges (uniformly) to f. □

Now we are ready to define the Cauchy integral of a regulated function.

24.3.4. Definition. Recall that $\mathcal{S} = \mathcal{S}([a,b], E)$ is a subset of the Banach space $\mathcal{B}([a,b], E)$. In the preceding section we defined the integral of a step function. The map

$$\int \colon \mathcal{S} \to E$$

was shown to be bounded and linear (Proposition 24.2.10); therefore, it is uniformly continuous (Proposition 24.1.9). Thus by Theorem 24.1.15, it has a unique continuous extension to $\overline{\mathcal{S}}$. This extension, which we denote also by $\int$, and which is, in fact, uniformly continuous, is the CAUCHY (or CAUCHY–BOCHNER) INTEGRAL. For f in $\overline{\mathcal{S}}$ we call $\int f$ the INTEGRAL OF f (over $[a,b]$). As with step functions we may wish to emphasize the domain of f or the role of a particular variable, in which case we may write $\int_a^b f$ or $\int_a^b f(t)\,dt$ for $\int f$.

24.3.5. Problem. Use the *definition* of the Cauchy integral to show that $\int_0^1 x^2 \, dx = 1/3$. *Hint.* Start by finding a sequence of step functions that converges uniformly to the function $x \mapsto x^2$. The proof of Proposition 24.3.2 may help; so also may Problem I.1.15.

Most of the properties of the Cauchy integral are derived from the corresponding properties of the integral of step functions by taking limits. In the remainder of this section it is well to keep in mind one aspect of Theorem 24.1.15 (and its proof): When a uniformly continuous function f is extended from a set $\mathcal{S}$ to a function g on its closure $\overline{\mathcal{S}}$, the value of g at a point a in $\overline{\mathcal{S}}$ is the limit of the values $f(x_n)$, where (x_n) is any sequence in $\mathcal{S}$ that converges to a. What does this say in the present context? If h is a regulated function, then there exists a sequence (σ_n) of step functions converging uniformly to h and, furthermore,

$$\int h = \lim_{n \to \infty} \int \sigma_n .$$

We use this fact repeatedly without explicit reference.

One more simple fact is worthy of notice. Following Lemma 24.2.9, we remarked that changing the value of a step function at finitely many points does not affect the value of its integral. The same is true of regulated functions. (*Proof.* Certainly, it suffices to show that changing the value of a regulated function f at a single point c does not alter the value of $\int f$. Suppose (σ_n) is a sequence of step functions converging uniformly to f. Also suppose that g differs from f only at c. Replace each step function σ_n by a function τ_n that is equal to σ_n at each point other than c and whose value at c is $g(c)$. Then $\tau_n \to g$ (unif), and (by the comment in the preceding paragraph) $\int g = \lim \int \tau_n = \lim \int \sigma_n = \int f$.)

The following theorem (and Proposition 24.3.10) generalizes Proposition 24.2.10.

24.3.6. Theorem. *The Cauchy integral is a bounded linear transformation that maps the space of E-valued regulated functions on $[a, b]$ into the space E.*

Proof. Exercise.

Next we show that for real valued functions the Cauchy integral is a positive linear functional. This generalizes Proposition 24.2.11.

24.3.7. Proposition. *Let f be a regulated real valued function on $[a, b]$. If $f(t) \geq 0$ for all t in $[a, b]$, then $\int f \geq 0$.*

Proof. Problem. *Hint.* Suppose that $f(t) \geq 0$ for all t in $[a, b]$ and that (σ_n) is a sequence of real valued step functions converging uniformly to f. For each $n \in N$, define
$$\sigma_n^+ : [a, b] \to \mathbb{R} : t \mapsto \max\{\sigma_n(t), 0\}.$$
Show that (σ_n^+) is a sequence of step functions converging uniformly to f. Then use Proposition 24.2.11.

24.3.8. Corollary. *If f and g are regulated real valued functions on $[a, b]$ and if $f(t) \leq g(t)$ for all t in $[a, b]$, then $\int f \leq \int g$.*

Proof. Problem.

24.3.9. Proposition. *Let $f : [a, b] \to E$ be a regulated function, and let (σ_n) be a sequence of E-valued step functions converging uniformly to f. Then*
$$\left\| \int f \right\| = \lim \left\| \int \sigma_n \right\|.$$
Furthermore, if $g(t) = \|f(t)\|$ for all t in $[a, b]$ and $\tau_n(t) = \|\sigma_n(t)\|$ for all n in $\mathbb{N}$ and t in $[a, b]$, then (τ_n) is a sequence of real valued step functions that converges uniformly to g and $\int g = \lim \int \tau_n$.

Proof. Problem.

24.3.10. Proposition. *Let f be a regulated E-valued function on $[a, b]$. Then*
$$\left\| \int f \right\| \leq \int \|f(t)\| \, dt$$
and, therefore,
$$\left\| \int f \right\| \leq (b - a)\|f\|_u.$$
Thus,
$$\left\| \int \right\| = b - a,$$

where $\int\colon \overline{\mathcal{S}} \to E$ is the Cauchy integral and $\overline{\mathcal{S}}$ is the family of regulated E-valued functions on the interval $[a, b]$.

Proof. Problem.

24.3.11. Problem. Explain in one or two brief sentences why the following is obvious: If (f_n) is a sequence of E-valued regulated functions on $[a, b]$ that converges uniformly to g, then g is regulated and $\int g = \lim \int f_n$.

24.3.12. Problem. Let $\sigma, \tau\colon [0, 3] \to \mathbb{R}$ be the step functions defined by

$$\sigma = \chi_{[0,2]} \qquad \text{and} \qquad \tau = \chi_{[0,2]} + 2\chi_{(2,3]}\,.$$

Recall from Appendix N that the function $(\sigma, \tau)\colon [0, 3] \to \mathbb{R}^2$ is defined by $(\sigma, \tau)(t) = \big(\sigma(t), \tau(t)\big)$ for $0 \le t \le 3$. It is clear that (σ, τ) is a step function.

(a) Find the partition R associated with (σ, τ). Find $(\sigma, \tau)_R$. Make a careful sketch of (σ, τ).

(b) Find $\int(\sigma, \tau)$.

24.3.13. Problem. Same as preceding problem, except let $\sigma = \chi_{[0,1]}$.

We now generalize Proposition 24.2.13 to regulated functions on adjoining intervals.

24.3.14. Proposition. *Let c be an interior point of the interval $[a, b]$. If g and h are regulated E-valued functions on the intervals $[a, c]$ and $[c, b]$, respectively, define a function $f\colon [a, b] \to E$ by*

$$f(t) = \begin{cases} g(t) & \text{if } a \le c \le c, \\ h(t) & \text{if } c < t \le b. \end{cases}$$

Then f is a regulated function on $[a, b]$ and

$$\int_a^b f = \int_a^c g + \int_c^b h\,.$$

Proof. Problem.

As was remarked after Proposition 24.2.13, it is the usual practice to use the same name for a function and for its restriction to subintervals, thus the notation of the next corollary.

24.3.15. Corollary. *If $f\colon [a, b] \to E$ is a regulated function and c is an interior point of the interval $[a, b]$, then*

(31) $$\int_a^b f = \int_a^c f + \int_c^b f\,.$$

Proof. Let $g = f|_{[a,c]}$ and $h = f|_{[c,b]}$. Notice that g and h are regulated. (If, for example, (σ_n) is a sequence of step functions converging uniformly on $[a, b]$ to f, then the step functions $\sigma_n|_{[a,c]}$ converge uniformly on $[a, c]$ to g.) Then apply the preceding proposition. $\square$

It is convenient for formula (31) to be correct when a, b, and c are not in increasing order or, for that matter, even necessarily distinct. This can be achieved by means of a simple notational convention.

24.3.16. Definition. For a regulated function f on $[a, b]$ where $a < b$, define

$$\int_b^a f := -\int_a^b f .$$

Furthermore, if g is any function whose domain contains the point a, then

$$\int_a^a g := 0 .$$

24.3.17. Corollary. *If f is an E-valued regulated function whose domain contains an interval to which the points a, b, and c belong, then*

$$\int_a^b f = \int_a^c f + \int_c^b f .$$

Proof. The result is obvious if any two of the points a, b, and c coincide; so we suppose that they are distinct. There are six possible orderings. We check one of these. Suppose $c < a < b$. By Corollary 24.3.15

$$\int_c^b f = \int_c^a f + \int_a^b f .$$

Thus

$$\int_a^b f = -\int_c^a f + \int_c^b f = \int_a^c f + \int_c^b f .$$

The remaining five cases are similar. $\square$

Suppose f is an E-valued regulated function and T is a bounded linear map from E into another Banach space F. It is interesting and useful to know that integration of the composite function $T \circ f$ can be achieved simply by integrating f and then applying T. This fact can be expressed by means of the following commutative diagram. *Notation.* $\overline{\mathcal{S}}_E$ and $\overline{\mathcal{S}}_F$ denote, respectively, the E-valued and F-valued regulated functions on $[a, b]$, and C_T is the bounded linear transformation discussed in Problem 23.1.21. ($C_T(f) = T \circ f$ for every f.)

$$
\begin{array}{ccc}
\overline{\mathcal{S}}_E & \xrightarrow{\;\int\;} & E \\
\downarrow{\scriptstyle C_T} & & \downarrow{\scriptstyle T} \\
\overline{\mathcal{S}}_F & \xrightarrow{\;\int\;} & F
\end{array}
$$

Alternatively it may be expressed by a formula, as in the next proposition.

24.3.18. Proposition. *Let $T \colon E \to F$ be a bounded linear map between Banach spaces, and let f be a regulated E-valued function on the interval $[a, b]$. Then $T \circ f$ is a regulated F-valued function on $[a, b]$ and*

$$\int (T \circ f) = T\left(\int f \right) .$$

Proof. Exercise. *Hint.* Use Problem 24.2.14.

24.3.19. Corollary. *If E and F are Banach spaces and*

$$T\colon [a,b] \to \mathfrak{B}(E,F)\colon t \mapsto T_t$$

is continuous, then for every x in E

$$\int T_t(x)\,dt = \left(\int T\right)(x).$$

Proof. Problem. *Hint.* For x in E let $E_x\colon \mathfrak{B}(E,F) \to F$ be the map (evaluation at x) defined in Problem 23.1.19. Write $T_t(x)$ as $(E_x \circ T)(t)$ and apply Proposition 24.3.18.

24.3.20. Proposition. *Let $f\colon [a,b] \to \mathbb{R}$ be a regulated function, and let $x \in E$. For all t in $[a,b]$ let $g(t) = f(t)\,x$. Then g is a regulated E-valued function and $\int g = (\int f)x$.*

Proof. Problem. *Hint.* Prove the result first for the case where f is a step function. Then take limits. $\square$

24.3.21. Proposition. *If $f\colon [a,b] \to E$ and $g\colon [a,b] \to F$ are regulated functions whose ranges lie in Banach spaces, then the function*

$$(f,g)\colon [a,b] \to E \times F\colon t \mapsto \big(f(t),g(t)\big)$$

is regulated and

$$\int (f,g) = \left(\int f, \int g\right).$$

Proof. Problem. *Hint.* Write $\int f$ as $\int\big(\pi_1 \circ (f,g)\big)$ where $\pi_1\colon E \times F \to E$ is the usual coordinate projection. Write $\int g$ in a similar fashion. Use Proposition 24.3.18. Keep in mind that if p is a point in the product $E \times F$, then $p = \big(\pi_1(p),\pi_2(p)\big)$.

24.3.22. Proposition. *Suppose $h\colon [a,b] \to E \times F$ is a regulated function from $[a,b]$ into the product of two Banach spaces. Then the components h^1 and h^2 are regulated functions and*

$$\int h = \left(\int h_1, \int h_2\right).$$

Proof. Problem.

24.3.23. Problem. Suppose $f\colon [a,b] \to \mathbb{R}^n$ is a regulated function. Express the integral of f in terms of the integrals of its components $f^1,\dots,f^n$. Justify your answer carefully. *Hint.* When $\mathbb{R}^n$ appears (without further qualification), its norm is assumed to be the usual Euclidean norm. This is not the product norm. What needs to be done to ensure that the results of the preceding problem will continue to be true if the product norm on $E \times F$ is replaced by an equivalent norm?

24.3.24. Problem. Suppose $T\colon [0,1] \to \mathfrak{B}(\mathbb{R}^2, \mathbb{R}^2)\colon t \mapsto T_t$ is continuous, and suppose that for each t in $[0,1]$ the matrix representation of T_t is given by

$$[T_t] = \begin{bmatrix} 1 & t \\ t^2 & t^3 \end{bmatrix}.$$

Find $\left[\int T\right]$ the matrix representation of $\int T$. *Hint.* Use Corollary 24.3.19 and Propositions 24.3.20 and 24.3.23.

24.3.25. Problem. For every f in $\mathcal{C}([a,b], E)$ define

$$\|f\|_1 = \int_a^b \|f(t)\| \, dt.$$

(a) Show that $\| \ \ \|_1$ is a norm on $\mathcal{C}([a,b], E)$. *Hint.* In showing that $\|f\|_1 = 0$ implies $f = 0$, Proposition 3.3.21 may help.

(b) Are the norms $\| \ \ \|_1$ and $\| \ \ \|_u$ on $\mathcal{C}([a,b], E)$ equivalent?

(c) Let $\mathcal{C}_1$ be the vector space $\mathcal{C}([a,b], E)$ under the norm $\| \ \ \|_1$, and let $\mathcal{C}_u$ be the same vector space under its usual uniform norm. Does convergence of a sequence in $\mathcal{C}_u$ imply convergence in C_1? What about the converse?

24.3.26. Problem. Show that if $f \in \mathcal{C}([0,1], \mathbb{R})$ and $\int_0^1 x^n f(x) \, dx = 0$ for $n = 0, 1, 2, \ldots$, then $f = 0$.

24.3.27. Problem. Find $\lim \frac{\int_0^1 x^n f(x) \, dx}{\int_0^1 x^n \, dx}$ when f is a continuous real valued function on $[0,1]$. *Hint.* For each n in $\mathbb{N}$ let $L_n(f) = \frac{\int_0^1 x^n f(x) \, dx}{\int_0^1 x^n \, dx}$. Show that L_n is a continuous linear functional on the space $\mathcal{C}([0,1], \mathbb{R})$ of continuous real valued functions on $[0,1]$. What is $\lim_{n \to \infty} L_n(p)$ when p is a polynomial? Use the Weierstrass approximation theorem (Theorem 23.2.8).

Chapter 25

Differential calculus

In Chapter 8 we studied the calculus of real valued functions of a real variable. We now extend that study to vector valued functions of a vector variable. The first three sections of this chapter repeat almost exactly the material in Chapter 8. Absolute values are replaced by norms; linear maps by continuous linear maps. Other than that there are few differences. In fact, if you have done Chapter 8 carefully, you may wish just to glance over the first three sections of this chapter and move on to section 25.4.

> **Throughout this chapter V, W, and X (with or without subscripts) will be normed linear spaces.**

25.1. $\mathfrak{O}$ and $\mathfrak{o}$ functions

25.1.1. Notation. Let $a \in V$. We denote by $\mathcal{F}_a(V, W)$ the family of all functions defined on a neighborhood of a taking values in W. That is, f belongs to $\mathcal{F}_a(V, W)$ if there exists a set U such that $a \in U \overset{\circ}{\subseteq} \operatorname{dom} f \subseteq V$ and if the image of f is contained in W. We shorten $\mathcal{F}_a(V, W)$ to $\mathcal{F}_a$ when no confusion will result. Notice that for each $a \in V$, the set $\mathcal{F}_a$ is closed under addition and scalar multiplication. (As usual, we define the sum of two functions f and g in $\mathcal{F}_a$ to be the function $f + g$ whose value at x is $f(x) + g(x)$ whenever x belongs to $\operatorname{dom} f \cap \operatorname{dom} g$.) Despite the closure of $\mathcal{F}_a$ under these operations, $\mathcal{F}_a$ is *not* a vector space.

25.1.2. Problem. Let $a \in V \neq \{\mathbf{0}\}$. Prove that $\mathcal{F}_a(V, W)$ is not a vector space.

Among the functions defined on a neighborhood of the zero vector in V are two subfamilies of crucial importance: they are $\mathfrak{O}(V, W)$ (the family of "big-oh" functions) and $\mathfrak{o}(V, W)$ (the family of "little-oh" functions).

25.1.3. Definition. A function f in $\mathcal{F}_0(V,W)$ belongs to $\mathfrak{O}(V,W)$ if there exist numbers $c > 0$ and $\delta > 0$ such that

$$\|f(x)\| \le c\,\|x\|$$

whenever $\|x\| < \delta$.

A function f in $\mathcal{F}_0(V,W)$ belongs to $\mathfrak{o}(V,W)$ if for every $c > 0$ there exists $\delta > 0$ such that

$$\|f(x)\| \le c\,\|x\|$$

whenever $\|x\| < \delta$. Notice that f belongs to $\mathfrak{o}(V,W)$ if and only if $f(\mathbf{0}) = \mathbf{0}$ and

$$\lim_{h \to \mathbf{0}} \frac{\|f(h)\|}{\|h\|} = 0.$$

When no confusion seems likely, we will shorten $\mathfrak{O}(V,W)$ to $\mathfrak{O}$ and $\mathfrak{o}(V,W)$ to $\mathfrak{o}$.

The properties of the families $\mathfrak{O}$ and $\mathfrak{o}$ are given in Propositions 25.1.4–25.1.9, 25.1.11, and 25.1.12.

25.1.4. Proposition. *Every member of $\mathfrak{o}(V,W)$ belongs to $\mathfrak{O}(V,W)$; so does every member of $\mathfrak{B}(V,W)$. Every member of $\mathfrak{O}(V,W)$ is continuous at $\mathbf{0}$.*

Proof. Obvious from the definitions. $\qquad\qquad\square$

25.1.5. Proposition. *Other than the zero transformation, no bounded linear transformation belongs to $\mathfrak{o}$.*

Proof. Exercise.

25.1.6. Proposition. *The family $\mathfrak{O}$ is closed under addition and scalar multiplication.*

Proof. Exercise.

25.1.7. Proposition. *The family $\mathfrak{o}$ is closed under addition and scalar multiplication.*

Proof. Problem.

The next two propositions say that the composite of a function in $\mathfrak{O}$ with one in $\mathfrak{o}$ (in either order) ends up in $\mathfrak{o}$.

25.1.8. Proposition. *If $g \in \mathfrak{O}(V,W)$ and $f \in \mathfrak{o}(W,X)$, then $f \circ g \in \mathfrak{o}(V,X)$.*

Proof. Problem.

25.1.9. Proposition. *If $g \in \mathfrak{o}(V,W)$ and $f \in \mathfrak{O}(W,X)$, then $f \circ g \in \mathfrak{o}(V,X)$.*

Proof. Exercise.

25.1.10. Notation. In what follows it will be convenient to multiply vectors not only by scalars but also by scalar valued functions. If ϕ is a function in $\mathcal{F}_a(V, \mathbb{R})$ and $w \in W$, we define the function ϕw by

$$(\phi w)(x) = \phi(x) \cdot w$$

for all x in the domain of ϕ. Clearly, ϕw belongs to $\mathcal{F}_a(V, W)$.

Similarly, it is useful to multiply vector valued functions by scalar valued functions. If $\phi \in \mathcal{F}_a(V, \mathbb{R})$ and $f \in \mathcal{F}_a(V, W)$, we define the function ϕf by

$$(\phi f)(x) = \phi(x) \cdot f(x)$$

for all x in $\operatorname{dom} \phi \cap \operatorname{dom} f$. Then ϕf belongs to $\mathcal{F}_a(V, W)$.

25.1.11. Proposition. *If $\phi \in \mathfrak{o}(V, \mathbb{R})$ and $w \in W$, then $\phi w \in \mathfrak{o}(V, W)$.*

Proof. Exercise.

25.1.12. Proposition. *If $\phi \in \mathfrak{D}(V, \mathbb{R})$ and $f \in \mathfrak{D}(V, W)$, then $\phi f \in \mathfrak{o}(V, W)$.*

Proof. Exercise.

Remark. The list summarizing these facts is almost the same as the one in section 8.1. In (1) and (2) the linear maps $\mathfrak{L}$ have been replaced by continuous linear maps $\mathfrak{B}$; and (7) is new. (As before, $\mathcal{C}_\mathbf{0}$ is the set of all functions in $\mathcal{F}_\mathbf{0}$ that are continuous at $\mathbf{0}$.)

$$
\begin{aligned}
&(1) && \mathfrak{B} \cup \mathfrak{o} \subseteq \mathfrak{D} \subseteq \mathcal{C}_\mathbf{0}\,. \\
&(2) && \mathfrak{B} \cap \mathfrak{o} = \{\mathbf{0}\}\,. \\
&(3) && \mathfrak{D} + \mathfrak{D} \subseteq \mathfrak{D}\,; \qquad \alpha\,\mathfrak{D} \subseteq \mathfrak{D}\,. \\
&(4) && \mathfrak{o} + \mathfrak{o} \subseteq \mathfrak{o}\,; \qquad \alpha\,\mathfrak{o} \subseteq \mathfrak{o}\,. \\
&(5) && \mathfrak{o} \circ \mathfrak{D} \subseteq \mathfrak{o}\,. \\
&(6) && \mathfrak{D} \circ \mathfrak{o} \subseteq \mathfrak{o}\,. \\
&(7) && \mathfrak{o}(V, \mathbb{R}) \cdot W \subseteq \mathfrak{o}(V, W)\,. \\
&(8) && \mathfrak{D}(V, \mathbb{R}) \cdot \mathfrak{D}(V, W) \subseteq \mathfrak{o}(V, W)\,.
\end{aligned}
$$

25.1.13. Problem. Find a function in $\mathcal{C}_\mathbf{0}(\mathbb{R}^2, \mathbb{R}) \setminus \mathfrak{D}(\mathbb{R}^2, \mathbb{R})$. Also, find a function in $\mathfrak{D}(\mathbb{R}^2, \mathbb{R}) \setminus \mathfrak{o}(\mathbb{R}^2, \mathbb{R})$.

25.1.14. Problem. Show that $\mathfrak{D} \circ \mathfrak{D} \subseteq \mathfrak{D}$. That is, if $g \in \mathfrak{D}(V, W)$ and $f \in \mathfrak{D}(W, X)$, then $f \circ g \in \mathfrak{D}(V, X)$. (As usual, the domain of $f \circ g$ is taken to be $\{x \in V : g(x) \in \operatorname{dom} f\}$.)

25.1.15. Problem. If $f^1 \in \mathfrak{D}(V_1, W)$ and $f^2 \in \mathfrak{D}(V_2, W)$, then the function g defined on $\operatorname{dom} f^1 \times \operatorname{dom} f^2$ by

$$g(v^1, v^2) = f^1(v^1) + f^2(v^2)$$

belongs to $\mathfrak{D}(V_1 \times V_2, W)$. *Hint.* The simplest proof never mentions the domain elements v^1 and v^2. Instead write g in terms of the projection maps π_1 and π_2 on $V_1 \times V_2$, and apply Problem 25.1.14.

25.1.16. Problem. If $\phi \in \mathfrak{O}(V_1, \mathbb{R})$ and $f \in \mathfrak{O}(V_2, W)$, then the function h defined on dom $\phi \times$ dom f by

$$h(v^1, v^2) = \phi(v^1) \, f(v^2)$$

belongs to $\mathfrak{o}(V_1 \times V_2, W)$. (Use the hint given in the preceding problem.)

25.1.17. Problem. Show that the membership of the families $\mathfrak{O}(V, W)$ and $\mathfrak{o}(V, W)$ is not changed when the norms on the spaces V and W are replaced by equivalent norms.

25.2. Tangency

25.2.1. Definition. Two functions f and g in $\mathcal{F}_0(V, W)$ are TANGENT AT ZERO, in which case we write $f \simeq g$, if $f - g \in \mathfrak{o}(V, W)$.

25.2.2. Proposition. *The relation "tangency at zero" is an equivalence relation on $\mathcal{F}_0$.*

Proof. Exercise.

The next result shows that at most one bounded linear transformation can be tangent at zero to a given function.

25.2.3. Proposition. *Let S, $T \in \mathfrak{B}$ and $f \in \mathcal{F}_0$. If $S \simeq f$ and $T \simeq f$, then $S = T$.*

Proof. Exercise.

25.2.4. Proposition. *If $f \simeq g$ and $j \simeq k$, then $f + j \simeq g + k$, and furthermore, $\alpha f \simeq \alpha g$ for all $\alpha \in \mathbb{R}$.*

Proof. Problem.

In the next proposition we see that if two real valued functions are tangent at zero, multiplication by a vector does not disrupt this relationship. (For notation, see Notation 25.1.10.)

25.2.5. Proposition. *Let $\phi, \psi \in \mathcal{F}_0(V, \mathbb{R})$ and $w \in W$. If $\phi \simeq \psi$, then $\phi w \simeq \psi w$.*

Proof. Exercise.

25.2.6. Proposition. *Let f, $g \in \mathcal{F}_0(V, W)$ and $T \in \mathfrak{B}(W, X)$. If $f \simeq g$, then $T \circ f \simeq T \circ g$.*

Proof. Problem.

25.2.7. Proposition. *Let $h \in \mathfrak{O}(V, W)$ and f, $g \in \mathcal{F}_0(W, X)$. If $f \simeq g$, then $f \circ h \simeq g \circ h$.*

Proof. Problem.

25.2.8. Problem. Fix a vector x in V. Define a function $M_x \colon \mathbb{R} \to V$ by

$$M_x(\alpha) = \alpha x \qquad \text{for all } \alpha \in \mathbb{R}.$$

Notice that $M_x \in \mathfrak{B}(\mathbb{R}, V)$. If $f \in \mathcal{F}_0(\mathbb{R}, V)$ and $f \simeq M_x$, then

(a) $\dfrac{f(\alpha)}{\alpha} \to x$ as $\alpha \to 0$, and

(b) $f(\alpha) \to 0$ as $\alpha \to 0$.

25.2.9. Problem. Each of the following is an abbreviated version of a proposition. Formulate precisely and prove.

(a) $\mathcal{C}_0 + \mathfrak{D} \subseteq \mathcal{C}_0$.

(b) $\mathcal{C}_0 + \mathfrak{o} \subseteq \mathcal{C}_0$.

(c) $\mathfrak{D} + \mathfrak{o} \subseteq \mathfrak{D}$.

25.2.10. Problem. Suppose that $f \simeq g$. Then the following hold.

(a) If g is continuous at $\mathbf{0}$, so is f.

(b) If g belongs to $\mathfrak{D}$, so does f.

(c) If g belongs to $\mathfrak{o}$, so does f.

25.3. Differentiation

25.3.1. Definition. Let $f \in \mathcal{F}_a(V, W)$. Define the function Δf_a by
$$\Delta f_a(h) := f(a + h) - f(a)$$
for all h such that $a + h$ is in the domain of f. Notice that since f is defined in a neighborhood of a, the function Δf_a is defined in a neighborhood of $\mathbf{0}$; that is, Δf_a belongs to $\mathcal{F}_\mathbf{0}(V, W)$. Notice also that $\Delta f_a(\mathbf{0}) = \mathbf{0}$.

25.3.2. Proposition. *If $f \in \mathcal{F}_a(V, W)$ and $\alpha \in \mathbb{R}$, then*
$$\Delta(\alpha f)_a = \alpha \Delta f_a.$$

Proof. No changes need to be made in the proof of Proposition 8.3.3. □

25.3.3. Proposition. *If $f, g \in \mathcal{F}_a(V, W)$, then*
$$\Delta(f + g)_a = \Delta f_a + \Delta g_a.$$

Proof. The proof given (in the Solutions to Exercises online) for Proposition 8.3.4 needs no alteration. □

25.3.4. Proposition. *If $\phi \in \mathcal{F}_a(V, \mathbb{R})$ and $f \in \mathcal{F}_a(V, W)$, then*
$$\Delta(\phi f)_a = \phi(a) \cdot \Delta f_a + \Delta\phi_a \cdot f(a) + \Delta\phi_a \cdot \Delta f_a.$$

Proof. Problem.

25.3.5. Proposition. *If $f \in \mathcal{F}_a(V, W)$, $g \in \mathcal{F}_{f(a)}(W, X)$, and $g \circ f \in \mathcal{F}_a(V, X)$, then*
$$\Delta(g \circ f)_a = \Delta g_{f(a)} \circ \Delta f_a.$$

Proof. Use the proof given (in the Solutions to Exercises online) for Proposition 8.3.6. □

25.3.6. Proposition. *A function* $f\colon V \to W$ *is continuous at the point* a *in* V *if and only if* Δf_a *is continuous at* $\mathbf{0}$.

Proof. Problem.

25.3.7. Proposition. *If* $f\colon U \to U_1$ *is a bijection between subsets of arbitrary vector spaces, then for each* a *in* U *the function* $\Delta f_a\colon U - a \to U_1 - f(a)$ *is invertible and*

$$\left(\Delta f_a\right)^{-1} = \Delta\left(f^{-1}\right)_{f(a)}.$$

Proof. Problem.

25.3.8. Definition. Let $f \in \mathcal{F}_a(V,W)$. We say that f is DIFFERENTIABLE AT a if there exists a bounded linear map that is tangent at $\mathbf{0}$ to Δf_a. If such a map exists, it is called the DIFFERENTIAL of f at a and is denoted by df_a. Thus df_a is just a member of $\mathfrak{B}(V,W)$ that satisfies $df_a \simeq \Delta f_a$. We denote by $\mathcal{D}_a(V,W)$ the family of all functions in $\mathcal{F}_a(V,W)$ that are differentiable at a. We often shorten this to $\mathcal{D}_a$.

25.3.9. Proposition. *Let* $f \in \mathcal{F}_a(V,W)$. *If* f *is differentiable at* a, *then its differential is unique. (That is, there is at most one bounded linear map tangent at* 0 *to* Δf_a.)

Proof. Proposition 25.2.3. $\qquad\qquad\qquad\qquad\qquad\qquad\qquad\qquad\qquad\qquad\square$

Remark. If f is a function in $\mathcal{F}_a(V,W)$ that is differentiable at a, its differential df_a has three important properties:

(i) it is linear;

(ii) it is continuous (that is, bounded as a linear map);

(iii) $\displaystyle\lim_{h \to 0} \frac{\Delta f_a(h) - df_a(h)}{\|h\|} = \mathbf{0}$.

(An expression of the form $\frac{\Delta f_a(h) - df_a(h)}{\|h\|}$ is called a NEWTON QUOTIENT.)

25.3.10. Exercise. Let $f(x,y) = 3x^2 - xy + 4y^2$. Show that $df_{(1,-1)}(x,y) = 7x - 9y$. Interpret $df_{(1,-1)}$ geometrically.

25.3.11. Exercise. Let

$$f\colon \mathbb{R}^3 \to \mathbb{R}^2\colon (x,y,z) \mapsto (x^2y - 7, 3xz + 4y)$$

and $a = (1, -1, 0)$. Use the definition of "differential" to find df_a. *Hint.* Work with the matrix representation of df_a. Since the differential must belong to $\mathfrak{B}(\mathbb{R}^3, \mathbb{R}^2)$, its matrix representation is a 2×3 matrix $M = \begin{bmatrix} r & s & t \\ u & v & w \end{bmatrix}$. Use the requirement (condition (iii) of the preceding remark) that $\|h\|^{-1}\|\Delta f_a(h) - Mh\| \to \mathbf{0}$ as $h \to \mathbf{0}$ to discover the identity of the entries in M.

25.3.12. Proposition. *If* $f \in \mathcal{D}_a$, *then* $\Delta f_a \in \mathfrak{D}$.

Proof. Use the proof given (in the Solutions to Exercises online) for Proposition 8.4.8. Since we are working with bounded linear maps between normed spaces instead of linear maps on $\mathbb{R}$, we must change $\mathfrak{L}$ to $\mathfrak{B}$. $\qquad \square$

25.3.13. Corollary. *Every function that is differentiable at a point is continuous there.*

Proof. Use the proof given (in the Solutions to Exercises online) for Corollary 8.4.9. $\qquad \square$

25.3.14. Proposition. *If f is differentiable at a and $\alpha \in \mathbb{R}$, then αf is differentiable at a and*

$$d(\alpha f)_a = \alpha \, df_a \, .$$

Proof. The proof given (in the Solutions to Exercises online) for Proposition 8.4.10 works, but three references need to be changed. (What are the correct references?) $\qquad \square$

25.3.15. Proposition. *If f and g are differentiable at a, then $f+g$ is differentiable at a and*

$$d(f + g)_a = df_a + dg_a \, .$$

Proof. Problem.

25.3.16. Proposition (Leibniz's rule). *If $\phi \in \mathcal{D}_a(V, \mathbb{R})$ and $f \in \mathcal{D}_a(V, W)$, then $\phi f \in \mathcal{D}_a(V, W)$ and*

$$d(\phi f)_a = d\phi_a \cdot f(a) + \phi(a) \, df_a \, .$$

Proof. Exercise.

In the chapters of beginning calculus texts devoted to the differential calculus of several variables, the expression "chain rule" refers most frequently to a potentially infinite collection of related results concerning the differentiation of composite functions. Several examples such as the following are usually given:

Let $w = f(x, y, z)$, $x = x(u, v)$, $y = y(u, v)$, and $z = z(u, v)$. If these functions are all differentiable, then the function

$$w = f(x(u, v) \, , \, y(u, v) \, , \, z(u, v))$$

is differentiable and

$$(*) \qquad \begin{aligned} \frac{\partial w}{\partial u} &= \frac{\partial w}{\partial x}\frac{\partial x}{\partial u} + \frac{\partial w}{\partial y}\frac{\partial y}{\partial u} + \frac{\partial w}{\partial z}\frac{\partial z}{\partial u}, \\ \frac{\partial w}{\partial v} &= \frac{\partial w}{\partial x}\frac{\partial x}{\partial v} + \frac{\partial w}{\partial y}\frac{\partial y}{\partial v} + \frac{\partial w}{\partial z}\frac{\partial z}{\partial v}. \end{aligned}$$

Then the reader is encouraged to invent new "chain rules" for functions having different numbers of variables. Formulations such as $(*)$ have many shortcomings, the most serious of which is a nearly complete lack of discernible geometric content. The version of the chain rule that we will prove says that (after a suitable translation) the best linear approximation to the composite of two functions is the composite of the best linear approximations. In other words, the differential of the composite is the composite of the differentials. Notice that Proposition 25.3.17, where this is stated formally, is simpler than $(*)$, it has obvious geometric content, it is a single rule rather than a family of them, and it holds in arbitrary infinite-dimensional normed linear spaces as well as in finite-dimensional ones.

25.3.17. Proposition (The chain rule). *If $f \in \mathcal{D}_a(V,W)$ and $g \in \mathcal{D}_{f(a)}(W,X)$, then $g \circ f \in \mathcal{D}_a(V,X)$ and*

$$d(g \circ f)_a = dg_{f(a)} \circ df_a .$$

Proof. Exercise.

Each of the preceding propositions concerning differentiation is a direct consequence of a similar result concerning the map Δ. In particular, the linearity of the map $f \mapsto df_a$ (Propositions 25.3.14 and 25.3.15) follows from the fact that the function $f \mapsto \Delta f_a$ is linear (Propositions 25.3.2 and 25.3.3); Leibniz's rule (Proposition 25.3.16) is a consequence of Proposition 25.3.4; and the proof of the chain rule (Proposition 25.3.17) makes use of Proposition 25.3.5. It is reasonable to hope that the result given in Proposition 25.3.7 concerning $\Delta\left(f^{-1}\right)_{f(a)}$ will lead to useful information concerning the differential of the inverse function. This is indeed the case, but obtaining information about $d\left(f^{-1}\right)_{f(a)}$ from Proposition 25.3.7 is a rather involved undertaking. It turns out, happily enough, that (under mild hypotheses) the differential of f^{-1} is the inverse of the differential of f. This result is known as the inverse function theorem, and the whole of Chapter 29 is devoted to detailing its proof and to examining some of its consequences.

25.3.18. Problem. Show that Proposition 25.3.14 is actually a special case of Leibniz's rule, Proposition 25.3.16. Also, suppose that $\phi \in \mathcal{D}_a(V,\mathbb{R})$ and $w \in W$. Prove that $\phi w \in \mathcal{D}_a(V,W)$ and that $d(\phi w)_a = d\phi_a \cdot w$.

25.3.19. Problem. Let $f \colon \mathbb{R}^2 \to \mathbb{R}^2$ be defined by $f(x) = (3x_1 - x_2 + 7, x_1 + 4x_2)$. Show that $\left[df_{(1,0)}\right] = \begin{bmatrix} 3 & -1 \\ 1 & 4 \end{bmatrix}$. *Hint.* Let $M = \begin{bmatrix} 3 & -1 \\ 1 & 4 \end{bmatrix}$ and $a = (1,0)$. Show that the Newton quotient $\frac{\Delta f_a(h) - Mh}{\|h\|}$ approaches $\mathbf{0}$ as $h \to \mathbf{0}$. Use the uniqueness of differentials (Proposition 25.3.9) to conclude that $\left[df_a\right] = M$.

25.3.20. Problem. Let $F \colon \mathbb{R}^2 \to \mathbb{R}^4$ be defined by

$$F(x) = (x_2,\, x_1{}^2,\, 4 - x_1 x_2,\, 7x_1),$$

and let $a = (1,1)$. Use the definition of differential to show that

$$[dF_a] = \begin{bmatrix} 0 & 1 \\ 2 & 0 \\ -1 & -1 \\ 7 & 0 \end{bmatrix} .$$

25.3.21. Problem. Let $F\colon \mathbb{R}^3 \to \mathbb{R}^4\colon x \mapsto (x_1 + 2x_3, x_2 - x_3, 4x_2, 2x_1 - 5x_2)$ and $a = (1, 2, -5)$. Use the definition of differential to find $[dF_a]$. *Hint.* Use the technique suggested in Exercise 25.3.11.

25.3.22. Problem. Let $F\colon \mathbb{R}^3 \to \mathbb{R}^2$ be defined by $F(x, y, z) = (xy - 3, y + 2z^2)$. Use the definition of differential to find $[dF_{(1,-1,2)}]$.

25.3.23. Problem. Let $F\colon \mathbb{R}^3 \to \mathbb{R}^3 \colon x \mapsto (x_1x_2, x_2 - x_3{}^2, 2x_1x_3)$. Use the definition of differential to find $[dF_a]$ at a in $\mathbb{R}^3$.

25.3.24. Problem. Let $T \in \mathfrak{B}(V, W)$ and $a \in V$. Find dT_a.

25.4. Differentiation of curves

In the preceding section we have discussed the differentiability of functions mapping one normed linear space into another. Here we briefly consider the important and interesting special case that occurs when the domains of the functions are one dimensional.

25.4.1. Definition. A CURVE in a normed linear space V is a continuous mapping from an interval in $\mathbb{R}$ into V. If $c\colon J \to V$ is a curve, if 0 belongs to the interval J, and if $c(0) = a \in V$, then c is a CURVE AT a.

In classical terminology a curve $c\colon J \to V\colon t \mapsto c(t)$ is usually referred to as a PARAMETRIZED CURVE in V. The interval J is the PARAMETER INTERVAL and the variable t belonging to J is the PARAMETER. If you start with a subset A of V and find a continuous function c from an interval J onto A, then c is called a PARAMETRIZATION of A.

25.4.2. Example. Let

$$c_1\colon [0, 2\pi] \to \mathbb{R}^2\colon t \mapsto (\cos t, \sin t)$$

and

$$c_2\colon [0, 3\pi] \to \mathbb{R}^2\colon t \mapsto (\sin(2t + 1), \cos(2t + 1)).$$

Then c_1 and c_2 are two different parametrizations of the unit circle

$$\mathbb{S}^1 := \{(x, y) \in \mathbb{R}^2\colon x^2 + y^2 = 1\}.$$

Parameters need have no physical significance, but it is quite common to think of a curve in $\mathbb{R}^2$ or $\mathbb{R}^3$ as representing the motion of a particle in that space: the parameter t is taken to be time, and the range value $c(t)$ is the position of the particle at time t. With this interpretation we may view c_1 and c_2 as representing the motion of particles traveling around the unit circle. In the first case the particle starts (when $t = 0$) at the point $(1, 0)$ and makes one complete trip around the circle traveling counterclockwise. In the second case, the particle starts at the point $(\sin 1, \cos 1)$ and traverses $\mathbb{S}^1$ three times moving clockwise.

25.4.3. Example. Let a and $u \neq \mathbf{0}$ be vectors in V. The curve

$$\ell\colon \mathbb{R} \to V\colon t \mapsto a + tu$$

is the PARAMETRIZED LINE through a in the direction of u. Of course, infinitely many other parametrizations of the range of ℓ are possible, but this is the standard one and we adopt it.

25.4.4. Problem. Find a parametrization of the unit square
$$A := \{(x,y) \in \mathbb{R}^2 : d_u((x,y),(\tfrac{1}{2},\tfrac{1}{2})) = \tfrac{1}{2}\}$$
that starts at $(0,0)$ and traverses A once in a counterclockwise direction.

25.4.5. Definition. Let $c\colon J \to V$ be a curve in V and suppose that a is a point in the interior of the interval J. Then $Dc(a)$, the DERIVATIVE of c at a, is defined by the formula
$$Dc(a) := \lim_{h \to 0} \frac{\Delta c_a(h)}{h}$$
if the indicated limit exists. Notice that this is just the definition of "derivative" given in beginning calculus. The derivative at a is also called the TANGENT VECTOR to c at a or, in case we are thinking of the motion of a particle, the VELOCITY of c at a. If $Dc(a)$ exists and is not zero, then the parametrized line through $c(a)$ in the direction of $Dc(a)$ is the TANGENT LINE to the image of c at the point $c(a)$.

25.4.6. Exercise. Let $c\colon [0,2\pi] \to \mathbb{R}^2\colon t \mapsto (\cos t, \sin t)$.

(a) Find the tangent vector to c at $t = \frac{\pi}{3}$.

(b) Find the tangent line to the range of c at the point $(\frac{1}{2}, \frac{\sqrt{3}}{2})$.

(c) Write the equation of the range in $\mathbb{R}^2$ of the tangent line found in (b).

25.4.7. Proposition. *If a curve $c\colon J \to V$ is differentiable at a point a in the interior of the interval J, then it has a derivative at a and*
$$Dc(a) = dc_a(1).$$

Proof. Exercise. *Hint.* Start with the Newton quotient $\frac{\Delta c_a(h)}{h}$. Subtract and add $\frac{dc_a(h)}{h}$.)

The converse of Proposition 25.4.7 is also true. Every curve possessing a derivative is differentiable. This is our next proposition.

25.4.8. Proposition. *If a curve $c\colon J \to V$ has a derivative at a point a in J°, then it is differentiable at a and*
$$dc_a(h) = h\,Dc(a) \qquad \text{for all } h \in \mathbb{R}.$$

Proof. Problem. *Hint.* Define $T\colon \mathbb{R} \to V\colon h \mapsto h\,Dc(a)$. Show that $\Delta c_a \simeq T$.

25.4.9. Problem. Suppose that curves c_1 and c_2 in a normed linear space are defined and differentiable in some neighborhood of $a \in \mathbb{R}$. Then

(a) $D(c_1 + c_2)(a) = Dc_1(a) + Dc_2(a)$.

(b) $D(\alpha c_1)(a) = \alpha\,Dc_1(a)$ for all $\alpha \in \mathbb{R}$.

25.4.10. Problem. Let V be a normed linear space, and let $a \in V$.

(a) Suppose c is a differentiable curve at the zero vector in V. Then $c \simeq \mathbf{0}$ if and only if $Dc(0) = \mathbf{0}$.

(b) Suppose c_1 and c_2 are differentiable curves at a. Then $c_1 \simeq c_2$ if and only if $Dc_1(0) = Dc_2(0)$.

25.4.11. Proposition. *If $c \in \mathcal{D}_t(\mathbb{R}, V)$ and $f \in \mathcal{D}_a(V, W)$, where $a = c(t)$, then $f \circ c \in \mathcal{D}_t(\mathbb{R}, W)$ and*

$$D(f \circ c)(t) = df_a(Dc(t)).$$

Proof. Problem.

Thus far, integration and differentiation have been treated as if they belong to separate worlds. In the next theorem, known as the "fundamental theorem of calculus", we derive the most important link between these two topics.

25.4.12. Theorem (Fundamental theorem of calculus, Version I). *Let a belong to an open interval J in the real line, let E be a Banach space, and let $f: J \to E$ be a regulated curve. Define $F(x) = \int_a^x f$ for all $x \in J$. If f is continuous at $c \in J$, then F is differentiable at c and $DF(c) = f(c)$.*

Proof. Exercise. *Hint.* For every $\epsilon > 0$ there exists $\delta > 0$ such that $c + h \in J$ and $\|\Delta f_c(h)\| < \epsilon$ whenever $|h| < \delta$. (Why?) Use the (obvious) fact that $h\, f(c) = \int_c^{c+h} f(c)\, dt$ to show that $\|\Delta F_c(h) - h\, f(c)\| < \epsilon |h|$ whenever $0 < |h| < \delta$. From this conclude that $\lim_{h \to 0} \frac{1}{h} \Delta F_c(h) = f(c)$.

25.5. Directional derivatives

We now return to the study of maps between arbitrary normed linear spaces. Closely related to differentiability is the concept of "directional derivative", an examination of which provides some technical advantages and also throws light on the geometric aspect of differentiation.

25.5.1. Definition. Let f be a member of $\mathcal{F}_a(V, W)$, and let v be a nonzero vector in V. Then $D_v f(a)$, the DERIVATIVE OF f AT a IN THE DIRECTION OF v, is defined by

$$D_v f(a) := \lim_{t \to 0} \frac{1}{t} \Delta f_a(tv)$$

if this limit exists. This directional derivative is also called the GÂTEAUX DIFFERENTIAL (or GÂTEAUX VARIATION) of f, and is sometimes denoted by $\delta f(a; v)$. Many authors require that in the preceding definition v be a unit vector. We will *not* adopt this convention.

Recall that for $\mathbf{0} \neq v \in V$, the curve $\ell: \mathbb{R} \to V$ defined by $\ell(t) = a + tv$ is the parametrized line through a in the direction of v. In the following proposition, which helps illuminate our use of the adjective "directional", we understand the domain of $f \circ \ell$ to be the set of all numbers t for which the expression $f(\ell(t))$ makes sense; that is,

$$\operatorname{dom}(f \circ \ell) = \{t \in \mathbb{R} \colon a + tv \in \operatorname{dom} f\}.$$

Since a is an interior point of the domain of f, the domain of $f \circ \ell$ contains an open interval about 0.

25.5.2. Proposition. *If $f \in \mathcal{D}_a(V, W)$ and $\mathbf{0} \neq v \in V$, then the directional derivative $D_v f(a)$ exists and is the tangent vector to the curve $f \circ \ell$ at 0 (where ℓ is the parametrized line through a in the direction of v). That is,*

$$D_v f(a) = D(f \circ \ell)(0) \, .$$

Proof. Exercise.

25.5.3. Exercise. Let $f(x, y) = \ln(x^2 + y^2)^{\frac{1}{2}}$. Find $D_v f(a)$ when $a = (1, 1)$ and $v = (\frac{3}{5}, \frac{4}{5})$.

25.5.4. Notation. For $a < b$, let $\mathcal{C}^1([a, b], \mathbb{R})$ be the family of all functions f differentiable on some open subset of $\mathbb{R}$ containing $[a, b]$ whose derivative Df is continuous on $[a, b]$.

25.5.5. Exercise. For all f in $\mathcal{C}^1([0, \frac{\pi}{2}], \mathbb{R})$ define

$$\phi(f) = \int_0^{\frac{\pi}{2}} (\cos x + Df(x))^2 \, dx \, .$$

Compute $D_v \phi(a)$ when $a(x) = 1 + \sin x$ and $v(x) = 2 - \cos x$ for $0 \leq x \leq \frac{\pi}{2}$.

25.5.6. Problem. Let $f(x, y) = e^{x^2 + y^2}$, $a = (1, -1)$, and $v = (-1, 2)$. Find $D_v f(a)$.

25.5.7. Problem. Let $f(x, y) = (2xy, y^2 - x^2)$, $a = (-1, 2)$, and $v = \left(\frac{1}{\sqrt{2}}, \frac{1}{\sqrt{2}}\right)$. Find $D_v f(a)$.

25.5.8. Problem. Let $\phi \colon \mathcal{C}([0, 1], \mathbb{R}) \to \mathbb{R} \colon f \mapsto \int_0^1 \left(\sin^6 \pi x + (f(x))^2 \right) dx$, and for $0 \leq x \leq 1$, let $a(x) = e^{-x} - x + 3$ and $v(x) = e^x$. Find $D_v \phi(a)$.

According to Proposition 25.5.2, differentiability implies the existence of directional derivatives in all directions. (In Problem 25.5.11 you are asked to show that the converse is not true.) The next proposition makes explicit the relationship between differentials and directional derivatives for differentiable functions.

25.5.9. Proposition. *If $f \in \mathcal{D}_a(V, W)$, then for every nonzero v in V*

$$D_v f(a) = df_a(v) \, .$$

Proof. Exercise. *Hint.* Use Problem 25.4.11.

It is worth noticing that even though the domain of a curve is one-dimensional, it is still possible to take directional derivatives. The relationship between derivatives and directional derivatives is very simple: the derivative of a curve c at a point t is just the directional derivative at t in the direction of the unit vector 1 in $\mathbb{R}$. *Proof*.

$$\begin{aligned} Dc(t) &= dc_t(1) \quad \text{(by Proposition 25.4.7)} \\ &= D_1 c(t) \quad \text{(by Proposition 25.5.9)}. \end{aligned}$$

25.5.10. Problem. Let $\phi \colon \mathcal{C}([0, 1], \mathbb{R}) \to \mathcal{C}^1([0, 1], \mathbb{R})$ be defined by $\phi f(x) = \int_0^x f(s) \, ds$ for all f in $\mathcal{C}([0, 1], \mathbb{R})$ and x in $[0, 1]$. For arbitrary functions a and $v \neq 0$ in $\mathcal{C}([0, 1], \mathbb{R})$ compute $D_v \phi(a)$ using the following:

(a) the definition of directional derivative.

(b) Proposition 25.5.2.

(c) Proposition 25.5.9.

25.5.11. Problem. Show that the converse of Proposition 25.5.2 need not hold. A function may have directional derivatives in all directions but fail to be differentiable. *Hint.* Consider the function defined by $f(x,y) = \frac{x^3}{x^2+y^2}$ for $(x,y) \neq (0,0)$ and $f(0,0) = 0$.

25.6. Functions mapping into product spaces

In this short section we examine the differentiability of functions that map into the product of normed linear spaces. This turns out to be a very simple matter: a necessary and sufficient condition for a function whose codomain is a product to be differentiable is that each of its components be differentiable. The somewhat more complicated topic of differentiation of functions whose *domain* lies in a product of normed linear spaces is deferred until the next chapter when we have the mean value theorem at our disposal.

25.6.1. Proposition. *If $f^1 \in \mathcal{D}_a(V, W_1)$ and $f^2 \in \mathcal{D}_a(V, W_2)$, then the function $f = (f^1, f^2)$ belongs to $\mathcal{D}_a(V, W_1 \times W_2)$ and $df_a = \big(d(f^1)_a, d(f^2)_a\big)$.*

Proof. Exercise. *Hint.* The function f is defined by $f(x) = \big(f^1(x), f^2(x)\big)$ for all $x \in \operatorname{dom} f^1 \cap \operatorname{dom} f^2$. Check that $f = j_1 \circ f^1 + j_2 \circ f^2$, where $j_1 \colon W_1 \to W_1 \times W_2 \colon u \mapsto (u, 0)$ and $j_2 \colon W_2 \to W_1 \times W_2 \colon v \mapsto (0, v)$. Use Propositions 25.3.15 and 25.3.17, and Problem 25.3.24.

The preceding proposition says that a function is differentiable if its components are differentiable. The converse, which is the next proposition, is also true: the components of a differentiable function are differentiable.

25.6.2. Proposition. *If f belongs to $\mathcal{D}_a(V, W_1 \times W_2)$, then its components f^1 and f^2 belong to $\mathcal{D}_a(V, W_1)$ and $\mathcal{D}_a(V, W_2)$, respectively, and*

$$df_a = \big(d(f^1)_a, d(f^2)_a\big).$$

Proof. Problem. *Hint.* For $k = 1, 2$ write $f^k = \pi_k \circ f$ where, as usual, $\pi_k(x_1, x_2) = x_k$. Then use the chain rule (Proposition 25.3.17) and Problem 25.3.24.

An easy consequence of the two preceding propositions is that curves in product spaces have derivatives if and only if their components do.

25.6.3. Corollary. *Let c^1 and c^2 be curves in W_1 and W_2, respectively. If c^1 and c^2 are differentiable at t, then so is the curve $c = (c^1, c^2)$ in $W_1 \times W_2$ and $Dc(t) = \big(Dc^1(t), Dc^2(t)\big)$. Conversely, if c is a differentiable curve in the product $W_1 \times W_2$, then its components c^1 and c^2 are differentiable and $Dc = \big(Dc^1, Dc^2\big)$.*

Proof. Exercise.

It is easy to see how to generalize the three preceding results to functions whose codomains are the products of any finite collection of normed linear spaces. Differentiation is done componentwise.

25.6.4. Example. Consider the helix

$$c\colon \mathbb{R} \to \mathbb{R}^3 \colon t \mapsto (\cos t, \sin t, t)\,.$$

Its derivative at t is found by differentiating each component separately. That is, $Dc(t) = (-\sin t, \cos t, 1)$ for all $t \in \mathbb{R}$.

25.6.5. Problem. Let $f(x, y) = (\ln(xy), y^2 - x^2)$, $a = (1, 1)$, and $v = (\frac{3}{5}, \frac{4}{5})$. Find the directional derivative $D_v f(a)$.

Chapter 26

Partial derivatives and iterated integrals

In this chapter we consider questions that arise concerning a function f whose domain is the product $V_1 \times V_2$ of normed linear spaces. What relationship (if any) exists between the differentiability of f and the differentiability of the functions $x \mapsto f(x, b)$ and $y \mapsto f(a, y)$, where a and b are fixed points in V_1 and V_2, respectively? What happens in the special case $V_1 = V_2 = \mathbb{R}$ if we integrate f first with respect to y (that is, integrate, for arbitrary x, the function $y \mapsto f(x, y)$ over the interval $[c, d]$) and then integrate with respect to x (that is, integrate the function $x \mapsto \int_c^d f(x, y)\, dy$)? Does this produce the same result as integrating first with respect to x and then with respect to y? Before we can answer these and similar questions, we must develop a fundamental tool of analysis: the mean value theorem.

26.1. The mean value theorem(s)

Heretofore, we have discussed differentiability only of functions defined on *open* subsets of normed linear spaces. It is occasionally useful to consider differentiability of functions defined on other types of subsets. The business from beginning calculus of right and left differentiability at endpoints of intervals does not extend in any very natural fashion to functions with more complicated domains in higher-dimensional spaces. Recall that according to Definition 12.1.1 a NEIGHBORHOOD of a point in a metric space is an open set containing that point. It will be convenient to expand slightly our use of this word.

26.1.1. Definition. A NEIGHBORHOOD of a subset A of a metric space is any open set that contains A.

26.1.2. Definition. Let V and W be normed linear spaces, and let $A \subseteq V$. A W-valued function f is said to be DIFFERENTIABLE ON A if it is (defined and)

differentiable on some neighborhood of A. The function f is CONTINUOUSLY DIF-
FERENTIABLE ON A if it is differentiable on (a neighborhood of) A and if its differential $df \colon x \mapsto df_x$ is continuous at each point of A. The family of all W-valued
continuously differentiable functions on A is denoted by $\mathcal{C}^1(A, W)$. (Keep in mind
that in order for a function f to belong to $\mathcal{C}^1(A, W)$, its domain must contain a
neighborhood of A.)

Given a function f for which both "differential" and "derivative" make sense,
it is natural to ask if there is any difference between requiring df to be continuous
and requiring Df to be. It is the point of the next problem to show that there is
not.

26.1.3. Problem. Let A be a subset of $\mathbb{R}$ with nonempty interior, and let W be
a normed linear space. A function f mapping a neighborhood of A into W belongs
to $\mathcal{C}^1(A, W)$ if and only if its derivative Df (exists and) is continuous on some
neighborhood of A.

Not every function that is differentiable is continuously differentiable.

26.1.4. Example. Let $f(x) = x^2 \sin 1/x$ for $x \neq 0$ and $f(0) = 0$. Then f is
differentiable on $\mathbb{R}$ but does not belong to $\mathcal{C}^1(\mathbb{R}, \mathbb{R})$.

Proof. Problem.

We have already encountered one version of the mean value theorem (see Theorem 8.4.26): if $a < b$ and f is a real valued function continuous on the interval
$[a, b]$ and differentiable on the interior of that interval, then

$$(32) \qquad\qquad \frac{f(b) - f(a)}{b - a} = Df(c)$$

for at least one number c between a and b.

We consider the problem of generalizing this formula to SCALAR FIELDS (that is,
real valued functions of a vector variable), to CURVES (vector valued functions of a
real variable), and to VECTOR FIELDS (vector valued functions of a vector variable).
There is no difficulty in finding an entirely satisfactory variant of (32) that holds
for scalar fields whose domain lies in $\mathbb{R}^n$. This is done in Chapter 27 once we have
the notion of "gradient" (see Proposition 27.2.17). On the other hand we show in
Exercise 26.1.5 that for curves (a fortiori, vector fields) formula (32) does *not* hold.
Nevertheless, one useful aspect of the mean value theorem (that changes in f over
the interval $[a, b]$ cannot exceed the maximum value of $|Df|$ multiplied by the length
of the interval) does have a direct generalization to curves (see Theorem 26.1.6).
We will refer to this as the MEAN INEQUALITY.

26.1.5. Exercise. Show that the classical mean value theorem (Theorem 8.4.26)
fails for vector valued functions. That is, find an interval $[a, b]$, a Banach space
E, and a continuous function $f \colon [a, b] \to E$ differentiable on (a, b) such that the
equation $(b - a)Df(c) = f(b) - f(a)$ holds for *no* point c in (a, b). *Hint.* Consider
a parametrization of the unit circle.

Here is our first generalization of the mean value theorem. Others will occur
in Proposition 26.1.7, Corollary 26.1.8, and Proposition 26.1.14.

26.1.6. Theorem (Mean inequality for curves). *Let $a < b$, and let W be a normed linear space. If a continuous function $f \colon [a, b] \to W$ has a derivative at each point of (a, b) and if there is a constant M such that $\|Df(t)\| \leq M$ for all $t \in (a, b)$, then*

$$\|f(b) - f(a)\| \leq M(b - a)\,.$$

Proof. Exercise. *Hint.* Given $\epsilon > 0$ define $h(t) = \|f(t) - f(a)\| - (t - a)(M + \epsilon)$ for $a \leq t \leq b$. Let $A = h^{\leftarrow}(-\infty, \epsilon]$. Show the following:

(a) A has a least upper bound, say l;

(b) $l > a$;

(c) $l \in A$; and

(d) $l = b$.

To prove (d), argue by contradiction. Assume $l < b$. Show that

$$\|(t - l)^{-1}(f(t) - f(l))\| < M + \epsilon$$

for t sufficiently close to and greater than l. For such t show that $t \in A$ by considering the expression

$$\|f(t) - f(l)\| + \|f(l) - f(a)\| - (t - l)(M + \epsilon)\,.$$

Finally, show that the desired conclusion follows from (c) and (d).

Next we extend the mean inequality from curves to vector fields.

26.1.7. Proposition. *Let V and W be normed linear spaces, and let a and h be points in V. If a W-valued function f is continuously differentiable on the segment $[a, a + h]$, then*

$$\|\Delta f_a(h)\| \leq M\|h\|$$

whenever M is a number such that $\|df_z\| \leq M$ for all z in $[a, a + h]$.

Proof. Problem. *Hint.* Use $l \colon t \mapsto a + th$ (where $0 \leq t \leq 1$) to parametrize the segment $[a, a + h]$. Apply Theorem 26.1.6 to the function $g = f \circ l$.

26.1.8. Corollary. *Let V and W be normed linear spaces, let a and h be points of V, let the operator T belong to $\mathfrak{B}(V, W)$, and let g be a W-valued function continuously differentiable on the segment $[a, a + h]$. If M is any number such that $\|dg_z - T\| \leq M$ for all z in $[a, a + h]$, then*

$$\|\Delta g_a(h) - Th\| \leq M\|h\|\,.$$

Proof. Problem. *Hint.* Apply Proposition 26.1.7 to the function $g - T$.

The next proposition is an important application of the mean inequality.

26.1.9. Proposition. *Let V and W be normed linear spaces, let U be a nonempty connected open subset of V, and let $f \colon U \to W$ be differentiable. If $df_x = \mathbf{0}$ for every $x \in U$, then f is constant.*

Proof. Problem. *Hint.* Choose $a \in U$ and set $G = f^{\leftarrow}\{f(a)\}$. Show that G is both an open and closed subset of U. Then use Proposition 17.1.6 to conclude that $U = G$.

To prove that G is open in U, take an arbitrary point y in G, find an open ball B about y that is contained in U, and use Proposition 26.1.7 to show that $\Delta f_y(w - y) = 0$ for every w in B.

26.1.10. Corollary. *Let V, W, and U be as in the preceding proposition. If f, $g: U \to W$ are differentiable on U and have the same differentials at each point of U, then f and g differ by a constant.*

Proof. Problem.

26.1.11. Problem. Show that the hypothesis that U be connected cannot be deleted in Proposition 26.1.9.

Corollary 26.1.10 makes possible a proof of the version of the fundamental theorem of calculus that is the basis for the procedures of formal integration taught in beginning calculus.

26.1.12. Definition. A differentiable curve f in a Banach space whose domain is an open interval in $\mathbb{R}$ is an ANTIDERIVATIVE of a function g if $Df = g$.

26.1.13. Theorem (Fundamental theorem of calculus, Version II). *Let a and b be points in an open interval $J \subseteq \mathbb{R}$ with $a < b$. If $g: J \to E$ is a continuous map into a Banach space and f is an antiderivative of g on J, then*

$$(33) \qquad \int_a^b g = f(b) - f(a)\,.$$

Proof. Problem. *Hint.* Let $h(x) = \int_a^x g$ for x in J. Use Corollary 26.1.10 to show that h and f differ by a constant. Find the value of this constant by setting $x = a$.

It is interesting to note that this version of the fundamental theorem of calculus may also be regarded as a mean value theorem. It is conventional to define $(b - a)^{-1} \int_a^b g$ to be the MEAN VALUE (or AVERAGE VALUE) of a function g over the interval $[a, b]$. Thus (33) may be regarded as saying that the Newton quotient $(b - a)^{-1}(f(b) - f(a))$ is just the mean value of the derivative of f. Since functions between Banach spaces do not in general have derivatives, it is better for purposes of generalization to rewrite (33) in terms of differentials. For a curve f with continuous derivative

$$\int_a^b Df(t)\, dt = \int_a^b df_t(1)\, dt \quad \text{(by Proposition 25.4.7)}$$

so (33) becomes

$$(34) \qquad f(b) - f(a) = \int_a^b df_t(1)\, dt.$$

If we divide both sides by $b - a$, this result says, briefly, that the Newton quotient of f is the mean value of the differential of f over $[a, b]$.

There is one more thing to consider in seeking to generalize the mean value theorem to vector fields. In Chapter 24 we defined the integral only for vector valued functions of a real variable. So if a and b are points in a general Banach space and f is a vector field, the expression $\int_a^b df_t(1)\,dt$ in (34) is meaningless (and so, of course, is the Newton quotient $(f(b) - f(a))/(b-a)$). This turns out not to be a big problem. In order to integrate over a closed segment $[a, b]$ in a Banach space, parametrize it: Let $l(t) = (1-t)a + tb$ for $0 \le t \le 1$, and let $g = f \circ l$. Then

(35)
$$dg_t(1) = df_{l(t)}(dl_t(1))$$
$$= df_{l(t)}(b-a)\,.$$

Apply (34) to the function g over the interval $[0, 1]$ to obtain

(36)
$$g(1) - g(0) = \int_0^1 dg_t(1)\,dt\,.$$

Substituting (35) in (36) leads to

(37)
$$f(b) - f(a) = g(1) - g(0)$$
$$= \int_0^1 df_{l(t)}(b-a)\,dt.$$

Notice that if we let $h = b - a$, then (37) may be written

(38)
$$\Delta f_a(h) = \int_0^1 df_{l(t)}(h)\,dt$$
$$= \left(\int_0^1 df_{l(t)}\,dt \right)(h)$$

by Corollary 24.3.19. It is in this form (either (37) or (38)) that the mean value theorem holds for a function f, between Banach spaces, that is continuously differentiable on a segment $[a, b]$. It is worth noticing that this generalization is weaker in three respects than the classical mean value theorem: we are not able to conclude that there exists a particular point where the differential of f is equal to $f(b) - f(a)$; we assume differentiability at a and b; and we assume continuity of the differential. Nonetheless, this will be adequate for our purposes.

Having the second version of the fundamental theorem of calculus in hand, we are in a position to prove a reasonable version of the mean value theorem for vector fields.

26.1.14. Proposition. *Suppose that E and F are Banach spaces, and $a, h \in E$. If an F-valued function f is continuously differentiable on the segment $[a, a+h]$, then*

(39)
$$\Delta f_a(h) = \left(\int_0^1 df_{l(t)}\,dt \right)(h),$$

where $l(t) = a + th$ for $0 \le t \le 1$.

Proof. *Problem.* *Hint.* Let $g = f \circ l$. Show that $Dg(t) = df_{l(t)}(h)$. Apply Corollary 24.3.19 to the right-hand side of equation (39). Use Theorem 26.1.13.

It is conventional when we invoke any of the results Theorem 26.1.6, Proposition 26.1.7, Corollary 26.1.8, or Proposition 26.1.14 to say that we have used "the" mean value theorem.

26.1.15. Problem. Verify Proposition 26.1.14 directly for the function $f(x, y) = x^2 + 6xy - 2y^2$ by computing both sides of equation (39).

26.1.16. Problem. Let W be a normed linear space, and let $a < b$. For each f in $\mathcal{C}^1([a, b], W)$ define
$$\|\|f\|\| = \sup\{\|f(x)\| + \|Df(x)\| : a \leq x \leq b\}\,.$$

(a) Show that $\mathcal{C}^1([a, b], W)$ is a vector space and that the map $f \mapsto \|\|f\|\|$ is a norm on this space.

(b) Let $j(x) = \sqrt{1 + x^2}$ for all x in $\mathbb{R}$. Use the mean value theorem to show that $|\Delta j_x(y)| \leq |y|$ for all x, $y \in \mathbb{R}$.

(c) Let k be a continuous real valued function on the interval $[a, b]$. For each f in $\mathcal{C}^1([a, b], \mathbb{R})$ define
$$\Phi(f) = \int_a^b k(x)\sqrt{1 + (f'(x))^2}\, dx\,.$$

Show that the function $\Phi \colon \mathcal{C}^1([a, b], \mathbb{R}) \to \mathbb{R}$ is uniformly continuous.

26.1.17. Proposition (Change of variables). *If $f \in \mathcal{C}(J_2, E)$ and $g \in \mathcal{C}^1(J_1, J_2)$, where J_1 and J_2 are open intervals in $\mathbb{R}$ and E is a Banach space, and if a, $b \in J_1$ with $a < b$, then*
$$\int_a^b g'(f \circ g) = \int_{g(a)}^{g(b)} f\,.$$

Proof. Problem. *Hint.* Use Theorem 26.1.13. Let $F(x) = \int_{g(a)}^x f$ for all x in J_2. Compute $(F \circ g)'(t)$.

26.1.18. Proposition (Integration by parts). *If $f \in \mathcal{C}^1(J, \mathbb{R})$ and $g \in \mathcal{C}^1(J, E)$, where J is an open interval in $\mathbb{R}$ and E is a Banach space, and if a and b are points in J with $a < b$, then*
$$\int_a^b fg' = f(b)g(b) - f(a)g(a) - \int_a^b f'g\,.$$

Proof. Problem. *Hint.* Differentiate the function $t \mapsto f(t)g(t)$.

26.1.19. Proposition. *If $f \in \mathcal{C}(J, \mathbb{R})$, where J is an interval, and if a, $b \in J$ with $a < b$, then there exists $c \in (a, b)$ such that*
$$\int_a^b f = f(c)(b - a)\,.$$

Proof. Problem.

26.1.20. Problem. Show by example that Proposition 26.1.19 is no longer true if it is assumed that f is a continuous function from J into $\mathbb{R}^2$.

26.2. Partial derivatives

Suppose that $V_1, V_2, \ldots, V_n$, and W are normed linear spaces. Let $V = V_1 \times \cdots \times V_n$. In the following discussion we will need the CANONICAL INJECTION MAPS $j_1, \ldots, j_n$, which map the coordinate spaces $V_1, \ldots, V_n$ into the product space V. For $1 \leq k \leq n$, the map $j_k \colon V_k \to V$ is defined by $j_k(x) = (0, \ldots, 0, x, 0, \ldots, 0)$, where x appears in the kth coordinate. It is an easy exercise to verify that each j_k is a bounded linear transformation and that if V_k does not consist solely of the zero vector, then $\|j_k\| = 1$. Also worth noting is the relationship given in the following exercise between the injections j_k and the projections π_k.

26.2.1. Exercise. Let $V_1, \ldots, V_n$ be normed linear spaces, and let $V = V_1 \times \cdots \times V_n$. Then

(a) For $k = 1, \ldots, n$, the injection j_k is a right inverse of the projection π_k.

(b) $\sum_{k=1}^{n} (j_k \circ \pi_k) = I_V$.

26.2.2. Problem. Let $V = V_1 \times \cdots \times V_n$, where $V_1, \ldots, V_n$ are normed linear spaces. Also let $a \in V$, $r > 0$, and $1 \leq k \leq n$. Then the image under the projection map π_k of the open ball in V about a of radius r is the open ball in V_k about a_k of radius r.

26.2.3. Definition. A mapping $f \colon M_1 \to M_2$ between metric spaces is OPEN if $f^{\to}(U)$ is open in M_2 whenever U is an open in M_1, and it is CLOSED if $f^{\to}(C)$ is closed whenever C is closed.

26.2.4. Problem. Show that each projection mapping $\pi_k \colon V_1 \times \cdots \times V_n \to V_k$ on the product of normed linear spaces is an open mapping. Construct an example to show that projection mappings need not be closed.

Now suppose that f belongs to $\mathcal{F}_a(V, W)$ and that $1 \leq k \leq n$. Let B be an open ball about a that is contained in the domain of f and $B_k = (\pi_k^{\to}(B)) - a_k$. From Problem 26.2.2 and the fact that translation by a_k is an isometry (see Problem 22.3.14), we see that B_k is an open ball in V_k about the origin (whose radius is the same as the radius of B). Define

$$g \colon B_k \to W \colon x \mapsto f(a + j_k(x)).$$

Notice that as x changes, only the kth variable of the domain of f is changing; the other $k - 1$ variables are fixed. Also notice that we can write $g = f \circ T_a \circ j_k$, where T_a is translation by a (that is, $T_a \colon x \mapsto x + a$).

26.2.5. Exercise. With notation as above show that

$$\Delta g_0 = \Delta f_a \circ j_k$$

in some neighborhood of the origin in V_k.

26.2.6. Proposition. *Let f, g, and a be as above. If f is differentiable at a, then g is differentiable at $\mathbf{0}$ and*

$$dg_0 = df_a \circ j_k.$$

Proof. Problem. *Hint.* Use Exercise 26.2.5 and Proposition 25.2.7.

26.2.7. Notation. Suppose that the function $g: B_k \to W: x \mapsto f(a + j_k(x))$ is differentiable at $\mathbf{0}$. Since g depends only on the function f, the point a, and the index k, it is desirable to have a notation for $dg_\mathbf{0}$ that does not require the use of the extraneous letter "g". A fairly common convention is to write $d_k f_a$ for $dg_\mathbf{0}$; this bounded linear map is the kth PARTIAL DIFFERENTIAL of f at a. Thus $d_k f_a$ is the unique bounded linear map that is tangent to $\Delta f_a \circ j_k$. We restate the preceding proposition using this notation.

26.2.8. Corollary. *Let $V_1, V_2, \ldots, V_n$, and let W be normed linear spaces. If the function f belongs to $\mathcal{D}_a(V_1 \times \cdots \times V_n, W)$, then for each $k = 1, \ldots, n$, the kth partial differential of f at a exists and*

$$d_k f_a = df_a \circ j_k .$$

26.2.9. Corollary. *Let $V_1, V_2, \ldots, V_n$, and let W be normed linear spaces. If the function f belongs to $\mathcal{D}_a(V_1 \times \cdots \times V_n, W)$, then*

$$df_a(x) = \sum_{k=1}^{n} d_k f_a(x_k)$$

for each x in $V_1 \times \cdots \times V_n$.

Proof. Problem. *Hint.* Write df_a as $df_a \circ I$, where I is the identity map on $V_1 \times \cdots \times V_n$, and use Exercise 26.2.1(b).

Corollaries 26.2.8 and 26.2.9 assure us that if f is differentiable at a point, then it has partial differentials at that point. The converse is not true. (Consider the function defined on $\mathbb{R}^2$ whose value is 1 everywhere except on the coordinate axes, where its value is 0. This function has partial differentials at the origin, but it is certainly not differentiable—or even continuous—there.) The following proposition shows that if we assume continuity (as well as the existence) of the partial differentials of a function in an open set, then the function is differentiable (in fact, continuously differentiable) on that set. To avoid complicated notation, we state this only for the product of two spaces.

26.2.10. Proposition. *Let V_1, V_2, and W be normed linear spaces, and let $f: U \to W$, where $U \overset{\circ}{\subseteq} V_1 \times V_2$. If the partial differentials $d_1 f$ and $d_2 f$ exist and are continuous on U, then f is continuously differentiable on U and*

$$(40) \qquad\qquad df_{(a,b)} = d_1 f_{(a,b)} \circ \pi_1 + d_2 f_{(a,b)} \circ \pi_2$$

at each point (a, b) in U.

Proof. Exercise. *Hint.* To show that f is differentiable at a point (a, b) in U and that its differential there is the expression on the right-hand side of (40), we must establish that $\Delta f_{(a,b)}$ is tangent to $R = S \circ \pi_1 + T \circ \pi_2$ where $S = d_1 f_{(a,b)}$ and $T = d_2 f_{(a,b)}$. Let $g: t \mapsto f(a, b + t)$ for all t such that $(a, b + t) \in U$ and $h^t: s \mapsto f(a + s, b + t)$ for all s and t such that $(a + s, b + t) \in U$. Show that $\Delta f_{(a,b)}(s, t)$ is the sum of $\Delta \left(h^t\right)_\mathbf{0}(s)$ and $\Delta g_0(t)$. Conclude from this that it suffices to show that, given $\epsilon > 0$, there exists $\delta > 0$ such that if $\|(u, v)\|_1 < \delta$, then

$$(41) \qquad\qquad \left\| \Delta \left(h^v\right)_\mathbf{0}(u) - Su \right\| \le \epsilon \|u\|$$

and

(42) $$\left\| \Delta g_{\mathbf{0}}(v) - Tv \right\| \le \epsilon \left\| v \right\|.$$

Find $\delta_1 > 0$ so that (42) holds whenever $\|v\| < \delta_1$. (This is easy.) Then find $\delta_2 > 0$ so that (41) holds whenever $\|(u, v)\|_1 < \delta_2$. This requires a little thought. Notice that (41) follows from the mean inequality (Corollary 26.1.8) provided that we can verify that

$$\left\| d\left(h^v\right)_z - S \right\| \le \epsilon$$

for all z in the segment $[\mathbf{0}, u]$. To this end show that $d\left(h^v\right)_z = d_1 f_{(a+z, b+v)}$ and use the fact that $d_1 f$ is assumed to be continuous.

26.2.11. Notation. Up to this point we have had no occasion to consider the problem of the various ways in which a Euclidean space $\mathbb{R}^n$ may be regarded as a product. Take $\mathbb{R}^5$ for example. If nothing to the contrary is specified, it is natural to think of $\mathbb{R}^5$ as being the product of five copies of $\mathbb{R}$; that is, points of $\mathbb{R}^5$ are 5-tuples $x = (x_1, \dots, x_5)$ of real numbers. If, however, we wish to regard $\mathbb{R}^5$ as the product of $\mathbb{R}^2$ and $\mathbb{R}^3$, then a point in $\mathbb{R}^5$ is an ordered pair (x, y) with $x \in \mathbb{R}^2$ and $y \in \mathbb{R}^3$. One good way of informing a reader that you wish $\mathbb{R}^5$ to be considered as a product in this particular fashion is to write $\mathbb{R}^2 \times \mathbb{R}^3$; another is to write $\mathbb{R}^{2+3}$. (Note the distinction between $\mathbb{R}^{2+3}$ and $\mathbb{R}^{3+2}$.) In many concrete problems the names of variables are given in the statement of the problem: for example, suppose we encounter several equations involving the variables u, v, w, x, and y and wish to solve for the last three variables in terms of the first two. We are then thinking of $\mathbb{R}^5$ as the product of $\mathbb{R}^2$ (the space of independent variables) and $\mathbb{R}^3$ (the space of dependent variables). This particular factorization may be emphasized by writing a point (u, v, w, x, y) of the product as $((u, v), (w, x, y))$. And if you wish to avoid such an abundance of parentheses, you may choose to write $(u, v; w, x, y)$ instead. Ordinarily, when $\mathbb{R}^n$ appears, it is clear from context what factorization (if any) is intended. The preceding notational devices are merely reminders designed to ease the burden on the reader. In the next exercise a function f of four variables is given, and you are asked to compute $d_1 f_a$ in several different circumstances. It is important to realize that the value of $d_1 f_a$ *depends on the factorization of $\mathbb{R}^4$ that is assumed.*

26.2.12. Exercise. Let

(43) $$f(t, u, v, w) = tu^2 + 3\,vw$$

for all t, u, v, $w \in \mathbb{R}$, and let $a = (1, 1, 2, -1)$. Find $d_1 f_a$ assuming that the domain of f is

(a) $\mathbb{R}^4$.

(b) $\mathbb{R} \times \mathbb{R}^3$.

(c) $\mathbb{R}^2 \times \mathbb{R}^2$.

(d) $\mathbb{R}^2 \times \mathbb{R} \times \mathbb{R}$.

(e) $\mathbb{R}^3 \times \mathbb{R}$.

Hint. First compute df_a. Then use Corollary 26.2.8. *Note.* The default case is (a); that is, if we were given only equation (43), we would assume that the domain of f is $\mathbb{R} \times \mathbb{R} \times \mathbb{R} \times \mathbb{R}$. In this case it is possible to compute $d_k f_a$ for $k = 1, 2, 3, 4$. In cases (b), (c), and (e) only $d_1 f_a$ and $d_2 f_a$ make sense; and in (d) we can compute $d_k f_a$ for $k = 1,\, 2,\, 3$.

26.2.13. Problem. Let

$$f(t, u, v, w) = tuv - 4\,u^2 w$$

for all t, u, v, $w \in \mathbb{R}$, and let $a = (1, 2, -1, 3)$. Compute $d_k f_a$ for all k for which this expression makes sense, assuming that the domain of f is

(a) $\mathbb{R}^4$.

(b) $\mathbb{R} \times \mathbb{R}^3$.

(c) $\mathbb{R}^2 \times \mathbb{R}^2$.

(d) $\mathbb{R} \times \mathbb{R}^2 \times \mathbb{R}$.

(e) $\mathbb{R}^3 \times \mathbb{R}$.

26.2.14. Definition. We now consider the special case of a differentiable function f that maps an open subset of $\mathbb{R}^n$ into a normed linear space. For $1 \leq k \leq n$ the injection $j_k \colon \mathbb{R} \to \mathbb{R}^n$ takes the real number t to the vector $t\,e^k$. Thus the function $g \equiv f \circ T_a \circ j_k$ is just $f \circ l$ where as usual l is the parametrized line through a in the direction of e^k. (*Proof.* $g(t) = f\left(a + j_k(t)\right) = f(a + t\,e^k) = (f \circ l)(t)$.) We define $f_k(a)$ (or $\frac{\partial f}{\partial x_k}(a)$, or $D_k f(a)$), the kth PARTIAL DERIVATIVE of f at a, to be $d_k f_a(1)$. Using Propositions 25.4.7 and 25.5.2, we see that

$$\begin{aligned}
f_k(a) &= d_k f_a(1) \\
&= dg_0(1) \\
&= Dg(0) \\
&= D(f \circ l)(0) \\
&= D_{e^k} f(a).
\end{aligned}$$

That is, the kth partial derivative of f is the directional derivative of f in the direction of the kth coordinate axis of $\mathbb{R}^n$. Thus, the notation $D_k f$ for the kth partial derivative can be regarded as a slight abbreviation of the usual notation $D_{e^k} f$ used for directional derivatives of functions on $\mathbb{R}^n$.

It is also useful to note that

(44)
$$\begin{aligned}
f_k(a) &= Dg(0) \\
&= \lim_{t \to 0} \frac{g(t) - g(0)}{t} \\
&= \lim_{t \to 0} \frac{f(a + t\,e^k) - f(a)}{t}.
\end{aligned}$$

This is the usual definition given for partial derivatives in beginning calculus. The mechanics of computing partial derivatives is familiar and is justified by (44): pretend that a function of n variables is a function only of its kth variable, then take the ordinary derivative.

One more observation. If f is differentiable at a in $\mathbb{R}^n$, then by Proposition 25.5.9

$$f_k(a) = D_{e^k} f(a) = df_a(e^k)\,.$$

Let $f \in \mathcal{D}_a(\mathbb{R}^n, W)$ where $W = W_1 \times \cdots \times W_m$ is the product of m normed linear spaces. From Proposition 25.6.2 (and induction) we know that

$$df_a = \left(d(f^1)_a, \ldots, d(f^m)_a\right).$$

From this it is an easy step to the next proposition.

26.2.15. Proposition. *If $f \in \mathcal{D}_a(\mathbb{R}^n, W)$ where $W = W_1 \times \cdots \times W_m$ is the product of m normed linear spaces, then*

$$\left(f^j\right)_k = (f_k)^j$$

for $1 \le j \le m$ and $1 \le k \le n$.

Proof. Problem.

The point of the preceding proposition is that no ambiguity arises if we write the expression f_k^j. It may correctly be interpreted either as the kth partial derivative of the jth component of the function f or as the jth component of the kth partial derivative of f.

26.2.16. Example. Let $f(x, y) = (x^5 y^2, x^3 - y^3)$. To find $f_1(x, y)$ hold y fixed and differentiate with respect to x using Corollary 25.6.3. Then

$$f_1(x, y) = \left(5\,x^4 y^2, 3\,x^2\right).$$

Similarly,

$$f_2(x, y) = \left(2\,x^5 y, -3\,y^2\right).$$

26.2.17. Exercise. Let $f(x, y, z) = (x^3 y^2 \sin z, x^2 + y \cos z)$ and $a = (1, -2, \frac{\pi}{2})$. Find $f_1(a)$, $f_2(a)$, and $f_3(a)$.

26.2.18. Problem. Let $f(w, x, y, z) = (wxy^2 z^3, w^2 + x^2 + y^2, wx + xy + yz)$ and $a = (-3, 1, -2, 1)$. Find $f_k(a)$ for all appropriate k.

26.2.19. Problem. Let $V_1, \ldots, V_n, W$ be normed linear spaces, let $U \overset{\circ}{\subseteq} V = V_1 \times \cdots \times V_n$, $\alpha \in \mathbb{R}$, and let $f, g \colon U \to W$. If the kth partial derivatives of f and g exist at a point a in U, then so do the kth partial differentials of $f + g$ and αf, and

(a) $d_k(f + g)_a = d_k f_a + d_k g_a$;

(b) $d_k(\alpha f)_a = \alpha d_k f_a$;

(c) $(f + g)_k(a) = f_k(a) + g_k(a)$; and

(d) $(\alpha f)_k(a) = \alpha f_k(a)$.

Hint. In (a), consider the function $(f + g) \circ T_a \circ j_k$.

26.2.20. Problem. Let $f, g \in \mathcal{F}_a(V, W)$, $\alpha \in \mathbb{R}$, and $\mathbf{0} \ne v \in V$. Suppose that $D_v f(a)$ and $D_v g(a)$ exist.

(a) Show that $D_v(f + g)(a)$ exists and is the sum of $D_v f(a)$ and $D_v g(a)$. *Hint.* Use the definition of directional derivative and Proposition 25.3.3. We *cannot* use either Proposition 25.5.2 or 25.5.9 here because we are *not* assuming that f and g are differentiable at a.

(b) Show that $D_v(\alpha f)(a)$ exists and is equal to $\alpha D_v f(a)$.

(c) Use (a) and (b) to prove parts (c) and (d) of Problem 26.2.19.

26.3. Iterated integrals

In the preceding section we considered partial differentiation of functions defined on the product of two or more spaces. In this section we consider the integration of such functions. To take the partial derivative of a function $f\colon (x,y) \mapsto f(x,y)$ with respect to x, we hold y fixed and differentiate the function $x \mapsto f(x,y)$. Partial integration works in very much the same way. If f is a continuous function mapping a rectangular subset $[a,b] \times [c,d]$ of $\mathbb{R}^2$ into a Banach space E, we may, for each fixed y in $[c,d]$, integrate the function $x \mapsto f(x,y)$ over the interval $[a,b]$. (This function is continuous since it is the composite of the continuous functions $x \mapsto (x,y)$ and f.) The integration will result in a vector that depends on y, call it $h(y)$. We will show shortly that the function $y \mapsto h(y)$ is also continuous and so may be integrated over $[c,d]$. The resulting vector in E is denoted by $\int_c^d \left(\int_a^b f(x,y)\, dx \right) dy$ or just $\int_c^d \int_a^b f(x,y)\, dx\, dy$. The two integrals operating successively are called ITERATED INTEGRALS.

26.3.1. Notation. Let $f\colon (x,y) \mapsto f(x,y)$ be a continuous function defined on a subset of $\mathbb{R}^2$ containing the rectangle $[a,b] \times [c,d]$. Throughout this section we denote by $f^{\,y}$ the function of x that results from holding y fixed and by $^x f$ the function of y resulting from fixing x. That is, for each y

$$f^{\,y}\colon x \mapsto f(x,y),$$

and for each x

$$^x f\colon y \mapsto f(x,y)\,.$$

For each y we interpret $\int_a^b f(x,y)\, dx$ to mean $\int_a^b f^{\,y}$, and for each x we take $\int_c^d f(x,y)\, dy$ to be $\int_c^d {}^x f$. Thus

$$\int_c^d \int_a^b f(x,y)\, dx\, dy = \int_c^d g,$$

where $g(y) = \int_a^b f^{\,y}$ for all $y \in [c,d]$. In order for $\int_c^d g$ to make sense, we must know that g is a regulated function. It will suffice for our needs to show that if f is continuous, then so is g.

There is an alternative notation that many prefer: $f(\,\cdot\,, y)$ for $f^{\,y}$ and $f(x,\,\cdot\,)$ for $^x f$.

26.3.2. Lemma. *Let $f\colon [a,b] \times [c,d] \to E$ be a continuous function into a Banach space. For each $y \in [c,d]$, let*

$$g(y) = \int_a^b f^{\,y}\,.$$

Then g is uniformly continuous on $[c,d]$.

Proof. *Exercise. Hint.* Use Proposition 24.1.11.

Perhaps the most frequently used result concerning iterated integrals is that if f is continuous, then the order of integration does not matter.

26.3.3. Proposition. *If E is a Banach space, if $a < b$ and $c < d$, and if $f \colon [a,b] \times [c,d] \to E$ is continuous, then*

$$\int_a^b \int_c^d f(x,y)\, dy\, dx = \int_c^d \int_a^b f(x,y)\, dx\, dy \,.$$

Proof. Problem. *Hint.* Define functions j and k for all z in $[c,d]$ by the formulas

$$j(z) = \int_a^b \int_c^z f(x,y)\, dy\, dx$$

and

$$k(z) = \int_c^z \int_a^b f(x,y)\, dx\, dy \,.$$

It suffices to show that $j = k$. (Why?) One may accomplish this by showing that $j'(z) = k'(z)$ for all z and that $j(c) = k(c)$ (see Corollary 26.1.10). Finding $k'(z)$ is easy. To find $j'(z)$, derive the formulas

(45) $$\frac{1}{h}\Delta j_z(h) = \int_a^b \frac{1}{h}\int_z^{z+h} f(x,y)\, dy\, dx$$

and

(46) $$\int_a^b f(x,z)\, dx = \int_a^b \frac{1}{h}\int_z^{z+h} f(x,z)\, dy\, dx \,.$$

Subtract (46) from (45) to obtain a new equation. Show that the right-hand side of this new equation can be made arbitrarily small by choosing h sufficiently small. (For this use an argument similar to the one used in the proof of Lemma 26.3.2. Conclude that $j'(z) = \int_a^b f(x,z)\, dx$.

Now that we have a result (Proposition 26.3.3) allowing us to change the order of two integrals, we can prove a result that justifies changing the order of integration and differentiation. We show that if f is continuous and f_2 (exists and) is continuous, then

$$\frac{d}{dy}\int_a^b f(x,y)\, dx = \int_a^b \frac{\partial f}{\partial y}(x,y)\, dx \,.$$

26.3.4. Proposition. *Let E be a Banach space, let $a < b$, $c < d$, and let $f \colon [a,b] \times [c,d] \to E$. If f and f_2 are continuous, then the function g defined for all $y \in [c,d]$ by*

$$g(y) = \int_a^b f^{\,y}$$

is continuously differentiable in (c,d), and for $c < y < d$

$$g'(y) = \int_a^b f_2(x,y)\, dx \,.$$

Proof. Exercise. *Hint.* Let $h(y) = \int_a^b f_2(x,y)\, dx$ for $c \le y \le d$. Use Proposition 26.3.3 to reverse the order of integration in $\int_c^z h$ (where $c < z < d$). Use the version of the fundamental theorem of calculus given in Theorem 26.1.13 to obtain $\int_c^z h = g(z) - g(c)$. Differentiate.

26.3.5. Problem. Compute

$$\int_0^1 \int_0^1 \frac{x^2 - y^2}{(x^2 + y^2)^2} \, dx \, dy \quad \text{and} \quad \int_0^1 \int_0^1 \frac{x^2 - y^2}{(x^2 + y^2)^2} \, dy \, dx \, .$$

Why does the result not contradict the assertion made in Proposition 26.3.3?

26.3.6. Problem.

(a) Suppose that the functions $g \colon \mathbb{R}^n \to \mathbb{R}$ and $h \colon \mathbb{R} \to \mathbb{R}$ are differentiable and $f = h \circ g$. Show that

$$f_k(x) = g_k(x)\,(Dh)(g(x))$$

whenever $x \in \mathbb{R}^n$ and $1 \leq k \leq n$.

(b) Let $g \colon \mathbb{R}^n \to \mathbb{R}$ be differentiable, and let $j \colon \mathbb{R} \to \mathbb{R}$ be continuous. Prove that

$$\frac{\partial}{\partial x_k} \int_c^{g(x)} j(t)\,dt = g_k(x)\,j(g(x))$$

whenever $c \in \mathbb{R}$, $x \in \mathbb{R}^n$, and $1 \leq k \leq n$. *Hint.* The expression on the left-hand side denotes $f_k(x)$ where f is the function $x \mapsto \int_c^{g(x)} j(t)\,dt$.

(c) Use part (b) to compute

$$\frac{\partial}{\partial x} \int_{x^3 y}^{x^2+y^2} \frac{1}{1 + t^2 + \cos^2 t} \, dt \, .$$

26.3.7. Proposition (Leibniz's formula). *Let $f \colon [a,b] \times [c,d] \to \mathbb{R}$ and $h \colon [c,d] \to \mathbb{R}$. If f and f_2 are continuous, if h is continuously differentiable on (c,d), and if $h(y) \in [a,b]$ for every $y \in (c,d)$, then*

$$\frac{d}{dy} \int_a^{h(y)} f(x,y)\,dx = \int_a^{h(y)} f_2(x,y)\,dx + Dh(y)\,f(h(y),y)\,.$$

Proof. Problem.

Chapter 27

Computations in $\mathbb{R}^n$

In the preceding two chapters we have developed some fundamental facts concerning the differential calculus in arbitrary Banach spaces. In the present chapter we restrict our attention to the Euclidean spaces $\mathbb{R}^n$. Not only are these spaces very important historically, but fortunately there are available a variety of powerful yet relatively simple techniques that make possible explicit computations of many of the concepts introduced in Chapter 25. The usefulness of these spaces seems to be associated with the emphasis in classical physics on systems having a finite number of degrees of freedom. The computational simplicity stems from two facts: First, differentials of functions between Euclidean spaces are always continuous (see Proposition 23.1.18). Second, the usual norm on $\mathbb{R}^n$ is derivable from an inner product. In the first section of this chapter we derive some standard elementary facts about inner products. It is important to appreciate that despite their algebraic appearance, inner products are the source of much of the geometry in Euclidean spaces.

27.1. Inner products

27.1.1. Definition. Let $x = (x_1, \ldots, x_n)$ and $y = (y_1, \ldots, y_n)$ be vectors in $\mathbb{R}^n$. The INNER PRODUCT (or DOT PRODUCT) of x and y, denoted by $\langle x, y \rangle$, is defined by

$$\langle x, y \rangle := \sum_{k=1}^{n} x_k y_k \, .$$

As a first result we list the most important properties of the inner product.

27.1.2. Proposition. *Let x, y, and z be vectors in $\mathbb{R}^n$, and let α be a scalar. Then*

(a) $\langle x + y, z \rangle = \langle x, z \rangle + \langle y, z \rangle$;

(b) $\langle \alpha x, y \rangle = \alpha \langle x, y \rangle$;

(c) $\langle x, y \rangle = \langle y, x \rangle$;

(d) $\langle x, x \rangle \geq 0$;

(e) $\langle x, x \rangle = 0$ *only if* $x = \mathbf{0}$; *and*

(f) $\|x\| = \sqrt{\langle x, x \rangle}$.

Items (a) *and* (b) *say that the inner product is* linear *in its first variable;* (c) *says it is* symmetric; *and* (d) *and* (e) *say that it is* positive definite. *It is virtually obvious that an inner product is also linear in its second variable (see Exercise 27.1.4). Thus an inner product may be characterized as a positive definite, symmetric, bilinear functional on* $\mathbb{R}^n$.

Proof. Problem.

27.1.3. Proposition. *If* x *is in* $\mathbb{R}^n$, *then*

$$x = \sum_{k=1}^{n} \langle x, e^k \rangle e^k \, .$$

This result is used so frequently that it has been stated formally as a proposition. Its proof, however, is trivial. (It is clear from the definition of the inner product that $\langle x, e^k \rangle = x_k$, where, as usual, $\{e^1, \ldots, e^n\}$ is the standard basis for $\mathbb{R}^n$.)

27.1.4. Exercise. Use Proposition 27.1.2(a)–(c), but not the definition of inner product, to prove that

$$\langle x, y + z \rangle = \langle x, y \rangle + \langle x, z \rangle$$

and

$$\langle x, \alpha y \rangle = \alpha \langle x, y \rangle$$

for all x, y, $z \in \mathbb{R}^n$ and $\alpha \in \mathbb{R}$.

27.1.5. Proposition (The parallelogram law). *If* x, $y \in \mathbb{R}^n$, *then*

$$\|x + y\|^2 + \|x - y\|^2 = 2\|x\|^2 + 2\|y\|^2 \, .$$

Proof. Problem.

27.1.6. Proposition (Schwarz's inequality). *If* u, $v \in \mathbb{R}^n$, *then*

$$|\langle u, v \rangle| \leq \|u\| \|v\| \, .$$

Proof. This has been proved in Chapter 9. Notice that the left-hand side of the inequality given in Proposition 9.2.6 is $\left(\langle u, v \rangle \right)^2$ and the right-hand side is $\|u\|^2 \|v\|^2$.   □

27.1.7. Definition. If x and y are nonzero vectors in $\mathbb{R}^n$, define $\angle(x, y)$, the ANGLE between x and y, by

$$\angle(x, y) := \arccos \left(\frac{\langle x, y \rangle}{\|x\| \|y\|} \right) \, .$$

A version of this formula, which is perhaps somewhat more familiar, is

$$\langle x, y \rangle = \|x\| \|y\| \cos \angle(x, y) \, .$$

27.1.8. Exercise. How do we know that the preceding definition makes sense? (What is the domain of the arccosine function?)

27.1.9. Problem. Prove the law of cosines: if x and y are nonzero vectors in $\mathbb{R}^n$ and $\theta = \measuredangle(x, y)$, then

$$\|x - y\|^2 = \|x\|^2 + \|y\|^2 - 2\|x\|\|y\| \cos \theta .$$

27.1.10. Exercise. What is the angle between the vectors $(1, 0, 1)$ and $(0, -1, 1)$ in $\mathbb{R}^3$?

27.1.11. Problem. Find the angle between the vectors $(1, 0, -1, -2)$ and $(-1, 1, 0, 1)$ in $\mathbb{R}^4$.

27.1.12. Problem. The angle of intersection of two curves is by definition the angle between the tangent vectors to the curves at the point of intersection. Find the angle of intersection at the point $(1, -2, 3)$ of the curves C_1 and C_2, where

$$C_1(t) = (t, t^2 + t - 4, 3 + \ln t)$$

and

$$C_2(u) = (u^2 - 8, u^2 - 2u - 5, u^3 - 3u^2 - 3u + 12) .$$

27.1.13. Definition. Two vectors x and y in $\mathbb{R}^n$ are PERPENDICULAR (or OR-THOGONAL) if $\langle x, y \rangle = 0$. In this case we write $x \perp y$. Notice that the relationship between perpendicularity and angle is what we expect: if x and y are nonzero vectors, then $x \perp y$ if and only if $\measuredangle(x, y) = \pi/2$. The zero vector is perpendicular to all vectors but the angle it makes with another vector is not defined.

27.1.14. Problem. Find a linear combination of the vectors $(1, 0, 2)$ and $(2, -1, 1)$ that is perpendicular to the vector $(2, 2, 1)$ in $\mathbb{R}^3$.

27.1.15. Problem. Prove the Pythagorean theorem: if $x \perp y$ in $\mathbb{R}^n$, then

$$\|x + y\|^2 = \|x\|^2 + \|y\|^2 .$$

Does the converse hold?

27.1.16. Notation. Let $f \colon U \to \mathbb{R}^n$ and $g \colon V \to \mathbb{R}^n$, where U and V are subsets of a normed linear space that are not disjoint. Then we denote by $\langle f, g \rangle$ the real valued function on $U \cap V$ whose value at a point x in $U \cap V$ is $\langle f(x), g(x) \rangle$. That is,

$$\langle f, g \rangle \colon U \cap V \to \mathbb{R} \colon x \mapsto \langle f(x), g(x) \rangle .$$

The scalar field $\langle f, g \rangle$ is the INNER PRODUCT (or DOT PRODUCT of f and g.

27.1.17. Proposition. *Suppose that the functions $f \colon U \to \mathbb{R}^n$ and $g \colon V \to \mathbb{R}^n$, defined on subsets of a normed linear space W, are differentiable at a point a in the interior of $U \cap V$. Then $\langle f, g \rangle$ is differentiable at a and*

$$d\langle f, g \rangle_a = \langle f(a), dg_a \rangle + \langle df_a, g(a) \rangle .$$

Proof. Problem. *Hint.* Use Propositions 25.3.16 and 25.6.2.

27.1.18. Corollary. *If f and g are curves at a point a in $\mathbb{R}^n$ and are differentiable, then*

$$D\langle f, g \rangle(a) = \langle f(a), Dg(a) \rangle + \langle Df(a), g(a) \rangle .$$

Proof. Use Definition 27.1.17 and Proposition 25.4.7.

$$D\langle f, g\rangle(a) = d\langle f, g\rangle_a(1)$$
$$= \langle f(a), dg_a(1)\rangle + \langle df_a(1), g(a)\rangle$$
$$= \langle f(a), Dg(a)\rangle + \langle Df(a), g(a)\rangle. \qquad \square$$

27.1.19. Problem. Let $f = (f^1, f^2, f^3)$, where

$$f^1(t) = t^3 + 2t^2 - 4t + 1,$$
$$f^2(t) = t^4 - 2t^3 + t^2 + 3,$$
$$f^3(t) = t^3 - t^2 + t - 2,$$

and let $g(t) = \|f(t)\|^2$ for all t in $\mathbb{R}$. Find $Dg(1)$.

27.1.20. Problem. Let c be a differentiable curve in $\mathbb{R}^n$. Show that the point $c(t)$ moves on the surface of a sphere centered at the origin if and only if the tangent vector $Dc(t)$ at t is perpendicular to the position vector $c(t)$ at each t. *Hint.* Use Corollary 27.1.18.

27.2. The gradient

In beginning calculus texts the gradient of a real valued function of n variables is usually defined to be an n-tuple of partial derivatives. This definition, although convenient for computation, disguises the highly geometric nature of the gradient. Here we adopt a different definition: the gradient of a scalar field is the vector that represents, in a sense to be made precise below, the differential of the function. First we look at an important example of a bounded linear functional on $\mathbb{R}^n$.

27.2.1. Example. Let $b \in \mathbb{R}^n$. Define

$$\psi_b \colon \mathbb{R}^n \to \mathbb{R} \colon x \mapsto \langle x, b\rangle.$$

Then ψ_b is a bounded linear functional on $\mathbb{R}^n$ and $\|\psi_b\| = \|b\|$.

Proof. Exercise.

The reason for the importance of the preceding example is that functions of the form ψ_b turn out to be the *only* bounded linear functionals on $\mathbb{R}^n$. Since on $\mathbb{R}^n$ every linear functional is bounded (see Propositions 23.1.18 and 23.1.4), the functions ψ_b are in fact the only real valued linear maps on $\mathbb{R}^n$. Thus, we say that every linear functional on $\mathbb{R}^n$ can be represented in the form ψ_b for some vector b in $\mathbb{R}^n$. Furthermore, this representation is unique. These assertions are stated formally in the next theorem.

27.2.2. Theorem (Riesz–Fréchet theorem). *If $f \in (\mathbb{R}^n)^*$, then there exists a unique vector b in $\mathbb{R}^n$ such that*

$$f(x) = \langle x, b\rangle$$

for all x in $\mathbb{R}^n$.

Proof. Problem. *Hint.* For the existence part, do the following. First (a), write $\sum_{k=1}^{n} x_k e^k$ for x in the expression $f(x)$, and use the linearity of f. Then (b), write $\langle x, b \rangle$ as a sum. Comparing the results of (a) and (b), guess the identity of the desired vector b. The uniqueness part is easy: suppose $f(x) = \langle x, a \rangle = \langle x, b \rangle$ for all x in $\mathbb{R}^n$. Show $a = b$.

27.2.3. Definition. If a map $T \colon V \to W$ between two normed linear spaces is both an isometry and a vector space isomorphism, we say that it is an ISOMETRIC ISOMORPHISM and that the spaces V and W are ISOMETRICALLY ISOMORPHIC.

27.2.4. Proposition. *Each Euclidean space $\mathbb{R}^n$ is isometrically isomorphic to its dual $(\mathbb{R}^n)^*$.*

Proof. Problem. *Hint.* Consider the map

$$\psi \colon \mathbb{R}^n \to (\mathbb{R}^n)^* \colon b \mapsto \psi_b \,.$$

One thing that must be established is that ψ is linear; don't confuse this with showing that each ψ_b is linear—a task already accomplished in Example 27.2.1. Use Riesz–Fréchet Theorem 27.2.2 and Problem 22.3.12.

Riesz–Fréchet Theorem 27.2.2 is the crucial ingredient in our definition of the *gradient* of a scalar field.

27.2.5. Definition. Let $U \subseteq \mathbb{R}^n$, and let $\phi \colon U \to \mathbb{R}$ be a scalar field. If ϕ is differentiable at a point a in U°, then its differential $d\phi_a$ is a bounded linear map from $\mathbb{R}^n$ into $\mathbb{R}$. That is, $d\phi_a \in (\mathbb{R}^n)^*$. Thus according to Riesz–Fréchet Theorem 27.2.2 there exists a unique vector, which we denote by $\nabla \phi(a)$, representing the linear functional $d\phi_a$. That is, $\nabla \phi(a)$ is the unique vector in $\mathbb{R}^n$ such that

$$d\phi_a(x) = \langle x, \nabla \phi(a) \rangle$$

for all x in $\mathbb{R}^n$. The vector $\nabla \phi(a)$ is the GRADIENT of ϕ at a. If U is an open subset of $\mathbb{R}^n$ and ϕ is differentiable at each point of U, then the function

$$\nabla \phi \colon U \to \mathbb{R}^n \colon u \mapsto \nabla \phi(u)$$

is the GRADIENT of ϕ. Notice two things: first, the gradient of a scalar field is a vector field; and second, the differential $d\phi_a$ is the zero linear functional if and only if the gradient at a, $\nabla \phi(a)$, is the zero vector in $\mathbb{R}^n$.

Perhaps the most useful fact about the gradient of a scalar field ϕ at a point a in $\mathbb{R}^n$ is that it is the vector at a that points in the direction of the most rapid increase of ϕ.

27.2.6. Proposition. *Let $\phi \colon U \to \mathbb{R}$ be a scalar field on a subset U of $\mathbb{R}^n$. If ϕ is differentiable at a point a in U and $d\phi_a$ is not the zero functional, then the maximum value of the directional derivative $D_u \phi(a)$, taken over all unit vectors u in $\mathbb{R}^n$, is achieved when u points in the direction of the gradient $\nabla \phi(a)$. The minimum value is achieved when u points in the opposite direction $-\nabla \phi(a)$.*

Proof. Exercise. *Hint.* Use Proposition 25.5.9 and recall that

$$\langle x, y \rangle = \|x\| \|y\| \cos \angle(x, y).$$

When a curve c is composed with a scalar field ϕ, we obtain a real valued function of a single variable. An easy but useful special case of the chain rule says that the derivative of the composite $\phi \circ c$ is the dot product of the derivative of c with the gradient of ϕ.

27.2.7. Proposition. *Suppose that c is a curve in $\mathbb{R}^n$ that is differentiable at a point t in $\mathbb{R}$ and that ϕ belongs to $\mathcal{D}_{c(t)}(\mathbb{R}^n, \mathbb{R})$. Then $\phi \circ c$ is differentiable at t and*

$$D(\phi \circ c)(t) = \langle Dc(t), (\nabla\phi)(c(t)) \rangle.$$

Proof. Problem. *Hint.* Use Proposition 25.4.7 and the chain rule, Proposition 25.3.17.

27.2.8. Problem. If $\phi \colon \mathbb{R}^n \to \mathbb{R} \colon x \mapsto \|x\|^2$, then ϕ is differentiable at each point b of $\mathbb{R}^n$ and

$$d\phi_b = 2\psi_b.$$

Furthermore, $\nabla\phi = 2I$, where I is the identity function on $\mathbb{R}^n$. *Hint.* The definition of ψ_b is given in Example 27.2.1. Write $\phi = \langle I, I \rangle$ and use Proposition 27.1.17.

27.2.9. Problem. For the function ϕ given in the preceding problem, verify by direct computation the formula for the mean value theorem (Proposition 26.1.14).

27.2.10. Definition. A linear transformation $T \colon \mathbb{R}^n \to \mathbb{R}^n$ is SELF-ADJOINT if $\langle Tx, y \rangle = \langle x, Ty \rangle$ for all $x, y \in \mathbb{R}^n$.

27.2.11. Problem. Let $T \in \mathfrak{B}(\mathbb{R}^n, \mathbb{R}^n)$ be self-adjoint, and let

$$\mu \colon \mathbb{R}^n \to \mathbb{R} \colon x \mapsto \langle Tx, x \rangle.$$

(a) Show that μ is differentiable at each point b in $\mathbb{R}^n$ and find $d\mu_b$.

(b) Find $\nabla\mu$.

27.2.12. Problem. Repeat Problem 27.2.9, this time using the function μ given in Problem 27.2.11.

27.2.13. Exercise (Conservation of energy). Consider a particle P moving in $\mathbb{R}^3$ under the influence of a force F. Suppose that the position of P at time t is $x(t)$ where $x \colon \mathbb{R} \to \mathbb{R}^3$ is at least twice differentiable. Let $v := Dx$ be the velocity of P, and let $a := Dv$ be its acceleration. Assume Newton's second law: $F \circ x = ma$, where $F \colon \mathbb{R}^3 \to \mathbb{R}^3$ is the force acting on P and m is the mass of P. Suppose further that the force field F is CONSERVATIVE; that is, there exists a scalar field $\phi \colon \mathbb{R}^3 \to \mathbb{R}$ such that $F = -\nabla\phi$. (Such a scalar field is a POTENTIAL FUNCTION for F.) The KINETIC ENERGY of P is defined by

$$KE := \tfrac{1}{2}m\|v\|^2,$$

its POTENTIAL ENERGY by

$$PE := \phi \circ x,$$

and its TOTAL ENERGY by

$$TE := KE + PE.$$

Prove, for this situation, the law of conservation of energy:

$$TE \text{ is constant.}$$

Hint. Use Propositions 26.1.9, 27.1.17, and 27.2.7.

In most circumstances the simplest way of computing the gradient of a scalar field ϕ on $\mathbb{R}^n$ is to calculate the n partial derivatives of ϕ. The n-tuple of these derivatives is the gradient. In most beginning calculus texts this is the definition of the gradient.

27.2.14. Proposition. *If ϕ is a scalar field on a subset of $\mathbb{R}^n$ and is differentiable at a point a, then*

$$\nabla\phi(a) = \sum_{k=1}^n \phi_k(a)e^k .$$

Proof. Exercise. *Hint.* Substitute $\nabla\phi(a)$ for x in Proposition 27.1.3. Use Proposition 25.5.9.

27.2.15. Exercise. Let $\phi(w,x,y,z) = wz - xy$, $u = (\frac{1}{2}, -\frac{1}{2}, \frac{1}{2}, -\frac{1}{2})$, and $a = (1,2,3,4)$. Find the directional derivative $D_u\phi(a)$. *Hint.* Use Proposition 25.5.9, the definition of gradient, and Proposition 27.2.14.

27.2.16. Exercise (Method of steepest descent). Let $\phi(x,y) = 2x^2 + 6y^2$ and $a = (2,-1)$. Find the steepest downhill path on the surface $z = \phi(x,y)$ starting at the point a and ending at the minimum point on the surface. *Hints.* (1) It is enough to find the equation of the projection of the curve onto the xy-plane; every curve $t \mapsto (x(t), y(t))$ in the xy-plane is the projection along the z-axis of a unique curve $t \mapsto (x(t), y(t), \phi(x(t), y(t)))$ on the surface $z = \phi(x,y)$. (2) If $c\colon t \mapsto (x(t), y(t))$ is the desired curve and we set $c(0) = a$, then according to Proposition 27.2.6 the unit vector u that minimizes the directional derivative $D_u\phi(b)$ at a point b in $\mathbb{R}^2$ is the one obtained by choosing u to point in the direction of $-\nabla\phi(b)$. Thus in order for the curve to point in the direction of the most rapid decrease of ϕ at each point $c(t)$, the tangent vector to the curve at $c(t)$ must be some positive multiple $p(t)$ of $-(\nabla\phi)(c(t))$. The function p will govern the speed of descent; since this is irrelevant in the present problem, set $p(t) = 1$ for all t. (3) Recall from beginning calculus that on an interval the only nonzero solution to an equation of the form $Dx(t) = kx(t)$ is of the form $x(t) = x(0)e^{kt}$. (4) The parameter t that we have introduced is artificial. Eliminate it to obtain an equation of the form $y = f(x)$.

27.2.17. Proposition (A mean value theorem for scalar fields). *Let ϕ be a differentiable scalar field on an open convex subset U of $\mathbb{R}^n$, and suppose that a and b are distinct points belonging to U. Then there exists a point c in the closed segment $[a,b]$ such that*

$$\phi(b) - \phi(a) = \langle b - a, \nabla\phi(c) \rangle .$$

Proof. Problem. *Hint.* Let $l(t) = (1-t)a + tb$ for $0 \le t \le 1$. Apply the mean value theorem for a real valued function of a single variable (Theorem 8.4.26) to the function $\phi \circ l$. Use Proposition 27.2.7.

27.2.18. Problem. Let $c(t) = (\cos t, \sin t, t)$ and $\phi(x,y,z) = x^2y - 3yz$. Find $D(\phi \circ c)(\pi/6)$. *Hint.* Use Propositions 27.2.7 and 27.2.14.

27.2.19. Problem. Let $\phi(x,y,z) = xz - 4y$, $u = (\frac{1}{2}, 0, \frac{1}{2}\sqrt{3})$, and $a = (1, 0, -\frac{\pi}{2})$. Find the directional derivative $D_u\phi(a)$.

27.2.20. Problem. Show that if V is a normed linear space, $f \in \mathcal{D}_a(\mathbb{R}^n, V)$, and v is a nonzero vector in $\mathbb{R}^n$, then

$$D_v f(a) = \sum_{k=1}^{n} v_k f_k(a).$$

Hint. Use Proposition 25.5.9.

27.2.21. Problem. Let $f: \mathbb{R}^2 \to \mathbb{R}^2: x \mapsto (x_1{}^2 - x_2{}^2, 3x_1 x_2)$, $a = (2, 1)$, and $v = (-1, 2)$. Use the preceding problem to find $D_v f(a)$.

27.2.22. Problem. Find the path of steepest descent on the surface $z = x^6 + 12y^4$ starting at the point whose x-coordinate is 1 and whose y-coordinate is $\frac{1}{2}$.

27.2.23. Problem. Suppose that the temperature $\phi(x, y)$ at points (x, y) on a flat surface is given by the formula

$$\phi(x, y) = x^2 - y^2.$$

Starting at a point (a, b) on the surface, what path should be followed so that the temperature will increase as rapidly as possible?

27.2.24. Problem. This (like Exercise 27.2.16 and Problem 27.2.22) is a steepest descent problem. But here we suppose that for some reason we are unable to solve explicitly the resulting differential equations. Instead we invoke an approximation technique. Let

$$\phi(x) = 13x_1{}^2 - 42x_1 + 13x_2{}^2 + 6x_2 + 10x_1 x_2 + 9$$

for all x in $\mathbb{R}^2$. The goal is to approximate the path of steepest descent. Start at an arbitrary point x^0 in $\mathbb{R}^2$ and choose a number $h > 0$. At x^0 compute the gradient of ϕ, take u^0 to be the unit vector pointing in the direction of $-\nabla\phi(x^0)$, and then move h units in the direction of u^0 arriving at a point x^1. Repeat the procedure: find the unit vector u^1 in the direction of $-\nabla\phi(x^1)$, then from x^1 move h units along u^1 to a point x^2. Continue in this fashion. In other words, $x^0 \in \mathbb{R}^2$ and $h > 0$ are arbitrary, and for $n \geq 0$

$$x^{n+1} = x^n + hu^n,$$

where $u^n = -\|\nabla\phi(x^n)\|^{-1}\nabla\phi(x^n)$.

(a) Start at the origin $x^0 = (0, 0)$ and choose $h = 1$. Compute 25 or 30 values of x^n. Explain geometrically what is happening here. Why is h "too large"? *Hint.* Don't attempt to do this by hand. Write a program for a computer or a programmable calculator. In writing your program, don't ignore the possibility that $\nabla\phi(x^n)$ may be zero for some n. (Keep in mind when you write this up that your reader probably has no idea how to read the language in which you write your program. Document it well enough that the reader can easily understand what you are doing at each step.)

(b) Describe what happens when h is "too small". Again start at the origin, take $h = 0.001$, and compute 25 or 30 values of x^n.

(c) By altering the values of h at appropriate times, find a succession of points $x^0, \ldots, x^n$ (starting with $x^0 = (0, 0)$) such that the distance between x^n and the point where ϕ assumes its minimum value is less than 0.001. (By examining the points $x^0, \ldots, x^n$ you should be able to guess, for this particular function, the exact location of the minimum.)

(d) Alter the program in part (a) to eliminate division by $\|\nabla\phi(x^n)\|$. (That is, let $x^{n+1} = x^n - h\nabla\phi(x^n)$.) Explain what happens in this case when h is "too large" (say $h = 1$). Explain why the altered program works better (provided that h is chosen appropriately) than the program in (a) for the present function ϕ.

27.2.25. Problem. Is it possible to find a differentiable scalar field ϕ on $\mathbb{R}^n$ and a point a in $\mathbb{R}^n$ such that $D_u\phi(a) > 0$ for every nonzero u in $\mathbb{R}^n$?

27.2.26. Problem. Is it possible to find a differentiable scalar field ϕ on $\mathbb{R}^n$ and a nonzero vector u in $\mathbb{R}^n$ such that $D_u\phi(a) > 0$ for every a in $\mathbb{R}^n$?

27.3. The Jacobian matrix

27.3.1. Definition. Let $U \overset{\circ}{\subseteq} \mathbb{R}^n$, and let $f\colon U \to \mathbb{R}^m$ be a differentiable function. Recall that the components $f^1, \ldots, f^m$ of f satisfy

$$f(x) = \left(f^1(x), \ldots, f^m(x)\right) = \sum_{j=1}^{m} f^j(x)e^j$$

for all x in U. More briefly, we may write

$$f = (f^1, \ldots, f^m) = \sum_{j=1}^{m} f^j e^j \, .$$

Recall also (from Proposition 26.2.15) that

$$(f^j)_k(a) = (f_k)^j(a)$$

whenever $1 \le j \le m$, $1 \le k \le n$, and $a \in U$. Consequently, the notation $f_k^j(a)$ is unambiguous; from now on we use it. As the differential of f at a point a in U is a (bounded) linear map from $\mathbb{R}^n$ into $\mathbb{R}^m$, it may be represented by an $m \times n$ matrix. This is called the JACOBIAN MATRIX of f at a. The entry in the jth row and kth column of this matrix is $f_k^j(a)$.

27.3.2. Proposition. *If $f \in \mathcal{D}_a(\mathbb{R}^n, \mathbb{R}^m)$, then*

$$[df_a] = [f_k^j(a)] \, .$$

Proof. Exercise. *Hint.* It helps to distinguish notationally between the standard basis vectors in $\mathbb{R}^n$ and those in $\mathbb{R}^m$. Denote the ones in $\mathbb{R}^n$ by $e^1, \ldots, e^n$ and those in $\mathbb{R}^m$ by $\hat{e}^1, \ldots, \hat{e}^m$. Use Proposition 21.3.11.

Note that the jth row of the Jacobian matrix is $f_1^j, \ldots, f_n^j$. Thought of as a vector in $\mathbb{R}^n$, this is just the gradient of the scalar field f^j. Thus we may think of the Jacobian matrix $[df_a]$ as being in the form

$$\begin{bmatrix} \nabla f^1(a) \\ \vdots \\ \nabla f^m(a) \end{bmatrix} \, .$$

27.3.3. Exercise. Let $f\colon \mathbb{R}^4 \to \mathbb{R}^3\colon (w,x,y,z) \mapsto (wxz, x^2+2y^2+3z^2, wy\arctan z)$, let $a=(1,1,1,1)$, and let $v=(0,2,-3,1)$.

(a) Find $[df_a]$.

(b) Find $df_a(v)$.

27.3.4. Problem. Let $f\colon \mathbb{R}^2 \to \mathbb{R}^2\colon x \mapsto (x_1{}^2 - x_2{}^2, 3x_1x_2)$, and let $a=(2,1)$.

(a) Find $[df_a]$.

(b) Use part (a) to find $df_a(-1,3)$.

27.3.5. Problem. Let $f\colon \mathbb{R}^3 \to \mathbb{R}^4\colon (x,y,z) \mapsto (xy, y-z^2, 2xz, y+3z)$, let $a=(1,-2,3)$, and let $v=(2,1,-1)$.

(a) Find $[df_a]$.

(b) Use part (a) to calculate $D_v f(a)$.

27.3.6. Problem. Let $f\colon \mathbb{R}^2 \to \mathbb{R}^2\colon (x,y) \mapsto (x^2y, 2xy^2)$, let $a=(2,-1)$, and let $u=(\frac{3}{5},\frac{4}{5})$. Compute $D_u f(a)$ in three ways:

(a) Use the definition of directional derivative.

(b) Use Proposition 25.5.2.

(c) Use Proposition 25.5.9.

27.3.7. Problem. Suppose that $f \in \mathcal{D}_a(\mathbb{R}^3, \mathbb{R}^4)$ and that the Jacobian matrix of f at a is

$$\begin{bmatrix} b & c & e \\ g & h & i \\ j & k & l \\ m & n & p \end{bmatrix}.$$

Find $f_1(a)$, $f_2(a)$, $f_3(a)$, $\nabla f^1(a)$, $\nabla f^2(a)$, $\nabla f^3(a)$, and $\nabla f^4(a)$.

27.3.8. Problem. Let $f \in \mathcal{D}_a(\mathbb{R}^n, \mathbb{R}^m)$ and $v \in \mathbb{R}^n$. Show that

(a) $df_a(v) = \sum_{j=1}^m \langle \nabla f^j(a), v \rangle e^j$, and

(b) $\|df_a\| \le \sum_{j=1}^m \|\nabla f^j(a)\|$.

27.4. The chain rule

In some respects it is convenient for scientists to work with variables rather than functions. Variables denote the physical quantities in which a scientist is ultimately interested. (In thermodynamics, for example, T is temperature, P pressure, S entropy, and so on.) Functions usually have no such standard associations. Furthermore, a problem that deals with only a small number of variables may turn out to involve a dauntingly large number of functions if they are specified. The simplification provided by the use of variables may, however, be more apparent than real, and the price paid in increased ambiguity for their suggestiveness is often substantial. Below are a few examples of ambiguities produced by the combined effects of excessive reliance on variables, inadequate (if conventional) notation, and the unfortunate mannerism of using the same name for a function and a dependent variable ("Suppose $x = x(s,t)\dots$").

(A) If $z = f(x,y)$, what does $\frac{\partial}{\partial x} z(y,x)$ mean? (Perhaps $f_1(y,x)$? Possibly $f_2(y,x)$?)

(B) If $z = f(x,t)$ where $x = x(t)$, then what is $\frac{\partial z}{\partial t}$? (Is it $f_2(t)$? Perhaps the derivative of $t \mapsto f(x(t),t)$ is intended?)

(C) Let $f(x,y)$ be a function of two variables. Does the expression $z = f(tx,ty)$ have three partial derivatives $\frac{\partial z}{\partial x}$, $\frac{\partial z}{\partial y}$, and $\frac{\partial z}{\partial t}$? Do $\frac{\partial z}{\partial x}$ and $\frac{\partial f}{\partial x}$ mean the same thing?

(D) Let $w = w(x,y,t)$ where $x = x(s,t)$ and $y = y(s,t)$. A direct application of the chain rule (as stated in most beginning calculus texts) produces

$$\frac{\partial w}{\partial t} = \frac{\partial w}{\partial x}\frac{\partial x}{\partial t} + \frac{\partial w}{\partial y}\frac{\partial y}{\partial t} + \frac{\partial w}{\partial t}.$$

Is this correct? Do the terms of the form $\frac{\partial w}{\partial t}$ cancel?

(E) Let $z = f(x,y) = g(r,\theta)$, where $x = r\cos\theta$ and $y = r\sin\theta$. Do $\frac{\partial z}{\partial r}$ and $\frac{\partial z}{\partial \theta}$ make sense? Do $\frac{\partial z}{\partial x}$ and $\frac{\partial z}{\partial y}$? How about z_1 and z_2? Are any of these equal?

(F) The formulas for changing polar to rectangular coordinates are $x = r\cos\theta$ and $y = r\sin\theta$. So if we compute the partial derivative of the variable r with respect to the variable x, we get

$$\frac{\partial r}{\partial x} = \frac{\partial}{\partial x}\sqrt{x^2+y^2} = \frac{x}{\sqrt{x^2+y^2}} = \frac{x}{r} = \cos\theta.$$

On the other hand, since $r = \dfrac{x}{\cos\theta}$, we use the chain rule to get

$$\frac{\partial r}{\partial x} = \frac{1}{\cos\theta} = \sec\theta.$$

Is something wrong here? What?

The principal goal of the present section is to provide a reliable formalism for dealing with partial derivatives of functions of several variables in such a way that questions like (A)–(F) can be avoided. The basic strategy is quite simple: when in doubt give names to the relevant functions (especially composite ones!), and then use the chain rule. Perhaps it should be remarked that one need not make a fetish of avoiding variables. Many problems stated in terms of variables can be solved quite simply without the intrusion of the names of functions; e.g., what is $\frac{\partial z}{\partial x}$ if $z = x^3 y^2$? This section is intended as a guide for the perplexed. Although its techniques are often useful in dissipating confusion generated by inadequate notation, it is neither necessary nor even particularly convenient to apply them routinely to every problem that arises. Let us start by writing the chain rule for functions between Euclidean spaces in terms of partial derivatives. Suppose that $f \in \mathcal{D}_a(\mathbb{R}^p, \mathbb{R}^n)$ and $g \in \mathcal{D}_{f(a)}(\mathbb{R}^n, \mathbb{R}^m)$. Then according to Proposition 25.3.17

$$d(g \circ f)_a = dg_{f(a)} \circ df_a.$$

Replacing these linear transformations by their matrix representations and using Proposition 21.5.12, we obtain

(47) $$\big[d(g \circ f)_a\big] = \big[dg_{f(a)}\big]\big[df_a\big].$$

27.4.1. Proposition. *If $f \in \mathcal{D}_a(\mathbb{R}^p, \mathbb{R}^n)$ and $g \in \mathcal{D}_{f(a)}(\mathbb{R}^n, \mathbb{R}^m)$, then $\big[d(g \circ f)_a\big]$ is the $m \times p$ matrix whose entry in the jth row and kth column is $\sum_{i=1}^{n} g_i^j(f(a))f_k^i(a)$. That is,*

$$(g \circ f)_k^j(a) = \sum_{i=1}^{n}(g_i^j \circ f)(a)f_k^i(a)$$

for $1 \le j \le m$ and $1 \le k \le p$.

Proof. Multiply the two matrices on the right-hand side of (47) and use Proposition 27.3.2. $\square$

It is occasionally useful to restate Proposition 27.4.1 in the following (clearly equivalent) way.

27.4.2. Corollary. *If $f \in \mathcal{D}_a(\mathbb{R}^p, \mathbb{R}^n)$ and $g \in \mathcal{D}_{f(a)}(\mathbb{R}^n, \mathbb{R}^m)$, then*

$$\big[d(g \circ f)_a\big] = \big[\langle \nabla g^j(f(a)), f_k(a) \rangle\big]_{j=1\,k=1}^{m\quad p}.$$

27.4.3. Exercise. This is an exercise in translation of notation. Suppose $y = y(u, v, w, x)$ and $z = z(u, v, w, x)$ where $u = u(s, t)$, $v = v(s, t)$, $w = w(s, t)$, and $x = x(s, t)$. Show that (under suitable hypotheses)

$$\frac{\partial z}{\partial t} = \frac{\partial z}{\partial u}\frac{\partial u}{\partial t} + \frac{\partial z}{\partial v}\frac{\partial v}{\partial t} + \frac{\partial z}{\partial w}\frac{\partial w}{\partial t} + \frac{\partial z}{\partial x}\frac{\partial x}{\partial t}.$$

27.4.4. Problem. Suppose that the variables x, y, and z are differentiable functions of the variables α, β, γ, δ, and ϵ, that in turn depend in a differentiable fashion on the variables r, s, and t. As in Exercise 27.4.3 use Proposition 27.4.1 to write $\frac{\partial z}{\partial r}$ in terms of quantities such as $\frac{\partial z}{\partial \alpha}$, $\frac{\partial \delta}{\partial r}$, etc.

27.4.5. Exercise. Let $f(x, y, z) = (xy^2, 3x - z^2, xyz, x^2 + y^2, 4xz + 5)$, $g(s, t, u, v, w) = (s^2 + u^2 + v^2, s^2v - 2tw^2)$, and $a = (1, 0, -1)$. Use the chain rule to find $\big[d(g \circ f)_a\big]$.

27.4.6. Problem. Let $f(x, y, z) = (x^3y^2\sin z, x^2 + y\cos z)$, $g(u, v) = (\sqrt{uv}, v^3)$, $k = g \circ f$, $a = (1, -2, \pi/2)$, and $h = (1, -1, 2)$. Use the chain rule to find $dk_a(h)$.

27.4.7. Problem. Let $f(x, y, z) = (x^2y + y^2z, xyz)$, $g(x, y) = (x^2y, 3xy, x - 2y, x^2 + 3)$, and $a = (1, -1, 2)$. Use the chain rule to find $\big[d(g \circ f)_a\big]$.

We now consider a slightly more complicated problem. Suppose that $w = w(x, y, t)$ where $x = x(s, t)$ and $y = y(s, t)$ and that all the functions mentioned are differentiable. (This is problem (D) at the beginning of this section.) It is perhaps tempting to write

(48)
$$\begin{aligned}
\frac{\partial w}{\partial t} &= \frac{\partial w}{\partial x}\frac{\partial x}{\partial t} + \frac{\partial w}{\partial y}\frac{\partial y}{\partial t} + \frac{\partial w}{\partial t}\frac{\partial t}{\partial t} \\
&= \frac{\partial w}{\partial x}\frac{\partial x}{\partial t} + \frac{\partial w}{\partial y}\frac{\partial y}{\partial t} + \frac{\partial w}{\partial t}
\end{aligned}$$

(since $\frac{\partial t}{\partial t} = 1$). The trouble with this is that the $\frac{\partial w}{\partial t}$ on the left-hand side is not the same as the one on the right-hand side. The $\frac{\partial w}{\partial t}$ on the right-hand side refers only to the rate of change of w with respect to t insofar as t appears *explicitly* in the formula for w; the one on the left-hand side takes into account the fact that, in addition, w depends *implicitly* on t *via* the variables x and y. What to do? Use functions. Relate the variables by functions as follows.

$$(49) \qquad \begin{array}{c} s \\ t \end{array} \xrightarrow{\ f\ } \begin{array}{c} x \\ y \\ t \end{array} \xrightarrow{\ g\ } w$$

Also let $h = g \circ f$. Notice that $f^3(s,t) = t$. Then according to the chain rule,

$$h_2 = \sum_{k=1}^{3} (g_k \circ f)\, f_2^k \,.$$

But $f_2^3 = 1$ (that is, $\frac{\partial t}{\partial t} = 1$). So

$$(50) \qquad h_2 = (g_1 \circ f)f_2^1 + (g_2 \circ f)f_2^2 + g_3 \circ f.$$

The ambiguity of (48) has been eliminated in (50). The $\frac{\partial w}{\partial t}$ on the left-hand side is seen to be the derivative with respect to t of the composite $h = g \circ f$, whereas the $\frac{\partial w}{\partial t}$ on the right-hand side is just the derivative with respect to t of the function g.

One last point. Many scientific workers adamantly refuse to give names to functions. What do they do? Look back at diagram (49) and remove the names of the functions.

$$(51) \qquad \begin{array}{c} s \\ t \end{array} \longrightarrow \begin{array}{c} x \\ y \\ t \end{array} \longrightarrow w$$

The problem is that the symbol "t" occurs twice. To specify differentiation of the composite function (our h) with respect to t, indicate that the "t" you are interested in is the one in the left column of (51). This may be done by listing everything else that appears in that column. That is, specify which variables are held constant. This specification conventionally appears as a subscript outside parentheses. Thus, the $\frac{\partial w}{\partial t}$ on the left-hand side of (48) (our h_2) is written as $\left(\frac{\partial w}{\partial t}\right)_s$ (and it is read "$\frac{\partial w}{\partial t}$ with s held constant"). Similarly, the $\frac{\partial w}{\partial t}$ on the right-hand side of (48)) (our g_3) involves differentiation with respect to t while x and y are fixed. So it is written $\left(\frac{\partial w}{\partial t}\right)_{x,y}$ (and it is read "$\frac{\partial w}{\partial t}$ with x and y held constant"). Thus, (48) becomes

$$(52) \qquad \left(\frac{\partial w}{\partial t}\right)_s = \frac{\partial w}{\partial x}\frac{\partial x}{\partial t} + \frac{\partial w}{\partial y}\frac{\partial y}{\partial t} + \left(\frac{\partial w}{\partial t}\right)_{x,y}.$$

It is not necessary to write, for example, an expression such as $\left(\frac{\partial w}{\partial x}\right)_{t,y}$ because there is no ambiguity; the symbol "x" occurs only once in (51). If you choose to use the convention just presented, it is best to use it only to avoid confusion; use it because you *must*, not because you *can*.

27.4.8. Exercise. Let $w = t^3 + 2yx^{-1}$ where $x = s^2 + t^2$ and $y = s\arctan t$. Use the chain rule to find $\left(\frac{\partial w}{\partial t}\right)_s$ at the point where $s = t = 1$.

We conclude this section with two more exercises on the use of the chain rule. Part of the difficulty here and in the problems at the end of the section is to interpret correctly what the problem says. The suggested solutions may seem longwinded, and they are. Nevertheless, these techniques prove valuable in situations complicated enough to be confusing. With practice, it is easy to do many of the indicated steps mentally.

27.4.9. Exercise. Show that if $z = xy + x\phi(yx^{-1})$, then $x\frac{\partial z}{\partial x} + y\frac{\partial z}{\partial y} = xy + z$. *Hint.* Start by restating the exercise in terms of functions. Add suitable hypotheses. In particular, suppose that $\phi\colon \mathbb{R} \to \mathbb{R}$ is differentiable. Let

$$j(x, y) = xy + x\phi(yx^{-1})$$

for $x, y \in \mathbb{R}$, $x \neq 0$. Then for each such x and y

(53) $$xj_1(x, y) + yj_2(x, y) = xy + j(x, y).$$

To prove this assertion proceed as follows.

(a) Let $g(x, y) = yx^{-1}$. Find $\left[dg_{(x,y)}\right]$.

(b) Find $\left[d(\phi \circ g)_{(x,y)}\right]$.

(c) Let $G(x, y) = \left(x, \phi(yx^{-1})\right)$. Use (b) to find $\left[dG_{(x,y)}\right]$.

(d) Let $m(x, y) = xy$. Find $\left[dm_{(x,y)}\right]$.

(e) Let $h(x, y) = x\phi(yx^{-1})$. Use (c) and (d) to find $\left[dh_{(x,y)}\right]$.

(f) Use (d) and (e) to find $\left[dj_{(x,y)}\right]$.

(g) Use (f) to prove (53).

27.4.10. Exercise. Show that if $f(u, v) = g(x, y)$ where f is a differentiable real valued function, $u = x^2 - y^2$, and $v = 2xy$, then

(54) $$y\frac{\partial g}{\partial x} - x\frac{\partial g}{\partial y} = 2v\frac{\partial f}{\partial u} - 2u\frac{\partial f}{\partial v}.$$

Hint. The equations $u = x^2 - y^2$ and $v = 2xy$ give u and v in terms of x and y. Think of the function $h\colon (x, y) \mapsto (u, v)$ as a change of variables in $\mathbb{R}^2$. That is, define

$$h\colon \mathbb{R}^2 \to \mathbb{R}^2\colon (x, y) \mapsto (x^2 - y^2, 2xy).$$

Then on the uv-plane (that is, the codomain of h) the function f is real valued and differentiable. The equation $f(u, v) = g(x, y)$ serves only to fix notation. It indicates that g is the composite function $f \circ h$. We may visualize the situation thus:

(55) $$\begin{array}{ccccc} x & \xrightarrow{\ h\ } & u & \xrightarrow{\ f\ } & w \\ y & & v & & \end{array}$$

where $g = f \circ h$.

Now what are we trying to prove? The conclusion (54) is clear enough if we evaluate the partial derivatives at the right place. Recalling that we have defined h so that $u = h^1(x, y)$ and $v = h^2(x, y)$, we may write (54) in the following form:

(56) $\quad yg_1(x, y) - xg_2(x, y) = 2h^2(x, y)f_1(h(x, y)) - 2h^1(x, y)f_2(h(x, y))$.

Alternatively, we may write

$$\pi_2 g_1 - \pi_1 g_2 = 2h^2 f_1 - 2h^1 f_2,$$

where π_1 and π_2 are the usual coordinate projections. To verify (56), use the chain rule to find $[dg_{(x,y)}]$.

27.4.11. Problem. Let $w = \frac{1}{2}x^2 y + \arctan(tx)$ where $x = t^2 - 3u^2$ and $y = 2tu$. Find $\left(\frac{\partial w}{\partial t}\right)_u$ when $t = 2$ and $u = -1$.

27.4.12. Problem. Let $z = \frac{1}{16}uw^2 xy$ where $w = t^2 - u^2 + v^2$, $x = 2tu + tv$, and $y = 3uv$. Find $\left(\frac{\partial z}{\partial u}\right)_{t,v}$ when $t = 1$, $u = -1$, and $v = -2$.

27.4.13. Problem. If $z = f\left(\frac{x-y}{y}\right)$, then $x\frac{\partial z}{\partial x} + y\frac{\partial z}{\partial y} = 0$. State this precisely and prove it.

27.4.14. Problem. If ϕ is a differentiable function on an open subset of $\mathbb{R}^2$ and $w = \phi(u^2 - t^2, t^2 - u^2)$, then $t\frac{\partial w}{\partial u} + u\frac{\partial w}{\partial t} = 0$. *Hint.* Let $h(t, u) = (u^2 - t^2, t^2 - u^2)$ and $w = \psi(t, u)$ where $\psi = \phi \circ h$. Compute $[dh_{(t,u)}]$. Use the chain rule to find $[d\psi_{(t,u)}]$. Then simplify $t\psi_2(t, u) + u\psi_1(t, u)$.

27.4.15. Problem. If $f(u, v) = g(x, y)$ where f is a differentiable real valued function on $\mathbb{R}^2$ and if $u = x^3 + y^3$ and $v = xy$, then

$$x\frac{\partial g}{\partial x} + y\frac{\partial g}{\partial y} = 3u\frac{\partial f}{\partial u} + 2v\frac{\partial f}{\partial v}.$$

27.4.16. Problem. Let $f(x, y) = g(r, \theta)$ where (x, y) are Cartesian coordinates and (r, θ) are polar coordinates in the plane. Suppose that f is differentiable at all (x, y) in $\mathbb{R}^2$.

(a) Show that except at the origin

$$\frac{\partial f}{\partial x} = (\cos\theta)\frac{\partial g}{\partial r} - \frac{1}{r}(\sin\theta)\frac{\partial g}{\partial \theta}.$$

(b) Find a similar expression for $\frac{\partial f}{\partial y}$.

Hint. Recall that Cartesian and polar coordinates are related by $x = r\cos\theta$ and $y = r\sin\theta$.

27.4.17. Problem. Let n be a fixed positive integer. A function $f\colon \mathbb{R}^2 \to \mathbb{R}^2$ is HOMOGENEOUS OF DEGREE n if $f(tx, ty) = t^n f(x, y)$ for all t, x, $y \in \mathbb{R}$. If such a function f is differentiable, then

(57) $$x\frac{\partial f}{\partial x} + y\frac{\partial f}{\partial y} = nf.$$

Hint. Try the following:

(a) Let $G(x, y, t) = (tx, ty)$. Find $[dG_{(t,x,y)}]$.
(b) Let $h = f \circ G$. Find $[dh_{(x,y,t)}]$.

(c) Let $H(x, y, t) = (t^n, f(x, y))$. Find $\left[dH_{(x,y,t)}\right]$.

(d) Let $k = m \circ H$ (where $m(u, v) = uv$). Find $\left[dk_{(x,y,t)}\right]$.

(e) By hypothesis $h = k$; so the answers to (b) and (d) must be the same. Use this fact to derive (57).

Chapter 28

Infinite series

It is perhaps tempting to think of an infinite series

$$\sum_{k=1}^{\infty} a_k = a_1 + a_2 + a_3 + \cdots$$

as being nothing but "addition performed infinitely often". This view of series is misleading, can lead to curious errors, and should not be taken seriously. Consider the infinite series

(58) $$1 - 1 + 1 - 1 + 1 - 1 + \cdots.$$

If we think of the "+" and "−" signs as functioning in essentially the same manner as the symbols we encounter in ordinary arithmetic, we might be led to the following "discovery":

(59)
$$
\begin{aligned}
0 &= (1-1) + (1-1) + (1-1) + \cdots \\
&= 1 - 1 + 1 - 1 + 1 - 1 + \cdots \\
&= 1 - (1-1) - (1-1) - \cdots \\
&= 1 - 0 - 0 - \cdots \\
&= 1.
\end{aligned}
$$

One can be more inventive: If S is the sum of the series (58), then

$$1 - S = 1 - (1 - 1 + 1 - 1 + \cdots) = 1 - 1 + 1 - 1 + 1 - 1 + \cdots = S,$$

from which it follows that $S = \frac{1}{2}$. This last result, incidentally, was believed (for quite different reasons) by both Leibniz and Euler; see [**Moo32**]. The point here is that if an intuitive notion of infinite sums and plausible arguments lead us to conclude that $1 = 0 = \frac{1}{2}$, then it is surely crucial for us to exercise great care in defining and working with convergence of infinite series.

In the first section of this chapter we discuss convergence of series in arbitrary normed linear spaces. One reason for giving the definitions in this generality is that doing so is no more complicated than discussing convergence of series in $\mathbb{R}$. A second reason is that it displays with much greater clarity the role of completeness of the underlying space in questions of convergence. (See, in particular, Propositions 28.1.17 and 28.3.2.) A final reason is that this generality is actually needed. We use it, for example, in the proofs of the inverse and implicit function theorems in the next chapter.

28.1. Convergence of series

28.1.1. Definition. Let (a_k) be a sequence in a normed linear space. For each n in $\mathbb{N}$ let $s_n = \sum_{k=1}^{n} a_k$. The vector s_n is the nth PARTIAL SUM of the sequence (a_k). As is true of sequences, we permit variations of this definition. For example, in section 28.4 on power series we consider sequences $(a_k)_{k=0}^{n}$ whose first term has index zero. In this case, of course, the proper definition of s_n is $\sum_{k=0}^{n} a_k$.

28.1.2. Exercise. Let $a_k = (-1)^{k+1}$ for each k in $\mathbb{N}$. For $n \in \mathbb{N}$ compute the nth partial sum of the sequence (a_k).

28.1.3. Exercise. Let $a_k = 2^{-k}$ for each k in $\mathbb{N}$. For $n \in \mathbb{N}$ show that the nth partial sum of the sequence (a_k) is $1 - 2^{-n}$.

28.1.4. Definition. Let (a_k) be a sequence in a normed linear space. The INFINITE SERIES $\sum_{k=1}^{\infty} a_k$ is defined to be the sequence (s_n) of partial sums of the sequence (a_k). We may also write $a_1 + a_2 + a_3 + \cdots$ for $\sum_{k=1}^{\infty} a_k$. Again we permit variants of this definition. For example, the infinite series associated with the sequence $(a_k)_{k=0}^{\infty}$ is denoted by $\sum_{k=0}^{\infty} a_k$. Whenever the range of the summation index k is understood from context or is unimportant, we may denote a series simply by $\sum a_k$.

28.1.5. Exercise. What are the infinite series associated with the sequences given in Exercises 28.1.2 and 28.1.3?

28.1.6. Definition. Let (a_k) be a sequence in a normed linear space V. If the infinite series $\sum_{k=1}^{\infty} a_k$ (that is, the sequence of partial sums of (a_k)) converges to a vector b in V, then we say that the sequence (a_k) is SUMMABLE or, equivalently, that the series $\sum_{k=1}^{\infty} a_k$ is a CONVERGENT SERIES. The vector b is called the SUM of the series $\sum_{k=1}^{\infty} a_k$, and we write

$$\sum_{k=1}^{\infty} a_k = b\,.$$

It is clear that a necessary and sufficient condition for a series $\sum_{k=1}^{\infty} a_k$ to be convergent or, equivalently, for the sequence (a_k) to be summable, is that there exist a vector b in V such that

(60) $$\left\| b - \sum_{k=1}^{n} a_k \right\| \to 0 \qquad \text{as } n \to \infty.$$

If a series does not converge, we say that it is a DIVERGENT SERIES (or that it DIVERGES).

CAUTION. It is an old, if illogical, practice to use the same notation $\sum_{k=1}^{\infty} a_k$ for both the sum of a convergent series and the series itself. As a result of this convention, the statements "$\sum_{k=1}^{\infty} a_k$ converges to b" and "$\sum_{k=1}^{\infty} a_k = b$" are interchangeable. It is possible for this to cause confusion, although in practice it is usually clear from the context which use of the symbol $\sum_{k=1}^{\infty} a_k$ is intended. Notice however that since a divergent series has no sum, the symbol $\sum_{k=1}^{\infty} a_k$ for such a series is unambiguous; it can refer only to the series itself.

28.1.7. Exercise. Are the sequences (a_k) given in Exercises 28.1.2 and 28.1.3 summable? If (a_k) is summable, what is the sum of the corresponding series $\sum_{k=1}^{\infty} a_k$?

28.1.8. Problem (Geometric series). Let a and r be real numbers.

(a) Show that if $|r| < 1$, then $\sum_{k=0}^{\infty} ar^k$ converges and

$$\sum_{k=0}^{\infty} ar^k = \frac{a}{1-r} \, .$$

Hint. To compute the nth partial sum s_n, use a technique similar to the one used in Exercise 28.1.3. See also Problem I.1.12.

(b) Show that if $|r| \geq 1$ and $a \neq 0$, then $\sum_{k=0}^{\infty} ar^k$ diverges. *Hint.* Look at the cases $r \geq 1$ and $r \leq -1$ separately.

28.1.9. Problem. Let $\sum a_k$ and $\sum b_k$ be convergent series in a normed linear space.

(a) Show that the series $\sum (a_k + b_k)$ also converges and that

$$\sum (a_k + b_k) = \sum a_k + \sum b_k \, .$$

Hint. Problem 22.3.9(a).

(b) Show that for every $\alpha \in \mathbb{R}$, the series $\sum (\alpha a_k)$ converges and that

$$\sum (\alpha a_k) = \alpha \sum a_k \, .$$

One very easy way of seeing that certain series do not converge is to observe that its terms do not approach 0. (The proof is given in the next proposition.) It is important not to confuse this assertion with its converse. *The condition $a_k \to 0$ does not guarantee that $\sum_{k=1}^{\infty} a_k$ converges.* (An example is given in Example 28.1.11.)

28.1.10. Proposition. *If $\sum_{k=1}^{\infty} a_k$ is a convergent series in a normed linear space, then $a_k \to 0$ as $k \to \infty$.*

Proof. Exercise. *Hint.* Write a_n in terms of the partial sums s_n and s_{n-1}.

28.1.11. Example. It is possible for a series to diverge even though the terms a_k approach 0. A standard example of this situation in $\mathbb{R}$ is the harmonic series $\sum_{k=1}^{\infty} 1/k$. The harmonic series diverges.

Proof. Exercise. *Hint.* Show that the difference of the partial sums s_{2p} and s_p is at least $1/2$. Assume that (s_n) converges. Use Proposition 18.1.4.

28.1.12. Problem. Show that if $0 < p \leq 1$, then the series $\sum_{k=1}^{\infty} k^{-p}$ diverges. *Hint.* Modify the argument used in Example 28.1.11.

28.1.13. Problem. Show that the series $\sum_{k=1}^{\infty} \frac{1}{k^2+k}$ converges and find its sum. *Hint.* $\frac{1}{k^2+k} = \frac{1}{k} - \frac{1}{k+1}$.

28.1.14. Problem. Use the preceding problem to show that the series $\sum_{k=1}^{\infty} \frac{1}{k^2}$ converges and that its sum is no greater than 2.

28.1.15. Problem. Show that the series $\sum_{k=4}^{\infty} \frac{1}{k^2-1}$ converges and find its sum.

28.1.16. Problem. Let $p \in \mathbb{N}$. Find the sum of the series $\sum_{k=1}^{\infty} \frac{(k-1)!}{(k+p)!}$. *Hint.* If $a_k = \frac{k!}{(k+p)!}$, what can you say about $\sum_{k=1}^{n} (a_{k-1} - a_k)$?

In *complete* normed linear spaces the elementary fact that a sequence is Cauchy if and only if it converges may be rephrased to give a simple necessary and sufficient condition for the convergence of series in the space.

28.1.17. Proposition (The Cauchy criterion). *Let V be a normed linear space. If the series $\sum a_k$ converges in V, then for every $\epsilon > 0$ there exists $n_0 \in \mathbb{N}$ such that $\left\| \sum_{k=m+1}^{n} a_k \right\| < \epsilon$ whenever $n > m \geq n_0$. If V is a Banach space, then the converse of this implication also holds.*

Proof. Exercise.

The principal use of the preceding proposition is to shorten proofs. By invoking the *Cauchy criterion*, one frequently can avoid explicit reference to the partial sums of the series involved. See, for example, Proposition 28.3.5.

One obvious consequence of the *Cauchy criterion* is that the convergence of an infinite series is unaffected by changing any finite number of its terms. If $a_n = b_n$ for all n greater than some fixed integer n_0, then the series $\sum a_n$ converges if and only if the series $\sum b_n$ does.

The examples of infinite series we have looked at thus far are all series of real numbers. We now turn to series in the Banach space $\mathcal{B}(S, E)$ of bounded E valued functions on a set S (where E is a Banach space). Most of the examples we consider will be real valued functions on subsets of the real line.

First a word of caution: The notations $\sum_{k=1}^{\infty} f_k$ and $\sum_{k=1}^{\infty} f_k(x)$ can, depending on context, mean many different things. There are numerous ways in which sequences (and therefore series) of functions can converge. There are, among a host of others, uniform convergence, pointwise convergence, convergence in mean, and convergence in measure. Only the first two of these appear in this text. Since we regard $\mathcal{B}(S, E)$ as a Banach space under the uniform norm $\| \ \|_u$, it is not, strictly speaking, necessary for us to write "$f_n \to g$ (unif)" when we wish to indicate that the sequence (f_n) converges to g in the space $\mathcal{B}(S, E)$; writing "$f_n \to g$" is enough, because, unless the contrary is explicitly stated, uniform convergence is understood. Nevertheless, in what follows we will frequently add the redundant "(unif)" just as a reminder that in the space $\mathcal{B}(S, E)$ we are dealing with uniform (and not some other type of) convergence of sequences and series.

It is important to keep in mind that in the space $\mathcal{B}(S, E)$ the following assertions are equivalent:

(a) $\sum_{k=1}^{\infty} f_k$ converges uniformly to g;

(b) $g = \sum_{k=1}^{\infty} f_k$; and

(c) $\left\| g - \sum_{k=1}^{n} f_k \right\|_u \to 0$ as $n \to \infty$.

Since uniform convergence implies pointwise convergence, but not conversely, each of the preceding three conditions implies—*but is not implied by*—the following three (which are also equivalent):

(a′) $\sum_{k=1}^{\infty} f_k(x)$ converges to $g(x)$ for every x in S;

(b′) $g(x) = \sum_{k=1}^{\infty} f_k(x)$ for every x in S; and

(c′) for every x in S

$$\left\| g(x) - \sum_{k=1}^{n} f_k(x) \right\| \to 0 \quad \text{as } n \to \infty .$$

One easy consequence of the Cauchy criterion (Proposition 28.1.17) is called the WEIERSTRASS M-TEST. The rather silly name that is attached to this result derives from the fact that in the statement of the proposition, the constants that appear are usually named M_n.

28.1.18. Proposition (Weierstrass M-test). *Let (f_n) be a sequence of functions in $\mathcal{B}(S, E)$ where S is a nonempty set and E is a Banach space. If there is a summable sequence of positive constants M_n such that $\|f_n\|_u \leq M_n$ for every n in $\mathbb{N}$, then the series $\sum f_k$ converges uniformly on S. Furthermore, if the underlying set S is a metric space and each f_n is continuous, then $\sum f_k$ is continuous.*

Proof. Problem.

28.1.19. Exercise. Let $0 < \delta < 1$. Show that the series $\sum_{k=1}^{\infty} \frac{x^k}{1+x^k}$ converges uniformly on the interval $[-\delta, \delta]$. *Hint.* Use Problem 28.1.8.

28.1.20. Problem. Show that $\sum_{n=1}^{\infty} \frac{1}{1+n^2 x}$ converges uniformly on $[\delta, \infty)$ for any δ such that $0 < \delta < 1$.

28.1.21. Problem. Let $M > 0$. Show that $\sum_{n=1}^{\infty} \frac{n^2 x^3}{n^4 + x^4}$ converges uniformly on $[-M, M]$.

28.1.22. Problem. Show that $\sum_{n=1}^{\infty} \frac{nx}{n^4 + x^4}$ converges uniformly on $\mathbb{R}$.

We conclude this section with a generalization of the alternating series test, familiar from beginning calculus. Recall that an alternating series in $\mathbb{R}$ is a series of the form $\sum_{k=1}^{\infty} (-1)^{k+1} \alpha_k$ where each $\alpha_k > 0$. The generalization here will not require that the multipliers of the α_k's be $+1$ and -1 in strict alternation. Indeed they need not even be real numbers; they may be the terms of any sequence of vectors in a Banach space for which the corresponding sequence of partial sums is bounded. You are asked in Problem 28.1.25 to show that the alternating series test actually follows from the next proposition.

28.1.23. Proposition. *Let (α_k) be a decreasing sequence of real numbers, each greater than or equal to zero, that converges to zero. Let (x_k) be a sequence of vectors in a Banach space for which there exists $M > 0$ such that $\|\sum_{k=1}^{n} x_k\| \leq M$ for all n in $\mathbb{N}$. Then $\sum_{k=1}^{\infty} \alpha_k x_k$ converges and $\|\sum_{k=1}^{\infty} \alpha_k x_k\| \leq M\alpha_1$.*

Proof. Problem. *Hint.* This is a bit complicated. Start by proving the following very simple geometrical fact about a normed linear space V: if $[x, y]$ is a closed segment in V, then one of its endpoints is at least as far from the origin as every other point in the segment. Use this to derive the fact that if x and y are vectors in V and $0 \leq t \leq 1$, then

$$\|x + ty\| \leq \max\{\|x\|, \|x + y\|\}.$$

Next prove the following result.

28.1.24. Lemma. *Let (α_k) be a decreasing sequence of real numbers with $\alpha_k \geq 0$ for every k, let $M > 0$, and let V be a normed linear space. If $x_1, \ldots, x_n \in V$ satisfy*

(61)
$$\left\| \sum_{k=1}^{m} x_k \right\| \leq M$$

for all $m \leq n$, then

$$\left\| \sum_{k=1}^{n} \alpha_k x_k \right\| \leq M\alpha_1 .$$

To prove this result, use mathematical induction. Supposing the lemma to be true for $n = p$, let $y_1, \ldots, y_{p+1}$ be vectors in V such that $\|\sum_{k=1}^{m} y_k\| \leq M$ for all $m \leq p + 1$. Let $x_k = y_k$ for $k = 1, \ldots, p - 1$ and let $x_p = y_p + (\alpha_{p+1}/\alpha_p)y_{p+1}$. Show that the vectors $x_1, \ldots, x_p$ satisfy (61) for all $m \leq p$ and invoke the inductive hypothesis.

Once the lemma is in hand, apply it to the sequence $(x_k)_{k=1}^{\infty}$ to obtain $\|\sum_{k=1}^{n} \alpha_k x_k\| \leq M\alpha_1$ for all $n \in \mathbb{N}$, and apply it to the sequence $(x_k)_{k=m+1}^{\infty}$ to obtain $\|\sum_{k=m+1}^{n} \alpha_k x_k\| \leq 2M\alpha_{m+1}$ for $0 < m < n$. Use this last result to prove that the sequence of partial sums of the series $\sum \alpha_k x_k$ is Cauchy.

28.1.25. Problem. Use Proposition 28.1.23 to derive the alternating series test: If (α_k) is a decreasing sequence of real numbers with $\alpha_k \geq 0$ for all k and if $\alpha_k \to 0$ as $k \to \infty$, then the alternating series $\sum_{k=1}^{\infty}(-1)^{k+1}\alpha_k$ converges. Furthermore, the absolute value of the difference between the sum of the series and its nth partial sum is no greater than α_{n+1}.

28.1.26. Problem. Show that the series $\sum_{k=1}^{\infty} k^{-1}\sin(k\pi/4)$ converges.

An important and interesting result in analysis is the TIETZE EXTENSION THE-OREM. For compact metric spaces it says that any continuous real valued function defined on a closed subset of the space can be extended to a continuous function on the whole space and that this process can be carried out in such a way that the (uniform) norm of the extension does not exceed the norm of the original function. One proof of this uses both the *M-test* and the approximation theorem of Weierstrass.

28.1.27. Theorem (Tietze extension theorem). *Let M be a compact metric space, let A be a closed subset of M, and let $g \colon A \to \mathbb{R}$ be continuous. Then there exists a continuous function $w \colon M \to \mathbb{R}$ such that $w\big|_A = g$ and $\|w\|_u = \|g\|_u$.*

Proof. Problem. *Hint.* First of all demonstrate that a continuous function can be truncated without disturbing its continuity. (Precisely, if $f \colon M \to \mathbb{R}$ is a continuous function on a metric space, if $A \subseteq M$, and if $f^{\to}(A) \subseteq [a,b]$, then there exists a continuous function $g\colon M \to \mathbb{R}$ that agrees with f on A and whose range is contained in $[a, b]$.) Let $\mathcal{F} = \{u\big|_A \colon u \in \mathcal{C}(M, \mathbb{R})\}$. Notice that the preceding comment reduces the proof of Theorem 28.1.27 to showing that $\mathcal{F} = \mathcal{C}(A, \mathbb{R})$. Use *Stone–Weierstrass Theorem* 23.2.6 to prove that $\mathcal{F}$ is dense in $\mathcal{C}(A, \mathbb{R})$. Next find a sequence (f_n) of functions in $\mathcal{F}$ such that

$$\left\| g - \sum_{k=1}^{n} f_k \right\|_u < \frac{1}{2^n}$$

for every n. Then for each k find a function u_k in $\mathcal{C}(M, \mathbb{R})$ whose restriction to A is f_k. Truncate each u_k (as above) to form a new function v_k that agrees with u_k (and therefore f_k) on A and that satisfies $\|v_k\|_u = \|f_k\|_u$. Use Weierstrass M-test 28.1.18 to show that $\sum_1^\infty v_k$ converges uniformly on M. Show that $w = \sum_1^\infty v_k$ is the desired extension.

Recall that in Problem 23.1.22 we showed that if $\phi\colon M \to N$ is a continuous map between compact metric spaces, then the induced map $T_\phi\colon \mathcal{C}(N, \mathbb{R}) \to \mathcal{C}(M, \mathbb{R})$ is injective if and only if ϕ is surjective, and ϕ is injective if T_ϕ is surjective. The missing part of this result (T_ϕ is surjective if ϕ is injective) happens also to be true, but it requires the Tietze extension theorem for its proof.

28.1.28. Problem. Let ϕ and T_ϕ be as in Problem 23.1.22. Show that if ϕ is injective, then T_ϕ is surjective.

28.2. Series of positive scalars

In this brief section we derive some of the standard tests of beginning calculus for convergence of series of positive numbers.

28.2.1. Definition. Let $S(n)$ be a statement in which the natural number n is a variable. We say that $S(n)$ holds for n SUFFICIENTLY LARGE if there exists N in $\mathbb{N}$ such that $S(n)$ is true whenever $n \geq N$.

28.2.2. Proposition (Comparison test). *Let (a_k) and (b_k) be sequences in $[0, \infty)$ and suppose that there exists $M > 0$ such that $a_k \leq M b_k$ for sufficiently large $k \in \mathbb{N}$. If $\sum b_k$ converges, then $\sum a_k$ converges. If $\sum a_k$ diverges, so does $\sum b_k$.*

Proof. Problem.

28.2.3. Proposition (Ratio test). *Let (a_k) be a sequence in $(0, \infty)$. If there exists $\delta \in (0, 1)$ such that $a_{k+1} \leq \delta a_k$ for k sufficiently large, then $\sum a_k$ converges. If there exists $M > 1$ such that $a_{k+1} \geq M a_k$ for k sufficiently large, then $\sum a_k$ diverges.*

Proof. Exercise.

28.2.4. Proposition (Integral test). *Let $f\colon [1, \infty) \to [0, \infty)$ be decreasing; that is, $f(x) \geq f(y)$ whenever $x < y$. If $\lim_{M\to\infty} \int_1^M f$ exists, then $\sum_1^\infty f(k)$ converges. If $\lim_{M\to\infty} \int_1^M f$ does not exist, then $\sum_1^\infty f(k)$ diverges.*

Proof. Problem. *Hint.* Show that $\int_k^{k+1} f \leq f(k) \leq \int_{k-1}^k f$ for $k \geq 2$.

28.2.5. Proposition (The root test). *Let $\sum a_k$ be a series of numbers in $[0, \infty)$. Suppose that the limit $L = \lim_{k \to \infty} (a_k)^{1/k}$ exists.*

(a) *If $L < 1$, then $\sum a_k$ converges.*

(b) *If $L > 1$, then $\sum a_k$ diverges.*

Proof. Problem.

28.2.6. Problem. Let (a_k) and (b_k) be decreasing sequences in $(0, \infty)$, and let $c_k = \min\{a_k, b_k\}$ for each k. If $\sum a_k$ and $\sum b_k$ both diverge, must $\sum c_k$ also diverge?

28.3. Absolute convergence

It is a familiar fact from beginning calculus that absolute convergence of a series of real numbers implies convergence of the series. The proof of this depends in a crucial way on the completeness of $\mathbb{R}$. We show in the first proposition of this section that for series in a normed linear space V absolute convergence implies convergence if *and only if* V is complete.

28.3.1. Definition. Let (a_k) be a sequence in a normed linear space V. We say that (a_k) is ABSOLUTELY SUMMABLE or, equivalently, that the series $\sum a_k$ CONVERGES ABSOLUTELY if the series $\sum \|a_k\|$ converges in $\mathbb{R}$.

28.3.2. Proposition. *A normed linear space V is complete if and only if every absolutely summable sequence in V is summable.*

Proof. Exercise. *Hint.* If V is complete, the Cauchy criterion, Theorem 28.1.17, may be used. For the converse, suppose that every absolutely summable sequence is summable. Let (a_k) be a Cauchy sequence in V. Find a subsequence (a_{n_k}) such that $\|a_{n_{k+1}} - a_{n_k}\| < 2^{-k}$ for each k. Consider the sequence (y_k) where $y_k := a_{n_{k+1}} - a_{n_k}$ for all k.

One of the most useful consequences of absolute convergence of a series is that the terms of the series may be rearranged without affecting the sum of the series. This is not true of CONDITIONALLY CONVERGENT series (that is, series that converge but do not converge absolutely). One can show, in fact, that a conditionally convergent series of real numbers can, by rearrangement, be made to converge to any real number whatever, or, for that matter, to diverge. We will not demonstrate this here, but a nice proof can be found in [**Apo74**].

28.3.3. Definition. A series $\sum_{k=1}^{\infty} b_k$ is said to be a REARRANGEMENT of the series $\sum_{k=1}^{\infty} a_k$ if there exists a bijection $\phi \colon \mathbb{N} \to \mathbb{N}$ such that $b_k = a_{\phi(k)}$ for all k in $\mathbb{N}$.

28.3.4. Proposition. *If $\sum b_k$ is a rearrangement of an absolutely convergent series $\sum a_k$ in a Banach space, then $\sum b_k$ is itself absolutely convergent, and it converges to the same sum as $\sum a_k$.*

Proof. Problem. *Hint.* Let $\beta_n := \sum_{k=1}^{n} \|b_k\|$. Show that the sequence (β_n) is increasing and bounded. Conclude that $\sum b_k$ is absolutely convergent. The hard part of the proof is showing that if $\sum_1^{\infty} a_k$ converges to a vector A, then so does

$\sum_1^\infty b_k$. Define partial sums as usual: $s_n := \sum_1^n a_k$ and $t_n := \sum_1^n b_k$. Given $\epsilon > 0$, you want to show that $\|t_n - A\| < \epsilon$ for sufficiently large n. Prove that there exists a positive N such that $\|s_n - A\| < \frac{1}{2}\epsilon$ and $\sum_n^\infty \|a_k\| \leq \frac{1}{2}\epsilon$ whenever $n \geq N$. Write

$$\|t_n - A\| \leq \|t_n - s_N\| + \|s_N - A\|.$$

Showing that $\|t_n - s_N\| \leq \frac{1}{2}\epsilon$ for n sufficiently large takes a little thought. For an appropriate function ϕ write $b_k = a_{\phi(k)}$. Notice that

$$\|t_n - s_N\| = \left\| \sum_1^n a_{\phi(k)} - \sum_1^N a_k \right\|.$$

The idea of the proof is to choose n so large that there are enough terms $a_{\phi(j)}$ to cancel *all* the terms a_k ($1 \leq k \leq N$).

If you have difficulty in dealing with sums like $\sum_{k=1}^n a_{\phi(k)}$ whose terms are not consecutive ($a_1, a_2, \ldots$ are consecutive terms of the sequence (a_k); $a_{\phi(1)}, a_{\phi(2)}, \ldots$ in general are not), a notational trick may prove useful. For P a *finite* subset of $\mathbb{N}$, write $\sum_P a_k$ for the sum of all the terms a_k such that k belongs to P. This notation is easy to work with. It should be easy to convince yourself that, for example, if P and Q are finite subsets of $\mathbb{N}$ and if they are disjoint, then $\sum_{P \cup Q} a_k = \sum_P a_k + \sum_Q a_k$. (What happens if $P \cap Q \neq \emptyset$?) In the present problem, let $C := \{1, \ldots, N\}$ (where N is the integer chosen above). Give a careful proof that there exists an integer p such that the set $\{\phi(1), \ldots, \phi(p)\}$ contains C. Now suppose n is any integer greater than p. Let $F := \{\phi(1), \ldots, \phi(n)\}$ and show that

$$\|t_n - s_N\| \leq \sum_G \|a_k\|,$$

where $G := F \setminus C$.

28.3.5. Proposition. *If (α_n) is an absolutely summable sequence in $\mathbb{R}$ and (x_n) is a bounded sequence in a Banach space E, then the sequence $(\alpha_n x_n)$ is summable in E.*

Proof. Problem. *Hint.* Use the Cauchy criterion, Theorem 28.1.17.

28.3.6. Problem. What happens in the previous proposition if the sequence (α_n) is assumed only to be bounded and the sequence (x_n) is absolutely summable?

28.3.7. Problem. Show that if the sequence (a_n) of real numbers is SQUARE SUMMABLE (that is, if the sequence $(a_n{}^2)$ is summable), then the series $\sum n^{-1} a_n$ converges absolutely. *Hint.* Use the Schwarz inequality, Proposition 27.1.6.

28.4. Power series

According to Problem 28.1.8, we may express the reciprocal of the real number $1 - r$ as the sum of a power series $\sum_0^\infty r^k$ provided that $|r| < 1$. One may reasonably ask if anything like this is true in Banach spaces other than $\mathbb{R}$, in the space of bounded linear maps from some Banach space into itself, for example. If T is such a map and if $\|T\| < 1$, is it necessarily true that $I - T$ is invertible? And if it is, can the inverse of $I - T$ be expressed as the sum of the power series $\sum_0^\infty T^k$? It turns out that the answer to both questions is "yes". Our interest in pursuing this matter is not limited to the fact that it provides an interesting generalization of facts concerning $\mathbb{R}$ to spaces with richer structure. In the next chapter we will need exactly this result for the proof we give of the inverse function theorem.

Of course it is not possible to study power series in arbitrary Banach spaces; there are, in general, no powers of vectors because there is no multiplication. Thus we restrict our attention to those Banach spaces that (like the space of bounded linear maps) are equipped with an additional operation $(x, y) \mapsto xy$ (we call it MULTIPLICATION) under which they become linear associative algebras (see Definition 21.2.6 for its definition) and on which the norm is SUBMULTIPLICATIVE (that is, $\|xy\| \leq \|x\| \|y\|$ for all x and y). We will, for simplicity, insist further that these algebras be unital and that the multiplicative identity $\mathbf{1}$ have norm one. Any Banach space thus endowed is a (UNITAL) BANACH ALGEBRA. It is clear that Banach algebras have many properties in common with $\mathbb{R}$. It is important, however, to keep firmly in mind those properties that are *not* shared with $\mathbb{R}$. Certainly there is, in general, no linear ordering $<$ of the elements of a Banach algebra (or for that matter of a Banach space). Another crucial difference is that in $\mathbb{R}$ every nonzero element has a multiplicative inverse (its reciprocal); this is not true in general Banach algebras (see, for example, Proposition 21.5.16). Furthermore, Banach algebras may have nonzero NILPOTENT elements (that is, elements $x \neq 0$ such that $x^n = 0$ for some natural number n). (*Example.* The 2×2 matrix $a = \begin{bmatrix} 0 & 1 \\ 0 & 0 \end{bmatrix}$ is not zero, but $a^2 = 0$.) This, of course, prevents us from requiring that the norm be multiplicative: while $|xy| = |x| \, |y|$ holds in $\mathbb{R}$, all that is true in general Banach algebras is $\|xy\| \leq \|x\| \, \|y\|$. Finally, multiplication in Banach algebras need not be commutative.

28.4.1. Definition. Let A be a normed linear space. Suppose there is an operation $(x, y) \mapsto xy$ from $A \times A$ into A satisfying the following: for all x, y, $z \in A$ and $\alpha \in \mathbb{R}$,

 (a) $(xy)z = x(yz)$,

 (b) $(x + y)z = xz + yz$,

 (c) $x(y + z) = xy + xz$,

 (d) $\alpha(xy) = (\alpha x)y = x(\alpha y)$, and

 (e) $\|xy\| \leq \|x\| \, \|y\|$.

Suppose additionally that there exists a vector $\mathbf{1}$ in A such that

 (f) $x\mathbf{1} = \mathbf{1}x = x$ for all $x \in A$, and

 (g) $\|\mathbf{1}\| = 1$.

Then A is a (UNITAL) NORMED ALGEBRA. If A is complete it is a (UNITAL) BANACH ALGEBRA. If all elements x and y in a normed (or Banach) algebra satisfy $xy = yx$, then the algebra A is COMMUTATIVE.

28.4.2. Example. The set $\mathbb{R}$ of real numbers is a commutative Banach algebra. The number 1 is the multiplicative identity.

28.4.3. Example. If S is a nonempty set, then, with pointwise multiplication

$$(fg)(x) := f(x)g(x) \qquad \text{for all } x \in S,$$

the Banach space $\mathcal{B}(S, \mathbb{R})$ becomes a commutative Banach algebra. The constant function $\mathbf{1} \colon x \mapsto 1$ is the multiplicative identity.

28.4.4. Exercise. Show that if f and g belong to $\mathcal{B}(S, \mathbb{R})$, then $\|fg\|_u \leq \|f\|_u \|g\|_u$. Show by example that equality need not hold.

28.4.5. Example. If M is a compact metric space, then (again with pointwise multiplication) $\mathcal{C}(M, \mathbb{R})$ is a commutative Banach algebra. It is a SUBALGEBRA of $\mathcal{B}(M, \mathbb{R})$ (that is, a subset of $\mathcal{B}(M, \mathbb{R})$ containing the multiplicative identity that is a Banach algebra under the induced operations).

28.4.6. Example. If E is a Banach space, then $\mathfrak{B}(E, E)$ is a Banach algebra (with composition as multiplication). We have proved in Problem 21.2.8 that the space of linear maps from E into E is a unital algebra. The same is easily seen to be true of $\mathfrak{B}(E, E)$, the corresponding space of bounded linear maps. The identity transformation I_E is the multiplicative identity. Its norm is 1 by Exercise 23.1.11(a). In Proposition 23.1.14 it was shown that $\mathfrak{B}(E, E)$ is a normed linear space; the submultiplicative property of the norm on this space was proved in Proposition 23.1.15. Completeness was proved in Proposition 23.3.6.

28.4.7. Proposition. *If A is a normed algebra, then the operation of multiplication*

$$M \colon A \times A \to A \colon (x, y) \mapsto xy$$

is continuous.

Proof. Problem. *Hint.* Try to adapt the proof of Example 14.1.9.

28.4.8. Corollary. *If $x_n \to a$ and $y_n \to b$ in a normed algebra, then $x_n y_n \to ab$.*

Proof. Problem.

28.4.9. Definition. An element x of a unital algebra A (with or without norm) is INVERTIBLE if there exists an element x^{-1} (called the MULTIPLICATIVE INVERSE of x) such that $xx^{-1} = x^{-1}x = \mathbf{1}$. The set of all invertible elements of A is denoted by Inv A. We list several almost obvious properties of inverses.

28.4.10. Proposition. *If A is a unital algebra, then the following hold:*

(a) *Each element of A has at most one multiplicative inverse.*

(b) *The multiplicative identity $\mathbf{1}$ of A is invertible and $\mathbf{1}^{-1} = \mathbf{1}$.*

(c) *If x is invertible, then so is x^{-1} and $\left(x^{-1}\right)^{-1} = x$.*

(d) *If x and y are invertible, then so is xy and $(xy)^{-1} = y^{-1}x^{-1}$.*

(e) *If x and y are invertible, then $x^{-1} - y^{-1} = x^{-1}(y - x)y^{-1}$.*

Proof. Let $\mathbf{1}$ be the multiplicative identity of A.

(a) If y and z are multiplicative inverses of x, then $y = y\mathbf{1} = y(xz) = (yx)z = \mathbf{1}z = z$.

(b) $\mathbf{1} \cdot \mathbf{1} = \mathbf{1}$ implies $\mathbf{1}^{-1} = \mathbf{1}$ by (a).

(c) $x^{-1}x = xx^{-1} = \mathbf{1}$ implies x is the inverse of x^{-1} by (a).

(d) $(xy)(y^{-1}x^{-1}) = xx^{-1} = \mathbf{1}$ and $(y^{-1}x^{-1})(xy) = y^{-1}y = \mathbf{1}$ imply $y^{-1}x^{-1}$ is the inverse of xy (again by (a)).

(e) $x^{-1}(y-x)y^{-1} = (x^{-1}y - 1)y^{-1} = x^{-1} - y^{-1}$.   □

28.4.11. Proposition. *If x is an element of a unital Banach algebra and $\|x\| < 1$, then $\mathbf{1} - x$ is invertible and $(\mathbf{1} - x)^{-1} = \sum_{k=0}^{\infty} x^k$.*

Proof. Exercise. *Hint.* First show that the geometric series $\sum_{k=0}^{\infty} x^k$ converges absolutely. Next evaluate $(\mathbf{1} - x)s_n$ and $s_n(\mathbf{1} - x)$ where $s_n = \sum_{k=0}^{n} x^k$. Then take limits as $n \to \infty$.

28.4.12. Corollary. *If x is an element of a unital Banach algebra and $\|x\| < 1$, then*

$$\|(\mathbf{1} - x)^{-1} - \mathbf{1}\| \le \frac{\|x\|}{1 - \|x\|}.$$

Proof. Problem.

Proposition 28.4.11 says that anything close to $\mathbf{1}$ in a Banach algebra A is invertible. In other words $\mathbf{1}$ is an interior point of Inv A. Corollary 28.4.12 says that if $\|x\|$ is small, that is, if $\mathbf{1} - x$ is close to $\mathbf{1}$, then $(\mathbf{1} - x)^{-1}$ is close to $\mathbf{1}^{-1}$ ($= \mathbf{1}$). In other words, the operation of inversion $x \mapsto x^{-1}$ is continuous at $\mathbf{1}$. These results are actually special cases of much more satisfying results: every point of Inv A is an interior point of that set (that is, Inv A is open); and the operation of inversion is continuous at every point of Inv A. We use the special cases above to prove the more general results.

28.4.13. Proposition. *If A is a Banach algebra, then* Inv $A \overset{\circ}{\subseteq} A$.

Proof. Problem. *Hint.* Let $a \in$ Inv A, $r = \|a^{-1}\|^{-1}$, $y \in B_r(a)$, and $x = a^{-1}(a-y)$. What can you say about $a(\mathbf{1} - x)$?

28.4.14. Proposition. *If A is a Banach algebra, then the operation of inversion*

$$r \colon \text{Inv } A \to \text{Inv } A \colon x \mapsto x^{-1}$$

is continuous.

Proof. Exercise. *Hint.* Show that r is continuous at an arbitrary point a in Inv A. Given $\epsilon > 0$ find $\delta > 0$ sufficiently small that if y belongs to $B_\delta(a)$ and $x = \mathbf{1} - a^{-1}y$, then

$$\|x\| < \|a^{-1}\|\delta < \tfrac{1}{2}$$

and

(62) $$\|r(y) - r(a)\| \le \frac{\|x\|\|a^{-1}\|}{1 - \|x\|} < \epsilon.$$

For (62) use Proposition 28.4.10(e), the fact that $y^{-1}a = (a^{-1}y)^{-1}$, and Corollary 28.4.12.

There are a number of ways of multiplying infinite series. The most common is the Cauchy product. And it is the only one we will consider.

28.4.15. Definition. If $\sum_{n=0}^{\infty} a_n$ and $\sum_{n=0}^{\infty} b_n$ are infinite series in a Banach algebra, then their CAUCHY PRODUCT is the series $\sum_{n=0}^{\infty} c_n$ where $c_n = \sum_{k=0}^{n} a_k b_{n-k}$. To see why this definition is rather natural, imagine trying to multiply two power series $\sum_{0}^{\infty} a_k x^k$ and $\sum_{0}^{\infty} b_k x^k$ just as if they were infinitely long polynomials. The result would be another power series. The coefficient of x^n in the resulting series would be the sum of all the products $a_i b_j$ where $i + j = n$. There are several ways of writing this sum

$$\sum_{i+j=n} a_i b_j = \sum_{k=0}^{n} a_k b_{n-k} = \sum_{k=0}^{n} a_{n-k} b_k .$$

If we just forget about the variable x, we have the preceding definition of the Cauchy product.

The first thing we observe about these products is that convergence of both the series $\sum a_k$ and $\sum b_k$ does *not* imply convergence of their Cauchy product.

28.4.16. Example. Let $a_k = b_k = (-1)^k (k+1)^{-1/2}$ for all $k \geq 0$. Then $\sum_{0}^{\infty} a_k$ and $\sum_{0}^{\infty} b_k$ converge by the alternating series test (Problem 28.1.25). The nth term of their Cauchy product is

$$c_n = \sum_{k=0}^{n} a_k b_{n-k} = (-1)^n \sum_{k=0}^{n} \frac{1}{\sqrt{k+1}\sqrt{n-k+1}} .$$

Since for $0 \leq k \leq n$

$$(k+1)(n-k+1) = k(n-k) + n + 1$$
$$\leq n^2 + 2n + 1$$
$$= (n+1)^2,$$

we see that

$$|c_n| = \sum_{k=0}^{n} \frac{1}{\sqrt{k+1}\sqrt{n-k+1}} \geq \sum_{k=0}^{n} \frac{1}{n+1} = 1 .$$

Since $c_n \nrightarrow 0$, the series $\sum_{0}^{\infty} c_k$ does not converge (see Proposition 28.1.10).

Fortunately, quite modest additional hypotheses do guarantee convergence of the Cauchy product. One very useful sufficient condition is that at least one of the series $\sum a_k$ or $\sum b_k$ converge absolutely.

28.4.17. Theorem (Mertens' theorem). *If in a Banach algebra $\sum_{0}^{\infty} a_k$ is absolutely convergent and has sum a and the series $\sum_{0}^{\infty} b_k$ is convergent with sum b, then the Cauchy product of these series converges and has sum ab.*

Proof. Exercise *Hint.* Although this exercise is slightly tricky, the difficulty has nothing whatever to do with Banach algebras. Anyone who can prove Mertens' theorem for series of real numbers can prove it for arbitrary Banach algebras.

For each $k \in \mathbb{N}$, let $c_k = \sum_{j=0}^{k} a_j b_{k-j}$. Let s_n, t_n, and u_n be the nth partial sums of the sequences (a_k), (b_k), and (c_k), respectively. First verify that for every

n in $\mathbb{N}$

(63)
$$u_n = \sum_{k=0}^{n} a_{n-k} t_k.$$

To this end define, for $0 \le j, k \le n$, the vector d_{jk} by

$$d_{jk} = \begin{cases} a_j b_{k-j} & \text{if } j \le k, \\ 0 & \text{if } j > k. \end{cases}$$

Notice that both the expression that defines u_n and the expression on the right-hand side of equation (63) involve only finding the sum of the elements of the matrix $[d_{jk}]$ but in different orders.

From (63) it is easy to see that for every n

$$u_n = s_n b + \sum_{k=0}^{n} a_{n-k}(t_k - b).$$

Since $s_n b \to ab$ as $n \to \infty$, the proof of the theorem is reduced to showing that

$$\left\| \sum_{k=0}^{n} a_{n-k}(t_k - b) \right\| \to 0 \qquad \text{as } n \to \infty.$$

Let $\alpha_k = \|a_k\|$ and $\beta_k = \|t_k - b\|$ for all k. Why does it suffice to prove that $\sum_{k=0}^{n} \alpha_{n-k} \beta_k \to 0$ as $n \to \infty$? In the finite sum

(64)
$$\alpha_n \beta_0 + \alpha_{n-1} \beta_1 + \cdots + \alpha_0 \beta_n,$$

the α_k's toward the left are small (for large n) and the β_k's toward the right are small (for large n). This suggests breaking the sum (64) into two pieces,

$$p = \sum_{k=0}^{n_1} \alpha_{n-k} \beta_k \qquad \text{and} \qquad q = \sum_{k=n_1+1}^{n} \alpha_{n-k} \beta_k,$$

and trying to make each piece small (smaller, say, than $\frac{1}{2}\epsilon$ for a preassigned ϵ).

For any positive number ϵ_1 it is possible (since $\beta_k \to 0$) to choose n_1 in $\mathbb{N}$ so that $\beta_k < \epsilon_1$ whenever $k \ge n_1$. What choice of ϵ_1 will ensure that $q < \frac{1}{2}\epsilon$?

For any $\epsilon_2 > 0$ it is possible (since $\alpha_k \to 0$) to choose n_2 in $\mathbb{N}$ so that $\alpha_k < \epsilon_2$ whenever $k \ge n_2$. Notice that $n - k \ge n_2$ for all $k \le n_1$ provided that $n \ge n_1 + n_2$. What choice of ϵ_2 will guarantee that $p < \frac{1}{2}\epsilon$?

28.4.18. Proposition. *On a Banach algebra A the operation of multiplication*

$$M \colon A \times A \to A \colon (x, y) \mapsto xy$$

is differentiable.

Proof. Problem. *Hint.* Fix (a, b) in $A \times A$. Compute the value of the function $\Delta M_{(a,b)}$ at the point (h, j) in $A \times A$. Show that $\|(h, j)\|^{-1} hj \to 0$ in A as $(h, j) \to 0$ in $A \times A$. How should $dM_{(a,b)}$ be chosen so that the Newton quotient

$$\frac{\Delta M_{(a,b)}(h, j) - dM_{(a,b)}(h, j)}{\|(h, j)\|}$$

approaches zero (in A) as $(h, j) \to 0$ (in $A \times A$)? Don't forget to show that your choice for $dM_{(a,b)}$ is a bounded linear map.

28.4.19. Proposition. *Let $c \in V$ where V is a normed linear space, and let A be a Banach algebra. If $f, g \in \mathcal{D}_c(V, A)$, then their product fg defined by*

$$(fg)(x) := f(x) g(x)$$

(for all x in some neighborhood of c) is differentiable at c and

$$d(fg)_c = f(c)dg_c + df_c \cdot g(c).$$

Proof. Problem. *Hint.* Use Proposition 28.4.18.

28.4.20. Proposition. *If A is a commutative Banach algebra and $n \in \mathbb{N}$, then the function $f : x \mapsto x^n$ is differentiable and $df_a(h) = na^{n-1}h$ for all $a, h \in A$.*

Proof. Problem. *Hint.* A simple induction proof works. Alternatively, you may choose to convince yourself that the usual form of the binomial theorem (see Problem I.1.17) holds in every commutative Banach algebra.

28.4.21. Problem. What happens in Proposition 28.4.20 if we do not assume that the Banach algebra is commutative? Is f differentiable? *Hint.* Try the cases $n = 2$, 3, and 4. Then generalize.

28.4.22. Problem. Generalize Proposition 28.3.5 to produce a theorem concerning the convergence of a series $\sum a_k b_k$ in a Banach algebra.

28.4.23. Definition. If (a_k) is a sequence in a Banach algebra A and $x \in A$, then a series of the form $\sum_{k=0}^{\infty} a_k x^k$ is a POWER SERIES in x.

28.4.24. Notation. It is a bad (but very common) habit to use the same notation for a polynomial function and for its value at a point x. One often encounters an expression such as "the function $x^2 - x + 5$" when clearly what is meant is "the function $x \mapsto x^2 - x + 5$". This abuse of language is carried over to power series. If (a_k) is a sequence in a Banach algebra A and $D = \{x \in A : \sum_{k=0}^{\infty} a_k x^k \text{ converges}\}$, then the function $x \mapsto \sum_{k=0}^{\infty} a_k x^k$ from D into A is usually denoted by $\sum_{k=0}^{\infty} a_k x^k$. Thus for example one may find the expression, "$\sum_{k=0}^{\infty} a_k x^k$ converges uniformly on the set U". What does this mean? *Answer.* If $s_n(x) = \sum_{k=0}^{n} a_k x^k$ for all $n \in \mathbb{N}$ and $f(x) = \sum_{k=0}^{\infty} a_k x^k$, then $s_n \to f$ (unif).

28.4.25. Proposition. *Let (a_k) be a sequence in a Banach algebra, and let $r > 0$. If the sequence $(\|a_k\| r^k)$ is bounded, then the power series $\sum_{k=0}^{\infty} a_k x^k$ converges uniformly on every open ball $B_s(0)$ such that $0 < s < r$. (And therefore it converges on the ball $B_r(0)$.)*

Proof. Exercise. *Hint.* Let $p = s/r$ where $0 < s < r$. Let $f_k(x) = a_k x^k$. Use the Weierstrass M-test, Proposition 28.1.18.

Since the uniform limit of a sequence of continuous functions is continuous (see Proposition 14.2.15), it follows easily from the preceding proposition that (under the hypotheses given) the function $f : x \mapsto \sum_{0}^{\infty} a_k x^k$ is continuous on $B_r(0)$. We will prove considerably more than this: the function is actually differentiable on $B_r(0)$, and furthermore, the correct formula for its differential is found by differentiating the power series term-by-term. That is, f behaves on $B_r(0)$ just like a polynomial with infinitely many terms. This is proved in Theorem 28.4.27. We need, however, a preliminary result.

28.4.26. Proposition. *Let V and W be normed linear spaces, let U be an open convex subset of V, and let (f_n) be a sequence of functions in $\mathcal{C}^1(U, W)$. If the sequence (f_n) converges pointwise to a function $F\colon U \to W$ and if the sequence $\big(d(f_n)\big)$ converges uniformly on U, then F is differentiable at each point a of U and*

$$dF_a = \lim_{n \to \infty} d(f_n)_a \,.$$

Proof. Exercise. *Hint.* Fix $a \in U$. Let $\phi\colon U \to \mathfrak{B}(V, W)$ be the function to which the sequence $\big(d(f_n)\big)$ converges uniformly. Given $\epsilon > 0$ show that there exists $N \in \mathbb{N}$ such that $\big\|d\big(f_n - f_N\big)_x\big\| < \epsilon/4$ for all $x \in U$ and $n \in \mathbb{N}$. Let $g_n = f_n - f_N$ for each n. Use one version of the mean value theorem to show that

(65) $$\big\|\Delta\big(g_n\big)_a(h) - d\big(g_n\big)_a(h)\big\| \leq \tfrac{1}{2}\epsilon\|h\|$$

whenever $n \geq N$ and h is a vector such that $a + h \in U$. In (65) take the limit as $n \to \infty$. Use the result of this together with the fact that $\Delta\big(f_N\big)_a \simeq d\big(f_N\big)_a$ to show that $\Delta F_a \simeq T$ when $T = \phi(a)$.

28.4.27. Theorem (Term-by-term differentiation of power series). *Suppose that (a_n) is a sequence in a commutative Banach algebra A and that $r > 0$. If the sequence $(\|a_k\|r^k)$ is bounded, then the function $F\colon B_r(0) \to A$ defined by*

$$F(x) = \sum_{k=0}^{\infty} a_k x^k$$

is differentiable and

$$dF_x(h) = \sum_{k=1}^{\infty} k a_k x^{k-1} h$$

for every $x \in B_r(0)$ and $h \in A$.

Proof. Problem. *Hint.* Let $f_n(x) = \sum_0^n a_k x^k$. Fix x in $B_r(0)$ and choose a number s satisfying $\|x\| < s < r$. Use Propositions 28.4.20 and 28.4.26 to show that F is differentiable at x and to compute its differential there. In the process you will need to show that the sequence $\big(d(f_n)\big)$ converges uniformly on $B_s(0)$. Use Proposition 28.4.25 and Example 4.2.8. If $s < t < r$, then there exists $N \in \mathbb{N}$ such that $k^{1/k} < r/t$ for all $k \geq N$. (Why?)

28.4.28. Problem. Give a definition of and develop the properties of the *exponential function* on a commutative Banach algebra A. Include at least the following:

(a) The series $\sum_{k=0}^{\infty} \frac{1}{k!}x^k$ converges absolutely for all x in A and uniformly on $B_r(0)$ for every $r > 0$.

(b) If $\exp(x) := \sum_{k=0}^{\infty} \frac{1}{k!}x^k$, then $\exp\colon A \to A$ is differentiable and

$$d\exp_x(h) = \exp(x) \cdot h \,.$$

This is the EXPONENTIAL FUNCTION on A.

(c) If x, $y \in A$, then
$$\exp(x) \cdot \exp(y) = \exp(x + y).$$

(d) If $x \in A$, then x is invertible and
$$\left(\exp(x)\right)^{-1} = \exp(-x).$$

28.4.29. Problem. Develop some trigonometry on a commutative Banach algebra A. (It will be convenient to be able to take the DERIVATIVE of a Banach algebra valued function. If $G\colon A \to A$ is a differentiable function, define $DG(a) := dG_a(1)$ for every $a \in A$.) Include at least the following:

(a) The series $\sum_{k=0}^{\infty} \frac{(-1)^k}{(2k)!} x^k$ converges absolutely for all x in A and uniformly on every open ball centered at the origin.

(b) The function $F\colon x \mapsto \sum_{k=0}^{\infty} \frac{(-1)^k}{(2k)!} x^k$ is differentiable at every $x \in A$. Find $dF_x(h)$.

(c) For every $x \in A$, let $\cos x := F(x^2)$. Let $\sin x := D \cos x$ for every x. Show that
$$\sin x = \sum_{k=0}^{\infty} \frac{(-1)^k}{(2k+1)!} x^{2k+1}$$
for every $x \in A$.

(d) Show that $D \sin x = \cos x$ for every $x \in A$.

(e) Show that $\sin^2 x + \cos^2 x = 1$ for every $x \in A$.

Chapter 29

The implicit function theorem

This chapter deals with the problem of solving equations and systems of equations for some of the variables that appear in terms of the others. In a few very simple cases this can be done explicitly. As an example, consider the equation

(66) $$x^2 + y^2 = 25$$

for the circle of radius 5 centered at the origin in $\mathbb{R}^2$. Although (66) *cannot* be solved *globally* for y in terms of x (that is, there is no function f such that $y = f(x)$ for all points (x, y) satisfying (66)), it nevertheless *is* possible at most points on the circle to solve *locally* for y in terms of x. For example, if (a, b) lies on the circle and $b > 0$, then there exist open intervals J_1 and J_2 containing a and b, respectively, and a function $f \colon J_1 \to J_2$ such that every point (x, y) in the rectangular region $J_1 \times J_2$ will satisfy $y = f(x)$ if and only if it satisfies equation (66). In particular, we could choose $J_1 = (-5, 5)$, $J_2 = (0, 6)$, and $f \colon x \mapsto \sqrt{25 - x^2}$. In case $b < 0$ the function f would be replaced by $f \colon x \mapsto -\sqrt{25 - x^2}$. If $b = 0$, then there is *no* local solution for y in terms of x: each rectangular region about either of the points $(5, 0)$ or $(-5, 0)$ will contain pairs of points symmetrically located with respect to the x-axis that satisfy (66); and it is not possible for the graph of a function to contain such a pair.

Our attention in this chapter is focused not on the relatively rare cases where it is possible to compute explicit (local) solutions for some of the variables in terms of the remaining ones but on the more typical situation where no such computation is possible. In this latter circumstance it is valuable to have information concerning the *existence* of (local) solutions and the differentiability of such solutions.

The simplest special case is a single equation of the form $y = f(x)$ where f is a continuously differentiable function. The inverse function theorem, derived in the first section of this chapter, provides conditions under which this equation can be solved locally for x in terms of y, say $x = g(y)$, and gives us a formula allowing us to compute the differential of g. More complicated equations and systems of equations require the implicit function theorem, which is the subject of the second section of the chapter.

29.1. The inverse function theorem

Recall that in Chapter 25 formulas concerning the function $f \mapsto \Delta f$ lead to corresponding formulas involving differentials. For example, $d(f + g)_a = df_a + dg_a$ followed from $\Delta(f + g)_a = \Delta f_a + \Delta g_a$ (see Proposition 25.3.15). It is natural to ask whether the formula

$$\Delta(f^{-1})_{f(x)} = (\Delta f_x)^{-1}$$

derived for bijective functions f in Proposition 25.3.7 leads to a corresponding formula

(67) $$d(f^{-1})_{f(x)} = (df_x)^{-1}$$

for differentials. Obviously, a necessary condition for (67) to hold for all x in some neighborhood of a point a is that the linear map df_a be invertible. The inverse function theorem states that for continuously differentiable (but not necessarily bijective) functions this is all that is required. The proof of the inverse function theorem is a fascinating application of the contractive mapping theorem (Theorem 19.1.5). First some terminology.

29.1.1. Definition. Let E and F be Banach spaces, and let $\emptyset \neq U \overset{\circ}{\subseteq} E$. A function f belonging to $\mathcal{C}^1(U, F)$ is $\mathcal{C}^1$-INVERTIBLE if f is a bijection between U and an open subset V of F and if f^{-1} belongs to $\mathcal{C}^1(V, E)$. Such a function is also called a $\mathcal{C}^1$-ISOMORPHISM between U and V.

29.1.2. Exercise. Find nonempty open subsets U and V of $\mathbb{R}$ and a continuously differentiable bijection $f \colon U \to V$ that is *not* a $\mathcal{C}^1$-isomorphism between U and V.

29.1.3. Definition. Let E and F be Banach spaces. A function f in $\mathcal{F}_a(E, F)$ is LOCALLY $\mathcal{C}^1$-INVERTIBLE (or a LOCAL $\mathcal{C}^1$-ISOMORPHISM) at a point a in E if there exists a neighborhood of a on which the restriction of f is $\mathcal{C}^1$-invertible. The inverse of this restriction is a LOCAL $\mathcal{C}^1$-INVERSE of f at a and is denoted by f_{loc}^{-1}.

29.1.4. Exercise. Let $f(x) = x^2 - 6x + 5$ for all x in $\mathbb{R}$. Find a local $\mathcal{C}^1$-inverse of f at $x = 1$.

29.1.5. Problem. Let $f(x) = x^6 - 2x^3 - 7$ for all x in $\mathbb{R}$. Find local $\mathcal{C}^1$-inverses for f at 0 and at 10.

29.1.6. Problem. Find a nonempty open subset U of $\mathbb{R}$ and a function f in $\mathcal{C}^1(U, \mathbb{R})$ that is *not* $\mathcal{C}^1$-invertible but is locally $\mathcal{C}^1$-invertible at every point in U.

Before embarking on a proof of the inverse function theorem it is worthwhile seeing why a naive proof of this result using the chain rule fails—even in the simple case of a real valued function of a real variable.

29.1.7. Exercise. If $f \in \mathcal{F}_a(\mathbb{R}, \mathbb{R})$ and if $Df(a) \neq 0$, then f is locally $\mathcal{C}^1$-invertible at a and

(68) $$Df_{\text{loc}}^{-1}(b) = \frac{1}{Df(a)},$$

where f_{loc}^{-1} is a local $\mathcal{C}^1$-inverse of f at a and $b = f(a)$.

This assertion is correct. Criticize the following "proof" of the result: Since f_{loc}^{-1} is a local $\mathcal{C}^1$-inverse of f at a,

$$f_{\text{loc}}^{-1}(f(x)) = x$$

for all x in some neighborhood U of a. Applying the chain rule (Proposition 8.4.19), we obtain

$$(Df_{\text{loc}}^{-1})(f(x)) \cdot Df(x) = 1$$

for all x in U. Letting $x = a$, we have

$$(Df_{\text{loc}}^{-1})(b)\, Df(a) = 1,$$

and since $Df(a) \neq 0$ equation (68) follows.

The inverse function theorem (29.1.16) deals with a continuously differentiable function f that is defined on a neighborhood of a point a in a Banach space E and that maps into a second Banach space F. We assume that the differential of f at a is invertible. Under these hypotheses we prove that f is locally $\mathcal{C}^1$-invertible at a and in some neighborhood of a equation (67) holds. To simplify the proof, we temporarily make some additional assumptions: we suppose that H is a continuously differentiable function that is defined on a neighborhood of 0 in a Banach space E and that maps into this same space E, that $H(\mathbf{0}) = \mathbf{0}$, and that the differential of H at 0 is the identity map on E. Once the conclusion of the inverse function theorem has been established in this restricted case, the more general version follows easily. The strategy we employ to attack the special case is straightforward, but there are numerous details that must be checked along the way. Recall that in Chapter 19 we were able to solve certain systems of simultaneous linear equations by putting them in the form $Ax = b$ where A is a square matrix and x and b are vectors in the Euclidean space of appropriate dimension. This equation was rewritten in the form $Tx = x$ where $Tx := x - b + Ax$, thereby reducing the problem to one of finding a fixed point of the mapping T. When T is contractive a simple application of the contractive mapping theorem (Theorem 19.1.5) is all that is required. We make use of exactly the same idea here. We want a local inverse of H. That is, we wish to solve the equation $H(x) = y$ for x in terms of y in some neighborhood of $\mathbf{0}$. Rewrite the equation $H(x) = y$ in the form $\phi_y(x) = x$, where for each y near $\mathbf{0}$ the function ϕ_y is defined by $\phi_y(x) := x - H(x) + y$. Thus, as before, the problem is to find for each y a unique fixed point of ϕ_y. In order to apply the contractive mapping theorem to ϕ_y, the domain of this function must be a *complete* metric space. For this reason we choose temporarily to take the domain of ϕ_y to be a *closed* ball about the origin in E.

In Lemma 29.1.8 we find such a closed ball C. It must satisfy two conditions: first, C must lie in the domain of H; and second, if u belongs to C, then dH_u must be close to the identity map on E, say, $\|dH_u - I\| < \frac{1}{2}$. (This latter condition turns out to be a crucial ingredient in proving that ϕ_y is contractive.) In Lemma 29.1.9 we show that (for y sufficiently small) ϕ_y maps the closed ball C into itself; and in Lemma 29.1.10 the basic task is to show that ϕ_y is contractive and therefore has a unique fixed point. The result of all this is that there exists a number $r > 0$ such that for every y in $B = B_r(\mathbf{0})$ there exists a unique x in the closed ball $C = C_{2r}(\mathbf{0})$ such that $y = H(x)$. Now this is not quite the end of the story. First of all we do *not* know that H restricted to C is injective: some points in C may be mapped to the region outside B, about which the preceding says nothing. Furthermore, the definition of local $\mathcal{C}^1$-invertibility requires a homeomorphism between *open* sets, and C is not open. This suggests we restrict our attention to points lying in the interior of C that map into B. So let $V = C^\circ \cap H^{\leftarrow}(B)$ and consider the restriction of H to V, which we denote by H_{loc}. In Lemma 29.1.11 we show that V is a neighborhood of 0 and that H_{loc} is injective. Thus the inverse function $H_{\mathrm{loc}}^{-1} : H^{\rightarrow}(V) \to V$ exists. The succeeding lemma is devoted to showing that this inverse is continuous.

In order to conclude that H is locally $\mathcal{C}^1$-invertible, we still need two things: we must know that H_{loc} is a homeomorphism between *open* sets and that H_{loc}^{-1} is continuously differentiable. Lemma 29.1.13 shows that $H^{\rightarrow}(V)$ is open in E. And in Lemma 29.1.14 we complete the proof of this special case of the inverse function theorem by showing that H_{loc}^{-1} is continuously differentiable and that, in the open set V, its differential is given by (67).

Corollary 29.1.15 shows that the conclusions of the preceding result remain true even when one of the hypotheses is eliminated and another weakened. Here we prove the inverse function theorem for a function G whose domain E and codomain F are *not* required to be identical. Of course, if $E \neq F$ we cannot assume that the differential of G at $\mathbf{0}$ is the identity map; we assume only that it is invertible.

Finally, in Theorem 29.1.16 we prove our final version of the inverse function theorem. Here we drop the requirement that the domain of the function in question be a neighborhood of the origin.

In Lemmas 29.1.8–29.1.14 the following hypotheses are in force:

$(1')$ $\mathbf{0} \in U_1 \overset{\circ}{\subseteq} E$, where E is a Banach space;

$(2')$ $H \in \mathcal{C}^1(U_1, E)$;

$(3')$ $H(\mathbf{0}) = \mathbf{0}$; and

$(4')$ $dH_{\mathbf{0}} = I$.

29.1.8. Lemma. *There exists $r > 0$ such that $B_{3r}(\mathbf{0}) \subseteq U_1$ and $\|dH_u - I\| < \frac{1}{2}$ whenever $\|u\| \leq 2r$.*

Proof. Problem.

29.1.9. Lemma. *For $\|y\| < r$, define a function ϕ_y by*

$$\phi_y(x) := x - H(x) + y$$

for all x such that $\|x\| \leq 2r$. Show that ϕ_y maps $C_{2r}(\mathbf{0})$ into itself.

Proof. Problem.

29.1.10. Lemma. *For every y in $B_r(\mathbf{0})$ there exists a unique x in $C_{2r}(\mathbf{0})$ such that $y = H(x)$.*

Proof. Problem. *Hint.* Show that the function ϕ_y defined in Lemma 29.1.9 is contractive on the metric space $C_{2r}(\mathbf{0})$ and has $\frac{1}{2}$ as a contraction constant. To find an appropriate inequality involving $\|\phi_y(u) - \phi_y(v)\|$, apply Corollary 26.1.8 to $\|H(v) - H(u) - dH_{\mathbf{0}}(v - u)\|$.

29.1.11. Lemma. *Show that if $V := \{x \in B_{2r}(\mathbf{0}) \colon \|H(x)\| < r\}$, then $\mathbf{0} \in V \overset{\circ}{\subseteq} E$. Let H_{loc} be the restriction of H to V. Show that H_{loc} is a bijection between V and $H^{\rightarrow}(V)$.*

Proof. Problem.

29.1.12. Lemma. *The function $H_{\mathrm{loc}}^{-1} \colon H^{\rightarrow}(V) \to V$ is continuous.*

Proof. Problem. *Hint.* Prove first that if $u,\ v \in C_{2r}(\mathbf{0})$, then $\|u - v\| \leq 2\|H(u) - H(v)\|$. In order to do this, look at

$$2\|\phi_{\mathbf{0}}(u) + H(u) - \phi_{\mathbf{0}}(v) - H(v)\| - \|u - v\|$$

and recall (from the proof of Lemma 29.1.10) that $\phi_{\mathbf{0}}$ has contraction constant $\frac{1}{2}$. Use this to conclude that if $w,\ z \in H^{\rightarrow}(V)$, then

$$\|H_{\mathrm{loc}}^{-1}(w) - H_{\mathrm{loc}}^{-1}(z)\| \leq 2\|w - z\|,$$

where H_{loc} is the restriction of H to V; (see Lemma 29.1.10.

29.1.13. Lemma. *Show that $H^{\rightarrow}(V) \overset{\circ}{\subseteq} E$.*

Proof. Problem. *Hint.* Show that if a point b belongs to $H^{\rightarrow}(V)$, then so does the open ball $B_{r - \|b\|}(b)$. Proceed as follows: Show that if a point y lies in this open ball, then $\|y\| < r$ and therefore $y = H(x)$ for some (unique) x in $C_{2r}(\mathbf{0})$. Prove that $y \in H^{\rightarrow}(V)$ by verifying $\|x\| < 2r$. To do this, look at

$$\|x - H_{\mathrm{loc}}^{-1}(b)\| + \|H_{\mathrm{loc}}^{-1}(b) - H_{\mathrm{loc}}^{-1}(\mathbf{0})\|$$

and use the first inequality given in the hint to the preceding problem.

29.1.14. Lemma. *The function H is locally $\mathcal{C}^1$-invertible at 0. Furthermore,*

$$d\big(H_{\mathrm{loc}}^{-1}\big)_{H(x)} = (dH_x)^{-1}$$

for every x in V.

Proof. Problem. *Hint.* First prove the differentiability of H_{loc}^{-1} on $H^{\rightarrow}(V)$. If $y \in H^{\rightarrow}(V)$, then there exists a unique x in V such that $y = H(x)$. By hypothesis $\Delta H_x \sim dH_x$, show that multiplication on the right-hand side by $\Delta\big(H_{\mathrm{loc}}^{-1}\big)_y$ preserves tangency. (For this it must be established that $\Delta\big(H_{\mathrm{loc}}^{-1}\big)_y$ belongs to $\mathfrak{O}(E, E)$.) Then show that multiplication on the left-hand side by $(dH_x)^{-1}$ preserves tangency. (How do we know that this inverse exists for all x in V?) Finally, show that the map $y \mapsto d\big(H_{\mathrm{loc}}^{-1}\big)_y$ is continuous on $H^{\rightarrow}(V)$ by using (67) to write it as the composite of H_{loc}^{-1}, dH, and the map $T \mapsto T^{-1}$ on $\operatorname{Inv}\mathfrak{B}(E, E)$ (see Proposition 28.4.14).

29.1.15. Corollary (Inverse function theorems—second version). *Let E and F be Banach spaces. If*

($1''$) $\mathbf{0} \in U_1 \overset{\circ}{\subseteq} E$,

($2''$) $G \in \mathcal{C}^1(U_1, F)$,

(3″) $G(\mathbf{0}) = \mathbf{0}$, and

(4″) $dG_{\mathbf{0}} \in \operatorname{Inv} \mathcal{B}(E, F)$,

then G is locally $\mathcal{C}^1$-invertible at $\mathbf{0}$. Furthermore,

$$d\big(G_{\mathrm{loc}}^{-1}\big)_{G(x)} = (dG_x)^{-1}$$

for all x in some neighborhood of $\mathbf{0}$, where G_{loc}^{-1} is a local $\mathcal{C}^1$-inverse of G at $\mathbf{0}$.

Proof. Problem. *Hint.* Let $H = (dG_{\mathbf{0}})^{-1} \circ G$. Apply Lemma 29.1.14.

29.1.16. Theorem (Inverse function theorem—third, and final, version). *Let E and F be Banach spaces. If*

(1) $a \in U \overset{\circ}{\subseteq} E$,

(2) $f \in \mathcal{C}^1(U, F)$, *and*

(3) $df_a \in \operatorname{Inv} \mathcal{B}(E, F)$,

then f is locally $\mathcal{C}^1$-invertible at a. Furthermore,

$$d\big(f_{\mathrm{loc}}^{-1}\big)_{f(x)} = (df_x)^{-1}$$

for all x in some neighborhood of a.

Proof. Problem. *Hint.* Let $U_1 = U - a$ and $G = \Delta f_a$. Write G as a composite of f with translation maps. Apply Corollary 29.1.15.

29.1.17. Problem. Let $P\colon \mathbb{R}^2 \to \mathbb{R}^2\colon (r, \theta) \mapsto (r\cos\theta, r\sin\theta)$, and let a be a point in $\mathbb{R}^2$ such that $P(a) = (1, \sqrt{3})$.

(a) Show that P is locally $\mathcal{C}^1$-invertible at a by finding a local $\mathcal{C}^1$-inverse P_{loc}^{-1} of P at a. For the inverse you have found, compute $d\big(P_{\mathrm{loc}}^{-1}\big)_{(1,\sqrt{3})}$.

(b) Use the inverse function theorem, Problem 29.1.16, to show that P is locally $\mathcal{C}^1$-invertible at a. Then use the formula given in that theorem to compute $d\big(P_{\mathrm{loc}}^{-1}\big)_{(1,\sqrt{3})}$, where P_{loc}^{-1} is a local $\mathcal{C}^1$-inverse of P at a. *Hint.* Use Proposition 21.5.16.

29.1.18. Problem. Let $U = \{(x, y, z) \in \mathbb{R}^3 \colon x, y, z > 0\}$, and let $g\colon U \to \mathbb{R}^3$ be defined by

$$g(x, y, z) = \left(\frac{x}{y^2 z^2}, \, yz, \, \ln y\right).$$

Calculate separately $\big[dg_{(x,y,z)}\big]^{-1}$ and $\big[d(g^{-1})_{g(x,y,z)}\big]$.

29.2. The implicit function theorem

In the preceding section we derived the inverse function theorem, which gives conditions under which an equation of the form $y = f(x)$ can be solved locally for x in terms of y. The implicit function theorem deals with the local solvability of equations that are not necessarily in the form $y = f(x)$ and of systems of equations. The inverse function theorem is actually a special case of the implicit function theorem. Interestingly, the special case can be used to prove the more general one.

This section consists principally of exercises and problems that illustrate how the inverse function theorem can be adapted to guarantee the existence of local solutions for various examples of equations and systems of equations. Once the computational procedure is well understood for these examples it is a simple matter to explain how it works in general; that is, to prove the implicit function theorem.

Suppose, given an equation of the form $y = f(x)$ and a point a in the domain of f, we are asked to show that *the equation can be solved for x in terms of y near a*. It is clear what we are being asked: to show that f is locally invertible at a. (Since the function f will usually satisfy some differentiability condition—continuous differentiability, for example—it is natural to ask for the local inverse to satisfy the same condition.)

As we have seen, local invertibility can be established by explicitly computing a local inverse, which can be done only rarely, or by invoking the inverse function theorem. Let us suppose we are given an equation of the form

(69) $$f(x, y) = 0$$

and a point (a, b) in $\mathbb{R}^2$ that satisfies (69).

Question. What does it mean to say that (69) can be solved for y near b in terms of x near a? (*Alternative wording.* What does it mean to say that (69) can be solved for y in terms of x near the point (a, b)?)

Answer. There exist a neighborhood V of a and a function $h : V \to R$ that satisfy

(i) $h(a) = b$,

(ii) $f(x, h(x)) = 0$

for all x in V.

29.2.1. Example. In the introduction to this chapter we discussed the problem of solving the equation

$$x^2 + y^2 = 25$$

for y in terms of x. This equation can be put in the form (69) by setting $f(x, y) = x^2 + y^2 - 25$. Suppose we are asked to show that (69) can be solved for y near 4 in terms of x near 3. As in the introduction, take $V = (-5, 5)$ and $h(x) = \sqrt{25 - x^2}$ for all x in V. Then $h(3) = 4$ and $f(x, h(x)) = x^2 + \left(\sqrt{25 - x^2}\right)^2 - 25 = 0$; so h is the desired local solution to (69). If we are asked to show that (69) can be solved for y near -4 in terms of x near 3, what changes? We choose $h(x) = -\sqrt{25 - x^2}$. Notice that condition (i) above dictates the choice of h. (Either choice will satisfy (ii).)

As was pointed out in the introduction, the preceding example is atypical in that it is possible to *specify* the solution. We can actually *solve* for y in terms of x. *Much* more common is the situation in which an explicit solution is not possible. What do we do then?

To see how this more complicated situation can be dealt with, let us pretend just for a moment that our computational skills have so totally deserted us that in the preceding example we are unable to specify the neighborhood V and the

function h required to solve (69). The problem is still the same: show that the
equation

(70) $$x^2 + y^2 - 25 = 0$$

can be solved for y near 4 in terms of x near 3. A good start is to define a function
$G : \mathbb{R}^2 \to \mathbb{R}^2$ by

(71) $$G(x, y) = (x, f(x, y)),$$

where as above $f(x, y) = x^2 + y^2 - 25$, and apply the inverse function theorem to G.
It is helpful to make a sketch here. Take the xy-plane to be the domain of G, and in
this plane sketch the circle $x^2 + y^2 = 25$. For the codomain of G take another plane,
letting the horizontal axis be called "x" and the vertical axis be called "z". Notice
that in this second plane the image under G of the circle drawn in the xy-plane
is the line segment $[-5, 5]$ along the x-axis and that the image of the x-axis is the
parabola $z = x^2 - 25$. Where do points in the interior of the circle go? What about
points outside the circle? If you think of the action of G as starting with a folding
of the xy-plane along the x-axis, you should be able to guess the identity of those
points where G is not locally invertible. In any case we will find these points using
the inverse function theorem.

The function G is continuously differentiable on $\mathbb{R}^2$ and (the matrix representation of) its differential at a point (x, y) in $\mathbb{R}^2$ is given by

$$\left[dG_{(x,y)} \right] = \begin{bmatrix} 1 & 0 \\ 2x & 2y \end{bmatrix}.$$

Thus according to the inverse function theorem, G is locally $\mathcal{C}^1$-invertible at every
point (x, y) where this matrix is invertible, that is, everywhere except on the x-
axis. In particular, at the point $(3, 4)$ the function G is locally $\mathcal{C}^1$-invertible. Thus
there exist a neighborhood W of $G(3, 4) = (3, 0)$ in $\mathbb{R}^2$ and a local $\mathcal{C}^1$-inverse
$H : W \to \mathbb{R}^2$ of G. Write H in terms of its component functions $H = \left(H^1, H^2 \right)$ and
set $h(x) = H^2(x, 0)$ for all x in $V := \{ x : (x, 0) \in W \}$. Then V is a neighborhood
of 3 in $\mathbb{R}$ and the function h is continuously differentiable (because H is). To show
that h is a solution of the equation (70) for y in terms of x we must show that

(i) $h(3) = 4$,   and

(ii) $f(x, h(x)) = 0$

for all x in V.

To obtain (i), equate the second components of the first and last terms of the
following computation:

$$\begin{aligned}
(3, 4) &= H(G(3, 4)) \\
&= H(3, f(3, 4)) \\
&= H(3, 0) \\
&= \left(H^1(3, 0), H^2(3, 0) \right) \\
&= \left(H^1(3, 0), h(3) \right).
\end{aligned}$$

To obtain (ii) notice that for all x in V

$$(x, 0) = G(H(x, 0))$$
$$= G(H^1(x, 0),\, H^2(x, 0))$$
$$= G(H^1(x, 0),\, h(x))$$
$$= (H^1(x, 0),\, f(H^1(x, 0), h(x)))\,.$$

Equating first components, we see that $H^1(x, 0) = x$. So the preceding can be written

$$(x, 0) = (x, f(x, h(x)))$$

from which (ii) follows by equating second components.

Although the preceding computations demonstrate the existence of a local solution $y = h(x)$ without specifying it, it is nevertheless possible to calculate the value of its derivative $\frac{dy}{dx}$ at the point $(3, 4)$, that is, to find $h'(3)$. Since $h'(3) = \left(H^2 \circ j_1\right)'(3) = d\left(H^2 \circ j_1\right)_3(1) = \left(dH^2{}_{(3,0)} \circ j_1\right)(1) = dH^2{}_{(3,0)}(1, 0) = \frac{\partial H^2}{\partial x}(3, 0)$, where j_1 is the inclusion map $x \mapsto (x, 0)$, and since H is a local inverse of G, the inverse function theorem tells us that

$$\begin{aligned}
\left[dH_{(3,0)}\right] &= \left[dH_{G(3,4)}\right] \\
&= \left[dG_{(3,4)}\right]^{-1} \\
&= \begin{bmatrix} 1 & 0 \\ f_1(3,4) & f_2(3,4) \end{bmatrix}^{-1} \\
&= \begin{bmatrix} 1 & 0 \\ 6 & 8 \end{bmatrix}^{-1} \\
&= \begin{bmatrix} 1 & 0 \\ -\frac{3}{4} & \frac{1}{8} \end{bmatrix}.
\end{aligned}$$

But the entry in the lower left corner is $\frac{\partial H^2}{\partial x}(3, 0)$. Therefore, $h'(3) = -\frac{3}{4}$.

In the next exercise we consider an equation whose solution cannot be easily calculated.

29.2.2. Exercise. Consider the equation

(72) $$x^2 y + \sin\left(\frac{\pi}{2} xy^2\right) = 2,$$

where $x,\, y \in R$.

(a) What does it mean to say that equation (72) can be solved for y in terms of x near the point $(1, 2)$?

(b) Show that it is possible to solve (72) as described in (a). *Hint.* Proceed as in the second solution to the preceding example.

(c) Use the inverse function theorem to find the value of $\frac{dy}{dx}$ at the point $(1, 2)$.

29.2.3. Problem. Consider the equation

(73) $$e^{xy^2} - x^2 y + 3x = 4,$$

where x, $y \in \mathbb{R}$.

(a) What does it mean to say that equation (73) can be solved for y near 0 in terms of x near 1?

(b) Show that such a solution does exist.

(c) Use the inverse function theorem to compute the value of $\frac{dy}{dx}$ at the point $(1,0)$.

    The preceding examples have all been equations involving only two variables. The technique used in dealing with these examples works just as well in cases where we are given an equation in an arbitrary number of variables and wish to demonstrate the existence of local solutions for one of the variables in terms of the remaining ones.

29.2.4. Exercise. Consider the equation

(74) $$x^2 z + yz^2 - 3z^3 = 8$$

for x, y, $z \in \mathbb{R}$.

(a) What does it mean to say that equation (74) can be solved for z near 1 in terms of x and y near $(3,2)$? (Alternatively, what does it mean to say that (74) can be solved for z in terms of x and y near the point $(x,y,z) = (3,2,1)$?)

(b) Show that such a solution does exist. *Hint.* Follow the preceding technique, but instead of using (71), define $G(x,y,z) := (x,y,f(x,y,z))$ for an appropriate function f.

(c) Use the inverse function theorem to find the values of $\left(\frac{\partial z}{\partial x}\right)_y$ and $\left(\frac{\partial z}{\partial y}\right)_x$ at the point $(3,2,1)$.

29.2.5. Problem. Let $f(x,y,z) = xz + xy + yz - 3$. By explicit computation find a neighborhood V of $(1,1)$ in $\mathbb{R}^2$ and a function $h\colon V \to \mathbb{R}$ such that $h(1,1) = 1$ and $f(x,y,h(x,y)) = 0$ for all x and y in V. Find $h_1(1,1)$ and $h_2(1,1)$.

29.2.6. Problem. Use the inverse function theorem, not direct calculation, to show that the equation

$$xz + xy + yz = 3$$

has a solution for z near 1 in terms of x and y near $(1,1)$ and to find $\left(\frac{\partial z}{\partial x}\right)_y$ and $\left(\frac{\partial z}{\partial y}\right)_x$ at the point $(1,1,1)$.

29.2.7. Problem.

(a) What does it mean to say that the equation

$$wx^2 y + \sqrt{w}y^2 z^4 = 3xz + 6x^3 z^2 + 7$$

can be solved for z in terms of w, x, and y near the point $(w,x,y,z) = (4,1,2,1)$?

(b) Show that such a solution exists.

(c) Use the inverse function theorem to find $\left(\frac{\partial z}{\partial w}\right)_{x,y}$, $\left(\frac{\partial z}{\partial x}\right)_{w,y}$, and $\left(\frac{\partial z}{\partial y}\right)_{w,x}$ at the point $(4,1,2,1)$.

29.2.8. Problem. Let $U \overset{\circ}{\subseteq} \mathbb{R}^{n-1} \times \mathbb{R}$ and $f \in \mathcal{C}^1(U,\mathbb{R})$. Suppose that the point $(a,b) = (a_1,\dots,a_{n-1};b)$ belongs to U, that $f(a,b) = 0$, and that $f_n(a,b) \neq 0$. Show that there exists a neighborhood V of a in $\mathbb{R}^{n-1}$ and a function h in $\mathcal{C}^1(V,\mathbb{R})$ such that $h(a) = b$ and $f(x,h(x)) = 0$ for all x in V. *Hint.* Show that the function

$$G \colon U \to \mathbb{R}^{n-1} \times \mathbb{R} \colon (x,y) = (x_1,\dots,x_{n-1};y) \mapsto (x, f(x,y))$$

has a local $\mathcal{C}^1$-inverse, say H, at (a,b).

29.2.9. Problem. Show that the equation

$$uvy^2 z + w\sqrt{x}z^{10} + v^2 e^y z^4 = 5 + uw^2 \cos\left(x^3 y^5\right)$$

can be solved for x near 4 in terms of u, v, w, y, and z near -3, 2, -1, 0, and 1, respectively. *Hint.* Use the preceding problem.

29.2.10. Problem.

(a) Use Problem 29.2.8 to make sense of the following "theorem": If $f(x,y,z) = 0$, then

$$\left(\frac{\partial z}{\partial x}\right)_y = -\frac{\left(\dfrac{\partial f}{\partial x}\right)_{y,z}}{\left(\dfrac{\partial f}{\partial z}\right)_{x,y}}.$$

Hint. After determining that there exists a function h that satisfies the equation $f(x,y,h(x,y)) = 0$ on an appropriate neighborhood of a point, use the chain rule to differentiate both sides of the equation with respect to x.

(b) Verify part (a) directly for the function f given in Problem 29.2.5 by computing each side independently.

29.2.11. Problem. In the very first problem of the first chapter of a classic thermodynamics text (cf. [**Som56**]) one is asked to show that if the variables x, y, and z are related so that $f(x,y,z) = 0$, then

$$\left(\frac{\partial x}{\partial y}\right)_z \left(\frac{\partial y}{\partial z}\right)_x \left(\frac{\partial z}{\partial x}\right)_y = -1.$$

Your old pal, Fred Dimm, is taking a class in thermodynamics and is baffled by the statement of the problem. Among other things he notices that there is a mysterious f in the hypothesis that does not appear in the conclusion. He wonders, not unreasonably, how assuming something about f is going to help him prove a result that makes no reference whatever to that quantity. He is also convinced that the answer is wrong: he thinks that the product of the partial derivatives should be $+1$. He claims it should be just like the single variable case: the chain rule says $\frac{dy}{dt} = \frac{dy}{dx}\frac{dx}{dt}$ because we can cancel the dx's.

(a) Help Fred by explaining the exercise to him. And comment on his "correction" of the problem.

(b) Unfortunately, once you have explained it all him, he still can't do the problem. So also show him how to solve it.

29.2.12. Problem. A commonly used formula in scientific work is

$$\left(\frac{\partial x}{\partial y}\right)_z = \frac{1}{\left(\dfrac{\partial y}{\partial x}\right)_z}.$$

Recast this as a carefully stated theorem. Then prove the theorem. *Hint.* Use Problem 29.2.8 twice to obtain appropriate functions h and j satisfying $f(h(y, z), y, z) = 0$ and $f(x, j(x, z), z) = 0$ on appropriate neighborhoods. Differentiate these equations using the chain rule. Evaluate at a particular point. Solve.

29.2.13. Problem.

(a) Make sense of the formula

$$\left(\frac{\partial x}{\partial z}\right)_y = -\frac{\left(\dfrac{\partial y}{\partial z}\right)_x}{\left(\dfrac{\partial y}{\partial x}\right)_z},$$

and prove it.

(b) Illustrate the result in part (a) by computing separately $\left(\frac{\partial P}{\partial V}\right)_T$ and $-\dfrac{\left(\frac{\partial T}{\partial V}\right)_P}{\left(\frac{\partial T}{\partial P}\right)_V}$ from the equation of state $PV = RT$ for an ideal gas. (Here R is a constant.)

We have dealt at some length with the problem of solving a single equation for one variable in terms of the remaining ones. It is pleasant to discover that the techniques used there can be adapted with only the most minor modifications to give local solutions for systems of n equations in p variables (where $p > n$) for n of the variables in terms of the remaining $p - n$ variables. We begin with some examples.

29.2.14. Exercise. Consider the following system of equations:

$$(75) \qquad \begin{cases} 2u^3vx^2 + v^2x^3y^2 - 3u^2y^4 = 0, \\ 2uv^2y^2 - uvx^2 + u^3xy = 2. \end{cases}$$

(a) What does it mean to say that the system (75) can be solved for x and y near (c, d) in terms of u and v near (a, b)?

(b) Show that the system (75) can be solved for x and y near $(1, 1)$ in terms of u and v near $(1, 1)$. *Hint.* Try to imitate the technique of Problem 29.2.8, except in this case define G on an appropriate subset of $\mathbb{R}^2 \times \mathbb{R}^2$.

29.2.15. Problem. Consider the following system of equations

$$(76) \qquad \begin{cases} 4x^2 + 4y^2 = z, \\ x^2 + y^2 = 5 - z. \end{cases}$$

(a) What does it mean to say that (76) can be solved for y and z near $(1, 4)$ in terms of x near 0?

(b) Show that such a solution exists by direct computation.

(c) Compute $\left(\frac{\partial y}{\partial x}\right)_z$ and $\left(\frac{\partial z}{\partial x}\right)_y$ at the point $x = 0$, $y = 1$, $z = 4$.

29.2.16. Problem.

(a) Repeat Problem 29.2.15(b), this time using the inverse function theorem instead of direct computation.

(b) Use the inverse function theorem to find $\left(\frac{\partial y}{\partial x}\right)_z$ and $\left(\frac{\partial z}{\partial x}\right)_y$ at the point $x = 0$, $y = 1$, $z = 4$.

29.2.17. Problem. Discuss the problem of solving the system of equations

$$(77) \qquad \begin{cases} ux^2 + vwy + u^2w & = 4, \\ uvy^3 + 2wx - x^2y^2 & = 3 \end{cases}$$

for x and y near $(1, 1)$ in terms of u, v, and w near $(1, 2, 1)$.

As the preceding examples indicate, the first step in solving a system of n equations in $n + k$ unknowns (where $k > 0$) for n of the variables in terms of the remaining k variables is to replace the system of equations by a function $f : U \to \mathbb{R}^n$ where $U \overset{\circ}{\subseteq} \mathbb{R}^k \times \mathbb{R}^n$. If the finite-dimensional spaces $\mathbb{R}^k$ and $\mathbb{R}^n$ are replaced by arbitrary Banach spaces, the subsequent calculations can be carried out in exactly the same fashion as in the examples we have just considered. The result of this computation is the implicit function theorem.

29.2.18. Theorem (The implicit function theorem). *Let E_1, E_2, and F be Banach spaces, let $(a, b) \in U \overset{\circ}{\subseteq} E_1 \times E_2$, and let $f \in \mathcal{C}^1(U, F)$. If $f(a, b) = \mathbf{0}$ and $d_2 f_{(a,b)}$ is invertible in $\mathfrak{B}(E_2, F)$, then there exist a neighborhood V of a in E_1 and a continuously differentiable function $h : V \to E_2$ such that $h(a) = b$ and $f(x, h(x)) = \mathbf{0}$ for all $x \in V$.*

Proof. Problem. *Hint.* Let $G(x, y) = (x, f(x, y))$ for all $(x, y) \in U$. Show that $dG_{(a,b)} = (x, Sx + Ty)$ where $S = d_1 f_{(a,b)}$ and $T = d_2 f_{(a,b)}$. Show that the map $(x, z) \mapsto (x, T^{-1}(z - Sx))$ from $E_1 \times F$ to $E_1 \times E_2$ is the inverse of $dG_{(a,b)}$. (One can guess what the inverse of $dG_{(a,b)}$ should be by regarding $dG_{(a,b)}$ as the matrix $\begin{bmatrix} I_{E_1} & 0 \\ S & T \end{bmatrix}$ acting on $E_1 \times E_2$.) Apply the inverse function theorem (Theorem 29.1.16) and proceed as in Exercise 29.2.14 and Problems 29.2.16 and 29.2.17.

29.2.19. Problem. Suppose we are given a system of n equations in p variables where $p > n$. What does the implicit function theorem, Theorem 29.2.18, say about the possibility of solving this system locally for n of the variables in terms of the remaining $p - n$ variables?

Higher order derivatives

Suppose that V and W are Banach spaces and that $U \overset{\circ}{\subseteq} V$. If a function $f \colon U \to W$ is differentiable, then at each point a in U the differential of f at a, denoted by df_a, is a bounded linear map from V into W. Thus df, the differential of f, is a map from U into $\mathfrak{B}(V, W)$. It is natural to inquire whether the function df is itself differentiable. If it is, its differential at a (which we denote by $d^2 f_a$) is a bounded linear map from V into $\mathfrak{B}(V, W)$; that is

$$d^2 f_a \in \mathfrak{B}(V, \mathfrak{B}(V, W)).$$

In the same vein, since $d^2 f$ maps U into $\mathfrak{B}(V, \mathfrak{B}(V, W))$, its differential (if it exists) belongs to $\mathfrak{B}(V, \mathfrak{B}(V, \mathfrak{B}(V, W)))$. It is moderately unpleasant to contemplate what an element of $\mathfrak{B}(V, \mathfrak{B}(V, W))$ or of $\mathfrak{B}(V, \mathfrak{B}(V, \mathfrak{B}(V, W)))$ might "look like". And clearly, as we pass to even higher order differentials, things look worse and worse. It is comforting to discover that an element of $\mathfrak{B}(V, \mathfrak{B}(V, W))$ may be regarded as a continuous map from V^2 into W that is linear in both of its variables, and that an element of $\mathfrak{B}(V, \mathfrak{B}(V, \mathfrak{B}(V, W)))$ may be thought of as a continuous map from V^3 into W linear in each of its three variables. Establishing the isometric isomorphisms that allow us to make these identifications is the principal goal of section 30.1. In section 30.2 higher order differentials and partial derivatives are introduced and some of their more important and useful properties are derived.

30.1. Multilinear functions

30.1.1. Definition. Let X and Y be vector spaces. A map $B \colon X \times X \to Y$ is BILINEAR if it is linear in each of its variables; that is, if

$$B(w + x, y) = B(w, y) + B(x, y),$$
$$B(\alpha x, y) = \alpha B(x, y),$$
$$B(w, x + y) = B(w, x) + B(w, y), \quad \text{and}$$
$$B(x, \alpha y) = \alpha B(x, y)$$

for all w, x, $y \in X$ and $\alpha \in \mathbb{R}$. Scalar valued bilinear mappings on X are called
BILINEAR FUNCTIONALS. Care should be taken not to confuse bilinear maps from
$X \times X$ into Y with linear maps between the same spaces. This is the point of the
first exercise.

30.1.2. Exercise. Let X and Y be vector spaces, let u, v, x, $y \in X$, and let $\alpha \in \mathbb{R}$.

(a) Expand $B(u+v, x+y)$ if B is a bilinear map from $X \times X$ into Y.

(b) Expand $B(u+v, x+y)$ if B is a linear map from $X \times X$ into Y.

(c) Write $B(\alpha x, \alpha y)$ in terms of α and $B(x,y)$ if B is a bilinear map from $X \times X$
into Y.

(d) Write $B(\alpha x, \alpha y)$ in terms of α and $B(x,y)$ if B is a linear map from $X \times X$
into Y.

CAUTION. If X and Y are vector spaces and $B \colon X \times X \to Y$ is a bilinear map,
it is very common to refer to B as a bilinear map on X, when, of course, one should
say that it is a bilinear map on $X \times X$. You will encounter many instances of this
laxity in these notes.

30.1.3. Definition. Let V and W be normed linear spaces. A bilinear map
$B \colon V \times V \to W$ is BOUNDED if there exists a constant $M > 0$ such that
$$\|B(u,v)\| \le M\|u\|\,\|v\|$$
for all u, $v \in V$. Let $\mathfrak{B}^2(V,W)$ denote the family of all bounded bilinear maps from
$V \times V$ into W.

30.1.4. Example. The inner product on $\mathbb{R}^n$
$$B \colon \mathbb{R}^n \times \mathbb{R}^n \to \mathbb{R} : (u,v) \mapsto \langle u, v \rangle$$
is a bounded bilinear functional.

Proof. Problem.

It is easy to make a useful extension of the preceding example.

30.1.5. Example. Let T be a linear operator on $\mathbb{R}^n$. For all u, $v \in \mathbb{R}^n$, define
$$\widehat{T}(u,v) = \langle u, Tv \rangle.$$
Then $\widehat{T}$ is a bounded bilinear functional on $\mathbb{R}^n$.

Proof. Problem.

The bounded bilinear map in the preceding example is called the BILINEAR
FUNCTIONAL ASSOCIATED WITH T.

30.1.6. Example. Composition of operators on a normed linear space V is a
bounded bilinear map on the space $\mathfrak{B}(V)$ of operators on V.

Proof. Problem.

30.1.7. Example. Under pointwise operations of addition and scalar multiplication, the set $\mathfrak{B}^2(V,W)$ is a vector space.

Proof. Problem.

30.1.8. Definition. If B is a bounded bilinear map on a normed linear space V, define
$$\|B\| = \inf\{M > 0 \colon \|B(u,v)\| \le M\|u\|\,\|v\| \text{ for all } u, v \in V\}.$$
We call this scalar quantity the "norm" of B.

Notice that if B is a bounded bilinear map, then $\|B(u,v)\| \le \|B\|\,\|u\|\,\|v\|$ for all $u, v \in V$. The basic facts concerning boundedness of bilinear maps on normed linear spaces are essentially the same as those for boundedness of linear maps on normed linear spaces: bilinear maps are bounded if and only if they are continuous (see Proposition 30.1.14); the function $B \mapsto \|B\|$ really is a norm on the vector space $\mathfrak{B}^2(V,W)$ (see Example 30.1.13); and $\mathfrak{B}^2(V,W)$ is complete under this norm if and only if W is complete (see Proposition 30.1.16). These help us to view members of $\mathfrak{B}^2(V,W)$ as rather familiar objects.

30.1.9. Proposition. *If $B \in \mathfrak{B}^2(V,W)$, where V and W are normed linear spaces, then $\|B\| = \sup\{\|B(u,v)\| \colon \|u\| = \|v\| = 1\}$.*

Proof. Problem.

30.1.10. Example. Considered as a bilinear functional on $\mathbb{R}^n$, the inner product on $\mathbb{R}^n$ has norm 1; see Example 30.1.4.

Proof. Problem.

30.1.11. Example. If T is a bounded linear map on $\mathbb{R}^n$, then the norm of its associated bilinear functional $\widehat{T}$ (see Example 30.1.5) has the same norm as T.

Proof. Problem.

30.1.12. Example. Let V be a normed linear space. Regarded as a bilinear map on the space $B(V,V)$, composition of operators has norm 1; see Example 30.1.6.

Proof. Problem.

30.1.13. Example. If V and W are normed linear spaces, then the map $B \mapsto \|B\|$ is a norm on $\mathfrak{B}^2(V,W)$.

Proof. Problem.

30.1.14. Proposition. *For a bilinear map $B \colon V \times V \to W$, the following are equivalent:*

(a) *B is continuous;*

(b) *B is continuous at $(\mathbf{0}, \mathbf{0})$; and*

(c) *B is bounded.*

Proof. Problem. *Hint.* Only minor modifications of the proof of Proposition 23.1.4 are needed.

At the beginning of this chapter, we promised an isometric isomorphism that will allow us to replace the somewhat unfriendly objects in the space $\mathfrak{B}(V, \mathfrak{B}(V, W))$ with the rather more agreeable bounded bilinear maps that live in $\mathfrak{B}^2(V, W)$. This will in fact be a natural isometric isomorphism. Now, *natural* is a technical term in category theory that we will not take the opportunity to discuss in detail here. We use the adjective only in an informal sense: we say that two vector spaces (with or without additional structure) are NATURALLY ISOMORPHIC if it is possible to show that these spaces are isomorphic without appealing to a basis (or what amounts to the same thing in finite-dimensional spaces, to dimension) in either space. Theorem 30.1.15 states that the spaces $\mathfrak{B}(V, \mathfrak{B}(V, W))$ and $\mathfrak{B}^2(V, W)$ are naturally (isometrically) isomorphic. (For an example of an isomorphism that *does* require the use of a basis, look at Theorem 21.5.8.) We adopt a convention based on this distinction: when two spaces are naturally isomorphic, we identify them. That is, if V and W are naturally isomorphic Banach spaces, we write $V = W$ rather than $V \cong W$. Thus, in particular, we will not notationally distinguish between the spaces $\mathfrak{B}(V, \mathfrak{B}(V, W))$ and $\mathfrak{B}^2(V, W)$. Experience indicates that such an identification, although technically an abuse of language, creates little confusion and reduces the notational burden on the reader. On the other hand, one should be cautious: identifying spaces that are isomorphic, but not naturally so, may well lead to serious misunderstandings.

30.1.15. Theorem. *Let V and W be normed linear spaces. Then the spaces $\mathfrak{B}(V, \mathfrak{B}(V, W))$ and $\mathfrak{B}^2(V, W)$ are naturally isometrically isomorphic. The isomorphism is implemented by the map*

$$(78) \qquad\qquad F \colon \mathfrak{B}(V, \mathfrak{B}(V, W)) \to \mathfrak{B}^2(V, W) \colon \phi \mapsto \widetilde{\phi},$$

where $\widetilde{\phi}(u, v) := \big(\phi(u)\big)(v)$ for all u, $v \in V$.

Proof. Exercise. *Hint.* This proof requires a great deal more writing than thinking. The only slightly tricky part is showing that F is surjective. Given a bounded bilinear map ψ, you must find a map ϕ in $\mathfrak{B}(V, \mathfrak{B}(V, W))$ such that $\widetilde{\phi} = \psi$. Keeping in mind that if you fix, say, the first variable of a bounded bilinear function from V to W, you end up with a bounded linear map from V to W, you might think of trying the map $\phi \colon u \mapsto {}^u\psi = \psi(u, \, \cdot \,)$.

30.1.16. Proposition. *Let V and W be normed linear spaces. Then $\mathfrak{B}^2(V, W)$ is complete if and only if W is.*

Proof. Problem. *Hint.* The best proof is *very* short. Use Theorem 30.1.15 and Corollary 23.3.6.

30.1.17. Proposition. *Every bilinear functional on $\mathbb{R}^n$ is bounded.*

Proof. Problem. *Hint.* Use Theorem 30.1.15, Example 23.1.18, and Theorem 21.5.8.

In Example 30.1.5 we saw that to every linear operator T on $\mathbb{R}^n$ there is an associated (bounded) bilinear form $\widehat{T}$ on $\mathbb{R}^n$ such that $\widehat{T}(u, v) = \langle u, Tv \rangle$. It is natural to ask if, for an arbitrary bilinear functional B on $\mathbb{R}^n$, there exists a linear operator T on $\mathbb{R}^n$ such that $B = \widehat{T}$. And, if so, is it unique? In short, given that every operator on $\mathbb{R}^n$ gives rise to a bilinear functional on $\mathbb{R}^n$, is it also true that every bilinear map on $\mathbb{R}^n$ comes from some unique linear operator? The next proposition gives an affirmative answer.

30.1.18. Proposition. *Let $B \colon \mathbb{R}^n \times \mathbb{R}^n \to \mathbb{R}$ be a bilinear functional on $\mathbb{R}^n$. Then there exists a unique linear operator T on $\mathbb{R}^n$ such that $B = \widehat{T}$.*

Proof. Problem. *Hint.* Let T be the linear operator whose matrix representation is

$$\left[B(e^i, e^j) \right]_{i=1\,j=1}^{n\quad n}.$$

Proposition 30.1.18 shows that the map $T \mapsto \widehat{T}$ establishes a one-to-one correspondence between $\mathfrak{B}(\mathbb{R}^n, \mathbb{R}^n)$ and $\mathfrak{B}^2(\mathbb{R}^n, \mathbb{R})$. It is not difficult to improve this result by verifying that this map is linear and norm preserving.

30.1.19. Proposition. *The map $T \mapsto \widehat{T}$ defined in Example 30.1.5 is an isometric isomorphism between $\mathfrak{B}(\mathbb{R}^n, \mathbb{R}^n)$ and $\mathfrak{B}^2(\mathbb{R}^n, \mathbb{R})$.*

Proof. Problem. *Hint.* To show that $\|\widehat{T}\| \geq \|T\|$, suppose that an arbitrary $\epsilon > 0$ has been specified. Let e be a unit vector not in the kernel of T such that $\|Te\| > \|T\| - \epsilon$. Consider the absolute value of $\widehat{T}\bigl(\|Te\|^{-1}Te, e\bigr)$.

30.1.20. Example. It is interesting to note that the matrix $[B] := \left[B(e^i, e^j) \right]_{i=1\,j=1}^{n\quad n}$ that occurs in the hint to the proof of Proposition 30.1.18 is not only the matrix representation of the linear operator T but it also represents B in the sense that for all $u,\, v \in \mathbb{R}^n$

$$B(u, v) = u[B]v^t,$$

where the intended operation on the right is matrix multiplication, u is regarded as an $1 \times n$ matrix (a ROW VECTOR), and v^t (the transpose of v) is regarded as a $n \times 1$ matrix (a COLUMN VECTOR). It is certainly appropriate to refer to $[B]$ as the MATRIX REPRESENTATION of B.

Proof. Problem.

30.1.21. Definition. A bilinear functional B on a normed linear space is SYMMETRIC if $B(u, v) = B(v, u)$ for all u and v in the space.

The adjective "symmetric" is also used to describe certain matrices.

30.1.22. Definition. A square matrix is SYMMETRIC if it is equal to its transpose.

In the present context these two uses of symmetric are intimately related.

30.1.23. Proposition. *A bilinear functional on $\mathbb{R}^n$ is symmetric if and only if its matrix representation is.*

Proof. Problem.

The objects we have been considering thus far in this section are bilinear maps. What about functions linear in three or even more variables? The definitions are almost the same; and so are the facts through Proposition 30.1.17. The proofs are much like the ones for the special bilinear case (although notationally messier) and are omitted.

30.1.24. Definition. Let X and Y be vector spaces, and let n be an integer strictly greater than one. A map $L\colon X^n \to Y$ is MULTILINEAR if it is linear in each of its n variables; that is, if for all k between 1 and n

$$L(x^1, \ldots, x^{k-1}, x^k + u, x^{k+1}, \ldots, x^n)$$
$$= L(x^1, \ldots, x^n) + L(x^1, \ldots, x^{k-1}, u, x^{k+1}, \ldots, x^n)$$

and

$$L(x^1, \ldots, x^{k-1}, \alpha x^k, x^{k+1}, \ldots, x^n) = \alpha L(x^1, \ldots, x^n),$$

whenever x^1, ..., x^n, $u \in X$, and $\alpha \in \mathbb{R}$. A scalar valued multilinear map is a MULTILINEAR FUNCTIONAL. When n is 2, as we have seen, the map is called "bilinear." When n is 3, the term "trilinear" is common. For larger n, the maps may be called "n-linear."

30.1.25. Definition. If V and W are normed linear spaces a multilinear map, $L\colon V^n \to W$ is BOUNDED if there exists a constant $M > 0$ such that

$$(79) \qquad \|L(v^1, \ldots, v^n)\| \leq M\|v^1\| \cdots \|v^n\|$$

for all v^1, ..., $v^n \in V$. We denote by $\mathfrak{B}^n(V, W)$ the family of all bounded n-linear maps from X^n into Y. The NORM of a bounded multilinear map L is the infimum of the set of all M for which (79) holds.

The elementary useful facts are just what you would expect. A multilinear map is continuous if and only if it is bounded. The family of all continuous multilinear maps from V^n into W is a normed linear space under pointwise operations and the map $L \mapsto \|L\|$, which is indeed a norm; it is complete if and only if W is. The analogue of Theorem 30.1.15 holds for $n > 2$. For example, $\mathfrak{B}(V, \mathfrak{B}(V, \mathfrak{B}(V, W)))$ is naturally isometrically isomorphic to the family $\mathfrak{B}^3(V, W)$ of all trilinear maps from V^3 into W. In general, the space

$$\mathfrak{B}(V, \mathfrak{B}(V, \ldots, \mathfrak{B}(V, W) \ldots)),$$

where n occurrences of the symbol $\mathfrak{B}$ are intended, is naturally isometrically isomorphic (therefore by our convention "equal") to $\mathfrak{B}^n(V, W)$.

30.1.26. Problem. A standard fact about determinants is that if an $n \times n$ matrix M has in its jth row the entries

$$a_1 + b_1, a_2 + b_2, \ldots, a_n + b_n,$$

then $\det M$ can be written as the sum $\det N + \det P$ of the determinants of two matrices N and P where the jth row of N is $a_1, \ldots, a_n$, the jth row of P is $b_1, \ldots, b_n$, and all other rows of N and P are the same as the corresponding rows of M.

A second fact is that if the jth row of M has as its entries

$$\alpha a_1, \alpha a_2, \ldots, \alpha a_n,$$

then $\det M = \alpha \det N$, where N is the same matrix as M except for its jth row where the entries are $a_1, \ldots, a_n$.

Express the preceding facts in the language of this section.

30.2. Second order differentials

> In this section V and W (with or without subscripts) are Banach spaces.

30.2.1. Definition. If $U \overset{\circ}{\subseteq} V$ and $f \colon U \to W$ is continuously differentiable, then the function $df : x \mapsto df_x$ is a map from U into the Banach space $\mathfrak{B}(V, W)$. It may be differentiable. If it is, we say that f is TWICE DIFFERENTIABLE on U. In this case the differential of the function df at a point a in U is denoted by $d^2 f_a$ (rather than $d(df)_a$) and belongs to $\mathfrak{B}(V, \mathfrak{B}(V, W))$. The result of evaluating $d^2 f_a$ at point h in V is a bounded linear map from V to W; that is, $d^2 f_a(h) \in \mathfrak{B}(V, W)$.

Evaluation of $d^2 f_a(h)$ at a point j in V produces a vector in W; that is, $\big(d^2 f_a(h)\big)(j) \in W$. In light of the convention we have made identifying $\mathfrak{B}^2(V, W)$ with $\mathfrak{B}(V, \mathfrak{B}(V, W))$, we can, and usually will, write $d^2 f_a(h, j)$ for $\big((d^2 f_a(h))\big)(j)$. In other words, ordinarily we will think of $d^2 f_a$, the SECOND DIFFERENTIAL of f at a, as a bounded bilinear map taking $V \times V$ into W.

If the function $d^2 f \colon U \to W \colon a \mapsto d^2 f_a$ (exists and) is continuous on U, then f is said to be TWICE CONTINUOUSLY DIFFERENTIABLE on U; the family of all such f is denoted by $\mathcal{C}^2(U, W)$.

The value of the linear map df_a at a vector h can be found by computing the value of the directional derivative $D_h f$ at a (see Proposition 25.5.9). As one might reasonably guess, the value of the bilinear map $d^2 f_a$ at the pair (h, j) can be calculated by taking directional derivatives in the directions first of h and then of j and then evaluating at a.

30.2.2. Proposition. *If $f \in \mathcal{C}^2(U, W)$ where $U \overset{\circ}{\subseteq} V$ and if $a \in U$, then each directional derivative $D_x f \colon U \to W$ is differentiable at a and*

$$(D_v(D_u f))(a) = d^2 f_a(v, u)$$

for all nonzero u and v in V.

Proof. Exercise. *Hint.* In order to take the differential of $D_u f$, write $D_u f$ as $E_u \circ df$ where E_u is the function "evaluate at u" defined on $\mathfrak{B}(V, W)$, then apply the chain rule.

30.2.3. Definition. Recall (from Definition 26.2.14) that for a function defined in $\mathbb{R}^n$, its partial derivatives are simply the directional derivatives along the coordinate axes (that is, in the directions of the standard basis vectors). The function $D_{e_j}\big(D_{e_i} f\big) = \frac{\partial}{\partial x_j}\left(\frac{\partial f}{\partial x_i}\right)$ is a SECOND ORDER PARTIAL DERIVATIVE. Among its (many) standard notations are

$$\frac{\partial^2 f}{\partial x_j \partial x_i}, \quad f_{ij}, \quad D_{x_j} D_{x_i} f, \quad \text{and} \quad D_j D_i f.$$

If we differentiate twice with respect to the same variable x_i, some possible notations are

$$\frac{\partial^2 f}{\partial x_i{}^2}, \quad f_{ii}, \quad D_{x_i}{}^2 f, \quad \text{and} \quad D_i{}^2 f.$$

30.2.4. Definition. Let f be a twice differentiable scalar field defined on an open subset U of $\mathbb{R}^n$ and $a \in U$. We regard $d^2 f_a$ as a bounded bilinear functional on $\mathbb{R}^n$. The matrix representation of the functional (see Example 30.1.20) is called the HESSIAN MATRIX of f at a and is denoted by $H(a)$.

The next proposition states that the value of $d^2 f_a$ at a pair (u, v) can be written as a linear combination of partial derivatives of f.

30.2.5. Proposition. *If f is a twice continuously differentiable scalar field on some open subset U of $\mathbb{R}^n$ and $a \in U$, then the second order partial derivatives f exist at a and*

$$d^2 f_a(e^i, e^j) = D_{e^i}\left(e_{e^j}(f)\right)(a)$$

for $1 \leq i, j \leq n$. It follows that

(80) $$d^2 f_a(u, v) = \sum_{i,j=1}^{n} u_i v_j f_{ji}(a)$$

for all u, $v \in \mathbb{R}^n$, and that the Hessian matrix of f is given at a by

$$H(a) = \left[\frac{\partial^2 f}{\partial x_i \partial x_j}(a) \right]_{i=1\,j=1}^{n \quad n}.$$

Proof. Problem.

According to Proposition 30.2.5, every twice continuously differentiable function has second order partial derivatives. Since it is often much easier to verify the existence of partial derivatives than the existence of differentials, it would be gratifying to have the converse of this assertion. Is it true? As in the single variable case, the answer is "yes" provided that the partial derivatives in question are continuous; cf. Proposition 26.2.10.

30.2.6. Proposition. *If $f \colon U \to W$ where $U \overset{\circ}{\subseteq} \mathbb{R}^n$ and f has continuous second partial derivatives on U, then f is twice continuously differentiable on U.*

Proof. Problem. *Hint.* Looking at the right function makes this proof quite easy. The function is $E = (E_{e^1}, \ldots, E_{e^n})$ where, for $1 \leq k \leq n$, the function E_{e^k} evaluates each member of $\mathfrak{B}(\mathbb{R}^n, W)$ at the standard basis vector e^k. In other words

$$E \colon \mathfrak{B}(\mathbb{R}^n, W) \to W^n \colon T \mapsto (Te^1, \ldots, Te^n).$$

Convince yourself that E is invertible, then use Proposition 25.6.1 to show that $E \circ df$ (and therefore df) is differentiable.

Just as the fundamental facts concerning the differential df_a are adumbrated by corresponding (purely algebraic) properties of the difference Δf_a, so properties of the second order differential $d^2 f_a$ arise from the algebraic behavior of the SECOND ORDER DIFFERENCE $\Delta^2 f_a := \Delta(\Delta f)_a$. In the remainder of this section we make some elementary observations about $\Delta^2 f_a$ and derive from them the basic facts concerning $d^2 f_a$.

First of all notice that if f belongs to $\mathcal{F}_a(V, W)$, then from the computations

$$\Delta^2 f_a(v) = \Delta(\Delta f)_a(v)$$
$$= \Delta f_{a+v} - \Delta f_a$$

and

(81)
$$\Delta^2 f_a(u)(v) = \Delta f_{a+u}(v) - \Delta f_a(v)$$
$$= f(a + u + v) - f(a + u) - f(a + v) + f(a),$$

it follows easily that

$$\Delta^2 f_a \in \mathcal{F}_0(V, \mathcal{F}_0(V, W)),$$
$$\Delta^2(f + g)_a = \Delta^2 f_a + \Delta^2 g_a,$$
$$\Delta^2(\alpha f)_a = \alpha \Delta^2 f_a,$$

and that

(82)
$$\Delta^2 f_a(u)(v) = \Delta^2 f_a(v)(u).$$

In what follows we make no notational distinction between functions in $\mathcal{F}_0(V, \mathcal{F}_0(V, W))$ and those in $\mathcal{F}_0(V^2, W)$. Thus, we will often regard $\Delta^2 f_a$ as a W-valued function of two variables and write $\Delta^2 f_a(u, v)$ for $\Delta^2 f_a(u)(v)$. Thus, in particular, (82) becomes

$$\Delta^2 f_a(u, v) = \Delta^2 f_a(v, u).$$

Recall that in Chapter 25 we defined the family $\mathfrak{o}(V, W)$ in such a way that a function $f \in \mathcal{F}_0(V, W)$ belongs to $\mathfrak{o}(V, W)$ if and only if $f(\mathbf{0}) = \mathbf{0}$ and

$$\lim_{h \to \vec{0}} \frac{\|f(h)\|}{h} = 0.$$

Tangency was then defined so that $f \simeq g$ if and only if $f - g \in \mathfrak{o}(V, W)$. What are the analogous concepts for a function of two variables? Just what you would guess.

30.2.7. Definition. A function $f \in \mathcal{F}_0(V^2, W)$ belongs to $\mathfrak{o}^2(V, W)$ if for every $\epsilon > 0$ there exists $\delta > 0$ such that

$$\|f(u, v)\| \le \epsilon \|u\| \|v\|$$

whenever $\|(u, v)\|_1 < \delta$. Equivalently, $f \in \mathfrak{o}^2(V, W)$ provided that $f(u, \mathbf{0}) = f(\mathbf{0}, v) = \mathbf{0}$ for all u and v in some neighborhood of $(\mathbf{0}, \mathbf{0})$ and

$$\lim_{\substack{(u,v) \to (\mathbf{0},\mathbf{0}) \\ u \ne \mathbf{0}, v \ne \mathbf{0}}} \frac{\|f(u, v)\|}{\|u\| \|v\|} = 0.$$

When there seems little likelihood of confusion, we often write $\mathfrak{o}^2$ for $\mathfrak{o}^2(V, W)$.

We say that two functions $f, g \in \mathcal{F}_0(V^2, W)$ are BITANGENT at $\mathbf{0}$ if $f - g \in \mathfrak{o}^2(V, W)$. In this case we write $f \overset{2}{\simeq} g$. (It is tempting to call this relation "second order tangency"; unfortunately this term already has a standard—and different—meaning.)

30.2.8. Proposition. *If V and W are Banach spaces, then $\mathfrak{o}^2(V,W) \subseteq \mathfrak{o}(V^2, W)$.*

Proof. Problem.

30.2.9. Example. Let $f(x,y) = 0$ if $x = 0$ or $y = 0$, and let

$$f(x,y) = xy \sin \frac{1}{xy}$$

otherwise. Then f belongs to $\mathfrak{o}(\mathbb{R}^2, \mathbb{R})$ but not to $\mathfrak{o}^2(\mathbb{R}, \mathbb{R})$.

Proof. Problem.

30.2.10. Example. Let $f(x,y) = 0$ if $x = 0$ or $y = 0$, and let

$$f(x,y) = x^2 y \sin \frac{1}{xy}$$

otherwise. Then $f \in \mathfrak{o}^2(\mathbb{R}, \mathbb{R})$.

Proof. Problem.

30.2.11. Proposition. *The space $\mathfrak{o}^2$ is closed under addition and scalar multiplication. In fact, $\mathfrak{o}^2 + \mathfrak{o}^2 = \mathfrak{o}^2$ and $\alpha \mathfrak{o}^2 = \mathfrak{o}^2$ for every nonzero scalar α.*

Proof. Problem.

30.2.12. Proposition. *The relation $\overset{2}{\simeq}$ of bitangency is an equivalence relation.*

Proof. Problem.

30.2.13. Proposition. *If $f \overset{2}{\simeq} g$ and $h \overset{2}{\simeq} k$, then $f + h \overset{2}{\simeq} g + k$.*

Proof. Problem.

Recall that according to Proposition 25.1.5 no bounded linear map other than zero can be tangent to zero; that is, $\mathfrak{B} \cap \mathfrak{o} = \{\mathbf{0}\}$. A corresponding fact is true for the two variable case: no bounded bilinear map other than zero is bitangent to zero.

30.2.14. Proposition. *If $B \in \mathfrak{B}^2(V,W) \cap \mathfrak{o}^2(V,W)$, then $B = \mathbf{0}$.*

Proof. Problem.

30.2.15. Exercise. Let $a \in \mathbb{R}$ and $f(x) = x^3 - 2x^2 + 5x - 7$ for all $x \in \mathbb{R}$.

(a) Compute $\Delta f_a(x)$.
(b) Compute $df_a(x)$ and show that $\Delta f_a \simeq df_a$.
(c) Compute $\Delta(df)_a(x)(y)$.
(d) Compute $\Delta^2 f_a(x)(y)$.
(e) Using (c) and (d) show that $\Delta(df)_a \overset{2}{\simeq} \Delta^2 f_a$.
(f) Compute $d^2 f_a(x)(y)$ and show that $\Delta^2 f_a \overset{2}{\simeq} d^2 f_a$.

Recall that in Proposition 26.1.14, our final version of the mean value theorem, the claim is that for a $\mathcal{C}^1$ function f the first order difference $\Delta f_a(h)$ can be expressed as the integral of the first order differential df over the line segment from a to $a+h$. It is both interesting and satisfying that there is a perfect second order analogue of this: For a $\mathcal{C}^2$ function f the second order difference $\Delta^2 f_a(x)(y)$ can be expressed as the integral of the second order differential $d^2 f$ over the parallelogram with vertices a, $a+x$, $a+y$, and $a+x+y$. This we call the SECOND ORDER MEAN VALUE THEOREM.

30.2.16. Theorem. *If U is an open convex subset of V and $f \in \mathcal{C}^2(U, W)$, then*

$$\Delta^2 f_a(x)(y) = \left(\int_0^1 \int_0^1 d^2 f_{a+sx+ty} \, ds \, dt \right)(x)(y)$$

whenever a, $a+x$, $a+y$, and $a+x+y$ all lie in U.

Proof. Exercise. *Hint.* Use the mean value theorem, Proposition 26.1.14, twice.

We know that for $\mathcal{C}^1$ functions (in fact, for differentiable ones) that the difference Δf_a is tangent to the differential df_a. We now establish the analogous fact for second order differences and differentials: if $f \in \mathcal{C}^2$, then $\Delta^2 f_a$ is bitangent to $d^2 f_a$.

30.2.17. Theorem. *If $f \in \mathcal{C}^2(U, W)$ where $U \overset{\circ}{\subseteq} V$, then*

$$\Delta^2 f_a \overset{2}{\simeq} d^2 f_a$$

for all $a \in U$.

Proof. Problem. *Hint.* Use the hypothesis that f is $\mathcal{C}^2$ to show that for every $\epsilon > 0$ there exists $\delta > 0$ such that $\|d^2 f_{a+sx+ty} - d^2 f_a\| < \epsilon$ whenever $\|x\|, \|y\| < \frac{1}{2}\delta$ and $s, t \in [0, 1]$. To show that $\|\Delta^2 f_a(x, y) - d^2 f_a(x, y)\| \leq \epsilon \|x\| \|y\|$ write $\Delta^2 f_a(x, y)$ as an integral (use the second order mean value theorem, Theorem 30.2.16) and $d^2 f_a$ as $\int_0^1 \int_0^1 d^2 f_a \, ds \, dt$, and then apply Proposition 24.3.10.

It is now an easy task to see that the second differential of a $\mathcal{C}^2$ function is symmetric.

30.2.18. Proposition. *If $f \in \mathcal{C}^2(U, W)$ where $U \overset{\circ}{\subseteq} V$, then*

$$d^2 f_a(x, y) = d^2 f_a(y, x)$$

for all $a \in U$ and all $x, y \in V$.

Proof. Problem. *Hint.* A notational convention makes this a very easy problem. If g is a function of two variables, let $g^\sharp$ be the function that results from switching the variables of g; that is, let $g^\sharp(x, y) = g(y, x)$. Use Theorem 30.2.16, equation (82), and Proposition 30.2.14.

30.2.19. Corollary. *If $f \in \mathcal{C}^2(U, W)$ where $U \overset{\circ}{\subseteq} \mathbb{R}^n$ and if $1 \leq i, j \leq n$, then*

$$\frac{\partial^2 f}{\partial x_j \partial x_i}(a) = \frac{\partial^2 f}{\partial x_i \partial x_j}(a)$$

for every $a \in U$.

Proof. Problem.

30.2.20. Example. Consider the function f defined by

$$f(x,y) = \frac{xy(x^2 - y^2)}{x^2 + y^2}$$

if x and y are not both zero and $f(0,0) = 0$. Compute $\dfrac{\partial^2 f}{\partial y \partial x}(0,0)$ and $\dfrac{\partial^2 f}{\partial x \partial y}(0,0)$. Explain why this example does not contradict Corollary 30.2.19.

30.3. Higher order differentials

The preceding material on second order differentials can be extended to differentials of all finite orders. For instance, if the function

$$d^2 f \colon V \to \mathfrak{B}^2(V,W) \colon v \mapsto d^2 f_v$$

is differentiable at a point a in V, then its differential, which we denote by $d^3 f_a$, belongs to $\mathfrak{B}(V, \mathfrak{B}(V, \mathfrak{B}(V,W)))$, which we may identify with the space $\mathfrak{B}^3(V,W)$ of all continuous maps from V^3 into W that are linear in each of the three variables.

The results in the preceding section generalize just as one would expect. For example, formula (80), which, in the case $V = \mathbb{R}^n$, allows us to write the third order differential at a point a in terms of partial derivatives, becomes

$$d^3 f_a(t, u, v) = \sum_{i,j,k=1}^{n} t_i u_j v_k f_{kji}(a)$$

for all t, u, $v \in \mathbb{R}^n$. (Here, of course, $f_{kji}(a) = \left(\frac{\partial}{\partial x_i}\left(\frac{\partial}{\partial x_j}\left(\frac{\partial f}{\partial x_k}\right)\right)\right)(a)$.) You will likely find little conceptual difficulty in seeing how to produce inductive proofs for the generalizations. However, it is entirely possible to be daunted by notational complexities that arise. Should you be interested in pursuing these matters further, you can find a pleasant introduction to the current "multi-index" notation in [**LS90**, Chapter 3, section 17], the use of which simplifies many formulas and proofs immensely.

Quantifiers

Certainly, $2 + 2 = 4$ and $2 + 2 = 5$ are statements—one true, the other false. On the other hand the appearance of the variable x prevents the expression $x + 2 = 5$ from being a statement. Such an expression we will call an OPEN SENTENCE; its truth is open to question since x is unidentified. There are three standard ways of converting open sentences into statements.

The first, and simplest, of these is to give the variable a particular value. If we "evaluate" the expression $x + 2 = 5$ at $x = 4$, we obtain the (false) statement $4 + 2 = 5$.

A second way of obtaining a statement from an expression involving a variable is UNIVERSAL QUANTIFICATION: we assert that the expression is true for all values of the variable. In the preceding example we get, "For all x, $x + 2 = 5$". This is now a statement (and again false). The expression "for all x" (or equivalently, "for every x") is often denoted symbolically by $(\forall x)$. Thus the preceding sentence may be written $(\forall x) x + 2 = 5$. (The parentheses are optional; they may be used in the interest of clarity.) We call $\forall$ a UNIVERSAL QUANTIFIER.

Frequently, there are several variables in an expression. They may all be universally quantified. For example

(83) $$(\forall x)(\forall y)\, x^2 - y^2 = (x - y)(x + y)$$

is a (true) statement, which says that for every x and for every y the expression $x^2 - y^2$ factors in the familiar way. The order of consecutive universal quantifiers is unimportant: the statement

$$(\forall y)(\forall x)\, x^2 - y^2 = (x - y)(x + y)$$

says exactly the same thing as (83). For this reason the notation may be contracted slightly to read

$$(\forall x, y)\, x^2 - y^2 = (x - y)(x + y)\,.$$

A third way of obtaining a statement from an open sentence $P(x)$ is EXISTENTIAL QUANTIFICATION. Here we assert that $P(x)$ is true for *at least one* value of x.

This is often written "$(\exists x)$ such that $P(x)$" or more briefly "$(\exists x)P(x)$", and is read "there exists an x such that $P(x)$" or "$P(x)$ is true for some x". For example, if we existentially quantify the expression "$x + 2 = 5$", we obtain "$(\exists x)$ such that $x + 2 = 5$" (which happens to be true). We call $\exists$ an EXISTENTIAL QUANTIFIER.

As is true for universal quantifiers, the order of consecutive existential quantifiers is immaterial.

CAUTION. It is absolutely essential to realize that the order of an existential and a universal quantifier may *not* in general be reversed. For example,

$$(\exists x)(\forall y)\, x < y$$

says that there is a number x with the property that no matter how y is chosen, x is less than y; that is, there is a smallest real number. (This is, of course, false.) On the other hand

$$(\forall y)(\exists x)\, x < y$$

says that for every y we can find a number x smaller than y. (This is true: take x to be $y - 1$ for example.) *The importance of getting quantifiers in the right order cannot be overemphasized.*

There is one frequently used convention concerning quantifiers that should be mentioned. In the statement of definitions, propositions, theorems, etc., missing quantifiers are assumed to be universal; furthermore, they are assumed to be the innermost quantifiers.

A.1.1. Example. Let f be a real valued function defined on the real line $\mathbb{R}$. Many texts give the following definition. The function f is *continuous* at a point a in $\mathbb{R}$ if for every $\epsilon > 0$ there exists $\delta > 0$ such that

$$|f(x) - f(a)| < \epsilon \quad \text{whenever } |x - a| < \delta.$$

Here ϵ and δ are quantified; the function f and the point a are fixed for the discussion, so they do not require quantifiers. What about x? According to the convention just mentioned, x is universally quantified and that quantifier is the innermost one. Thus the definition reads, for every $\epsilon > 0$ there exists $\delta > 0$ such that for every x

$$|f(x) - f(a)| < \epsilon \quad \text{whenever } |x - a| < \delta.$$

A.1.2. Example. Sometimes all quantifiers are missing. In this case the preceding convention dictates that all variables are universally quantified. Thus

Theorem. $\quad x^2 - y^2 = (x - y)(x + y)$

is interpreted to mean

Theorem. $\quad (\forall x)(\forall y)\, x^2 - y^2 = (x - y)(x + y)\,.$

Sets

In this text everything is defined ultimately in terms of two primitive (that is, undefined) concepts: set and set membership. We assume that these are already familiar to the reader. In particular, it is assumed to be understood that distinct elements (or members, or points) can be regarded collectively as a single set (or family, or class, or collection). To indicate that x belongs to a set A (or that x is a member of A), we write $x \in A$; to indicate that it does not belong to A, we write $x \notin A$.

We specify a set by listing its members between braces (for instance, $\{1, 2, 3, 4, 5\}$ is the set of the first five natural numbers), by listing some of its members between braces with ellipses (three dots) indicating the missing members (e.g., $\{1, 2, 3, \dots\}$ is the set of all natural numbers), or by writing $\{x \colon P(x)\}$ where $P(x)$ is an open sentence that specifies what property the variable x must satisfy in order to be included in the set (e.g., $\{x \colon 0 \le x \le 1\}$ is the closed unit interval $[0, 1]$).

B.1.1. Problem. Let $\mathbb{N}$ be the set of natural numbers 1, 2, 3, ..., and let

$$S = \{x \colon x < 30 \text{ and } x = n^2 \text{ for some } n \in \mathbb{N}\}.$$

List all the elements of S.

B.1.2. Problem. Let $\mathbb{N}$ be the set of natural numbers 1, 2, 3, ..., and let

$$S = \{x \colon x = n + 2 \text{ for some } n \in \mathbb{N} \text{ such that } n < 6\}.$$

List all the elements of S.

B.1.3. Problem. Suppose that

$$S = \{x \colon x = n^2 + 2 \text{ for some } n \in \mathbb{N}\}$$

and that

$$T = \{3, 6, 11, 18, 27, 33, 38, 51\}.$$

(a) Find an element of S that is not in T.

(b) Find an element of T that is not in S.

B.1.4. Problem. Suppose that
$$S = \{x \colon x = n^2 + 2n \text{ for some } n \in \mathbb{N}\}$$
and that
$$T = \{x \colon x = 5n - 1 \text{ for some } n \in \mathbb{N}\}.$$
Find an element that belongs to both S and T.

CAUTION. How many elements are in the set $\{3, 7, 4, 7, 4, 1, 3, 1, 7\}$? Exactly four. Because
$$\{3, 7, 4, 7, 4, 1, 3, 1, 7\} = \{1, 3, 4, 7\} = \{4, 7, 3, 1\}.$$
Repetitions don't increase the size of a set. The order in which the elements are written does not matter.

It is prudent to give some thought to the possibility of sets having repeated elements. This occurs quite naturally in many contexts. The fundamental theorem of algebra, for example, says *every nonzero, single-variable, polynomial of degree n with complex coefficients has, counting multiplicity, exactly n complex roots.* Now let's look at an example: Consider the polynomial $p(x) = x^3 - x^2$. According to the preceding theorem, it is reasonable to write, "Let $S = \{r_1, r_2, r_3\}$ be the set of roots of p". However, when we factor p, we get $(x - 0)(x - 0)(x - 1)$, so that $r_1 = r_2 = 0$ and $r_3 = 1$. (A root r of a polynomial p has MULTIPLICITY m if the factor $(x - r)$ occurs exactly m times in the factorization of p.) So, despite its notation, the set S has only two elements, not three. There are only two *distinct* roots: the root 0 has multiplicity two, and the root 1 has multiplicity 1. To avoid consideration of uninteresting special cases, the following convention is frequently made.

B.1.5. Convention. When a set is *defined* by listing its elements, it is assumed that the elements are unique. For example, "Let $S = \{a, b, c\}$" is intended as a short version of, "Let $S = \{a, b, c\}$ be a set of three elements" or "Let $S = \{a, b, c\}$, where a, b, and c are distinct".

Since all of our subsequent work depends on the notions of set and of set membership, it is not altogether satisfactory to rely on intuition and shared understanding to provide a foundation for these crucial concepts. It is possible in fact to arrive at paradoxes using a naive approach to sets. (For example, ask yourself the question, "If S is the set of all sets that do not contain themselves as members, then does S belong to S?" If the answer is "yes", then it must be "no", and vice versa.) One satisfactory alternative to our intuitive approach to sets is axiomatic set theory. There are many ways of axiomatizing set theory to provide a secure foundation for subsequent mathematical development. Unfortunately, each of these ways turns out to be extremely intricate, and it is generally felt that the abstract complexities of axiomatic set theory do not serve well as a beginning to an introductory course in advanced calculus.

Most of the paradoxes inherent in an intuitive approach to sets have to do with sets that are too "large". For example, the set S mentioned in the preceding paragraph is enormous. Thus in what follows we will assume that in each situation all the mathematical objects we are then considering (sets, functions, etc.) belong to some appropriate "universal" set that is "small" enough to avoid set theoretic paradoxes. (Think of "universal" in terms of "universe of discourse", not "all-encompassing".) In many cases an appropriate universal set is clear from the context. Previously, we considered a statement

$$(\forall y)(\exists x)\, x < y\,.$$

The appearance of the symbol "$<$" suggests to most readers that x and y are real numbers. Thus the universal set from which the variables are chosen is the set $\mathbb{R}$ of all real numbers. When there is doubt that the universal set will be properly interpreted, it may be specified. In the example just mentioned, we might write

$$(\forall y \in \mathbb{R})(\exists x \in \mathbb{R})\, x < y.$$

This makes explicit the intended restriction that x and y be real numbers.

As another example recall that in Appendix A, we defined a real valued function f to be continuous at a point $a \in \mathbb{R}$ if

$$(\forall \epsilon > 0)(\exists \delta > 0) \text{ such that } (\forall x)\, |f(x) - f(a)| < \epsilon \text{ whenever } |x - a| < \delta.$$

Here the first two variables, ϵ and δ, are restricted to lie in the open interval $(0, \infty)$. Thus we might rewrite the definition as

$$\forall \epsilon \in (0, \infty)\ \exists \delta \in (0, \infty)$$
$$\text{such that } (\forall x \in \mathbb{R})\, |f(x) - f(a)| < \epsilon \text{ whenever } |x - a| < \delta.$$

The expressions $\forall x \in \mathbb{R}$, $\exists \delta \in (0, \infty)$, etc., are called RESTRICTED QUANTIFIERS.

B.1.6. Definition. Let S and T be sets. We say that S is a SUBSET of T and write $S \subseteq T$ (or $T \supseteq S$) if every member of S belongs to T. If $S \subseteq T$ we also say that S is CONTAINED IN T or that T CONTAINS S. Notice that the relation $\subseteq$ is REFLEXIVE (that is, $S \subseteq S$ for all S) and TRANSITIVE (that is, if $S \subseteq T$ and $T \subseteq U$, then $S \subseteq U$). It is also ANTISYMMETRIC (that is, if $S \subseteq T$ and $T \subseteq S$, then $S = T$). If we wish to claim that S is *not* a subset of T, we may write $S \nsubseteq T$. In this case there is at least one member of S that does not belong to T.

B.1.7. Example. Since every number in the closed interval $[0, 1]$ also belongs to the interval $[0, 5]$, it is correct to write $[0, 1] \subseteq [0, 5]$. Since the number π belongs to $[0, 5]$ but not to $[0, 1]$, we may also write $[0, 5] \nsubseteq [0, 1]$.

B.1.8. Definition. If $S \subseteq T$ but $S \neq T$, then we say that S is a PROPER SUBSET of T (or that S is PROPERLY CONTAINED IN T, or that T PROPERLY CONTAINS S) and write $S \subsetneqq T$.

B.1.9. Problem. Suppose that $S = \{x \colon x = 2n + 3 \text{ for some } n \in \mathbb{N}\}$ and that T is the set of all odd natural numbers $1, 3, 5, \ldots$.

(a) Is $S \subseteq T$? If not, find an element of S that does not belong to T.

(b) Is $T \subseteq S$? If not, find an element of T that does not belong to S.

B.1.10. Definition. The EMPTY SET (or NULL SET), which is denoted by $\emptyset$, is defined to be the set that has no elements. (Or, if you like, define it to be $\{x \colon x \neq x\}$.) It is regarded as a subset of every set, so that $\emptyset \subseteq S$ is always true. (*Note.* "$\emptyset$" is a letter of the Danish alphabet, not the Greek letter "phi".)

B.1.11. Definition. If S is a set, then the POWER SET of S, which we denote by $\mathfrak{P}(S)$, is the set of all subsets of S.

B.1.12. Example. Let $S = \{a, b, c\}$. (See Convention B.1.5.) Then the members of the power set of S are the empty set, the three one-element subsets, the three two-element subsets, and the set S itself. That is,

$$\mathfrak{P}(S) = \{\emptyset, \{a\}, \{b\}, \{c\}, \{a, b\}, \{a, c\}, \{b, c\}, S\}.$$

B.1.13. Problem. In each of the words (a)–(d) below let S be the set of letters in the word. In each case find the number of members of S and the number of members of $\mathfrak{P}(S)$, the power set of S.

(a) lull

(b) appall

(c) attract

(d) calculus

CAUTION. In attempting to prove a theorem that has as a hypothesis "Let S be a set" do not include in your proof something like "Suppose $S = \{s_1, s_2, \ldots, s_n\}$" or "Suppose $S = \{s_1, s_2, \ldots\}$". In the first case you are tacitly assuming that S is finite and in the second that it is countable. Neither is justified by the hypothesis.

CAUTION. A single letter $\mathfrak{S}$ (an S in fraktur font) is an acceptable symbol in printed documents. Don't try to imitate it in handwritten work or on the blackboard. Use script letters instead.

Finally, a word on the use of the symbols $=$ and $:=$. In this text equality is used in the sense of identity. We write $x = y$ to indicate that x and y are two names for the same object. For example, $0.5 = 1/2 = 3/6 = 1/\sqrt{4}$ because 0.5, $1/2$, $3/6$, and $1/\sqrt{4}$ are different names for the same real number. You have probably encountered other uses of the term equality. In many high school geometry texts, for example, one finds statements to the effect that a triangle is isosceles if it has two equal sides (or two equal angles). What is meant of course is that a triangle is isosceles if it has two sides of *equal length* (or two angles of *equal angular measure*). We also make occasional use of the symbol $:=$ to indicate *equality by definition*. Thus when we write $a := b$ we are giving a new name a to an object b with which we are presumably already familiar.

Special subsets of $\mathbb{R}$

We denote by $\mathbb{R}$ the set of real numbers. Certain subsets of $\mathbb{R}$ have standard names. We list some of them here for reference. The set $\mathbb{P} = \{x \in \mathbb{R}\colon x > 0\}$ of strictly positive numbers is discussed in Appendix H. The set $\{1, 2, 3, \dots\}$ of all natural numbers is denoted by $\mathbb{N}$, initial segments $\{1, 2, 3, \dots, m\}$ of this set by $\mathbb{N}_m$, and the set $\{\dots, -3, -2, -1, 0, 1, 2, 3, \dots\}$ of all integers by $\mathbb{Z}$. The set of all rational numbers (numbers of the form p/q where $p, q \in \mathbb{Z}$ and $q \neq 0$) is denoted by $\mathbb{Q}$. There are the *open intervals*

$$(a, b) := \{x \in \mathbb{R}\colon a < x < b\},$$
$$(-\infty, b) := \{x \in \mathbb{R}\colon x < b\}, \quad \text{and}$$
$$(a, \infty) := \{x \in \mathbb{R}\colon x > a\}.$$

There are the *closed intervals*

$$[a, b] := \{x \in \mathbb{R}\colon a \le x \le b\},$$
$$(-\infty, b] := \{x \in \mathbb{R}\colon x \le b\}, \quad \text{and}$$
$$[a, \infty) := \{x \in \mathbb{R}\colon x \ge a\}.$$

And there are the intervals that (for $a < b$) are neither open nor closed:

$$[a, b) := \{x \in \mathbb{R}\colon a \le x < b\} \quad \text{and}$$
$$(a, b] := \{x \in \mathbb{R}\colon a < x \le b\}.$$

The set $\mathbb{R}$ of all real numbers may be written in interval notation as $(-\infty, \infty)$. (As an interval it is considered both open and closed. The reason for applying the words "open" and "closed" to intervals is discussed in Chapter 2.)

A subset A of $\mathbb{R}$ is BOUNDED if there is a positive number M such that $|a| \le M$ for all $a \in A$. Thus intervals of the form $[a, b]$, $(a, b]$, $[a, b)$, and (a, b) are bounded. The other intervals are UNBOUNDED.

If A is a subset of $\mathbb{R}$, then $A^+ := A \cap [0, \infty)$. These are the POSITIVE elements of A. Notice, in particular, that $\mathbb{Z}^+$ (the set of positive integers) contains 0, but $\mathbb{N}$ (the set of natural numbers) does not.

Logical connectives

D.1. Disjunction and conjunction

The word "or" in English has two distinct uses. Suppose you ask a friend how he intends to spend the evening, and he replies, "I'll walk home or I'll take in a film." If you find that he then walked home and on the way stopped to see a film, it would not be reasonable to accuse him of having lied. He was using the inclusive "or", which is true when one or both alternatives are. On the other hand suppose that while walking along a street you are accosted by an armed robber who says, "Put up your hands or I'll shoot." You obligingly raise your hands. If he then shoots you, you have every reason to feel ill-used. Convention and context dictate that he used the exclusive "or": either alternative, but not both.

Since it is undesirable for ambiguities to arise in mathematical discourse, the inclusive "or" has been adopted as standard for mathematical (and most scientific) purposes. A convenient way of defining logical connectives such as "or" is by means of a truth table. The formal definition of "or" looks like this.

P	Q	$P \vee Q$
T	T	T
T	F	T
F	T	T
F	F	F

Here P and Q are any sentences. In the columns labeled P and Q, we list all possible combinations of truth values for P and Q (T for true, F for false). In the third column appears the corresponding truth value for "P or Q". According to the table "P or Q" is true in all cases except when both P and Q are false. The notation "$P \vee Q$" is frequently used for "P or Q". The operation $\vee$ is called DISJUNCTION.

D.1.1. Exercise. Construct a truth table giving the formal definition of "and", frequently denoted by $\wedge$. The operation $\wedge$ is called CONJUNCTION.

We say that two sentences depending on variables $P, Q, \ldots$ are LOGICALLY EQUIVALENT if they have the same truth value no matter how truth values T and F are assigned to the variables $P, Q, \ldots$. It turns out that truth tables are quite helpful in deciding whether certain statements encountered in mathematical reasoning are logically equivalent to one another. (But do *not* clutter up mathematical proofs with truth tables. Everyone is supposed to argue logically. Truth tables are only scratch work for the perplexed.)

D.1.2. Example. Suppose that P, Q, and R are any sentences. To a person who habitually uses language carefully, it will certainly be clear that the following two assertions are equivalent:

(a) P is true and so is either Q or R.

(b) Either both P and Q are true or both P and R are true.

Suppose for a moment, however, that we are in doubt concerning the relation between (a) and (b). We may represent (a) symbolically by $P \wedge (Q \vee R)$ and (b) by $(P \wedge Q) \vee (P \wedge R)$. We conclude that they are indeed logically equivalent by examining the following truth table. (Keep in mind that since there are three variables, P, Q, and R, there are $2^3 = 8$ ways of assigning truth values to them; so we need eight lines in our truth table.)

(1)	(2)	(3)	(4)	(5)	(6)	(7)	(8)
P	Q	R	$Q \vee R$	$P \wedge (Q \vee R)$	$P \wedge Q$	$P \wedge R$	$(P \wedge Q) \vee (P \wedge R)$
T	T	T	T	T	T	T	T
T	T	F	T	T	T	F	T
T	F	T	T	T	F	T	T
T	F	F	F	F	F	F	F
F	T	T	T	F	F	F	F
F	T	F	T	F	F	F	F
F	F	T	T	F	F	F	F
F	F	F	F	F	F	F	F

Column (4) is obtained from (2) and (3), column (5) from (1) and (4), column (6) from (1) and (2), column (7) from (1) and (3), and column (8) from (6) and (7). Comparing the truth values in columns (5) and (8), we see that they are exactly the same. Thus $P \wedge (Q \vee R)$ is logically equivalent to $(P \wedge Q) \vee (P \wedge R)$. This result is a DISTRIBUTIVE LAW; it says that conjunction distributes over disjunction.

D.1.3. Problem. Use truth tables to show that the operation of disjunction is associative; that is, show that $(P \vee Q) \vee R$ and $P \vee (Q \vee R)$ are logically equivalent.

D.1.4. Problem. Use truth tables to show that disjunction distributes over conjunction; that is, show that $P \vee (Q \wedge R)$ is logically equivalent to $(P \vee Q) \wedge (P \vee R)$.

One final remark: Quantifiers may be "moved past" portions of disjunctions and conjunctions that do not contain the variable being quantified. For example,

$$(\exists y)(\exists x)\,[(y^2 \le 9) \wedge (2 < x < y)]$$

says the same thing as

$$(\exists y)\,[(y^2 \le 9) \wedge (\exists x)\,2 < x < y].$$

D.2. Implication

Consider the assertion, "If $1 = 2$, then $2 = 3$". If you ask a large number of people (not mathematically trained) about the truth of this, you will probably find some who think it is true, some who think it is false, and some who think it is meaningless (therefore neither true nor false). This is another example of the ambiguity of ordinary language. In order to avoid ambiguity and to ensure that "P implies Q" has a truth value whenever P and Q do, we define the operation of IMPLICATION, denoted by $\Rightarrow$, by means of the following truth table.

P	Q	$P \Rightarrow Q$
T	T	T
T	F	F
F	T	T
F	F	T

There are many ways of saying that P implies Q. The following assertions are all identical.

$P \Rightarrow Q$.

P implies Q.

If P, then Q.

P is sufficient (or a sufficient condition) for Q.

Whenever P, then Q.

$Q \Leftarrow P$

Q is implied by P.

Q is a consequence of P.

Q follows from P.

Q is necessary (or a necessary condition) for P.

Q whenever P.

The statement $Q \Rightarrow P$ is the CONVERSE of $P \Rightarrow Q$. It is a common mistake to confuse a statement with its converse; however, this is a grievous error. For example, it is correct to say that if a geometric figure is a square, then it is a quadrilateral; but it is *not* correct to say that if a figure is a quadrilateral, it must be a square.

D.2.1. Definition. If P and Q are sentences, we define the logical connective "iff" (read "if and only if") by the following truth table.

P	Q	P iff Q
T	T	T
T	F	F
F	T	F
F	F	T

Notice that the sentence "P iff Q" is true exactly in those cases where P and Q have the same truth values. That is, saying that "P iff Q" is a TAUTOLOGY (true for all truth values of P and Q) is the same as saying that P and Q are equivalent sentences. Thus the connective "iff" is called EQUIVALENCE. An alternative notation for "iff" is "$\Leftrightarrow$".

D.2.2. Example. By comparing columns (3) and (6) of the following truth table, we see that "P iff Q" is logically equivalent to "$(P \Rightarrow Q) \wedge (Q \Rightarrow P)$".

(1)	(2)	(3)	(4)	(5)	(6)
P	Q	P iff Q	$P \Rightarrow Q$	$Q \vee (\sim P)$	$(P \Rightarrow Q) \wedge (Q \Rightarrow P)$
T	T	T	T	T	T
T	F	F	F	T	F
F	T	F	T	F	F
F	F	T	T	T	T

This is a very important fact. Many theorems of the form P iff Q are most conveniently proved by verifying separately that $P \Rightarrow Q$ and that $Q \Rightarrow P$.

D.3. Restricted quantifiers

Now that we have the logical connectives $\Rightarrow$ and $\wedge$ at our disposal, it is possible to introduce restricted quantifiers formally in terms of unrestricted ones. This enables one to obtain properties of the former from corresponding facts about the latter.

(See Exercise D.3.2 and Problems D.4.8 and D.4.9.)

D.3.1. Definition (Of restricted quantifiers). Let S be a set, and let $P(x)$ be an open sentence. We define $(\forall x \in S)\, P(x)$ to be true if and only if $(\forall x)\big((x \in S) \Rightarrow P(x)\big)$ is true; and we define $(\exists x \in S)\, P(x)$ to be true if and only if $(\exists x)\big((x \in S) \wedge P(x)\big)$ is true.

D.3.2. Exercise. Use the preceding definition and the fact (mentioned in Appendix A) that the order of unrestricted existential quantifiers does not matter to show that the order of restricted existential quantifiers does not matter. That is, show that if S and T are sets and $P(x,y)$ is an open sentence, then $(\exists x \in S)\,(\exists y \in T)\, P(x,y)$ holds if and only if $(\exists y \in T)\,(\exists x \in S)\, P(x,y)$ does.

D.4. Negation

If P is a sentence, then $\sim P$ (read "the NEGATION of P" or "the DENIAL of P" or just "not P") is the sentence whose truth values are the opposite of P.

P	$\sim P$
T	F
F	T

D.4.1. Example. It should be clear that the denial of the disjunction of two sentences P and Q is logically equivalent to the conjunction of their denials. If we were in doubt about the correctness of this, however, we could appeal to a truth table to prove that $\sim (P \vee Q)$ is logically equivalent to $\sim P \wedge \sim Q$.

(1)	(2)	(3)	(4)	(5)	(6)	(7)
P	Q	$P \vee Q$	$\sim (P \vee Q)$	$\sim P$	$\sim Q$	$(\sim P) \wedge (\sim Q)$
T	T	T	F	F	F	F
T	F	T	F	F	T	F
F	T	T	F	T	F	F
F	F	F	T	T	T	T

Columns (4) and (7) have the same truth values; that is, the denial of the disjunction of P and Q is logically equivalent to the conjunction of their denials. This result is one of DE MORGAN'S LAWS. The other is given as Problem D.4.2.

D.4.2. Problem (De Morgan's law). Use a truth table to show that $\sim (P \wedge Q)$ is logically equivalent to $(\sim P) \vee (\sim Q)$.

D.4.3. Problem. Obtain the result in Problem D.4.2 without using truth tables. *Hint.* Use Example D.4.1 together with the fact that a proposition P is logically equivalent to $\sim\sim P$. Start by writing $(\sim P) \vee (\sim Q)$ iff $\sim\sim ((\sim P) \vee (\sim Q))$.

D.4.4. Exercise. Let P and Q be sentences. Then $P \Rightarrow Q$ is logically equivalent to $Q \vee (\sim P)$.

One very important matter is the process of taking the negation of a quantified statement. Let $P(x)$ be an open sentence. If it is not the case that $P(x)$ holds for every x, then it must fail for some x, and conversely. That is, $\sim (\forall x)P(x)$ is logically equivalent to $(\exists x) \sim P(x)$.

Similarly, $\sim (\exists x)P(x)$ is logically equivalent to $(\forall x) \sim P(x)$. (If it is not the case that $P(x)$ is true for some x, then it must fail for all x, and conversely.)

D.4.5. Example. In Chapter 3 we define a real valued function f on the real line to be continuous provided that

$$(\forall a)(\forall \epsilon)(\exists \delta)(\forall x)\, |x - a| < \delta \Rightarrow |f(x) - f(a)| < \epsilon.$$

How does one prove that a particular function f is *not* continuous? *Answer*. Find numbers a and ϵ such that for every δ it is possible to find an x such that $|f(x) - f(a)| \geq \epsilon$ and $|x - a| < \delta$. To see that this is in fact what we must do, notice that each pair of consecutive lines in the following argument are logically equivalent.

$$\sim [(\forall a)(\forall \epsilon)(\exists \delta)(\forall x) \ |x - a| < \delta \Rightarrow |f(x) - f(a)| < \epsilon];$$
$$(\exists a) \sim [(\forall \epsilon)(\exists \delta)(\forall x) \ |x - a| < \delta \Rightarrow |f(x) - f(a)| < \epsilon];$$
$$(\exists a)(\exists \epsilon) \sim [(\exists \delta)(\forall x) \ |x - a| < \delta \Rightarrow |f(x) - f(a)| < \epsilon];$$
$$(\exists a)(\exists \epsilon)(\forall \delta) \sim [(\forall x) \ |x - a| < \delta \Rightarrow |f(x) - f(a)| < \epsilon];$$
$$(\exists a)(\exists \epsilon)(\forall \delta)(\exists x) \sim [\ |x - a| < \delta \Rightarrow |f(x) - f(a)| < \epsilon];$$
$$(\exists a)(\exists \epsilon)(\forall \delta)(\exists x) \sim [(|f(x) - f(a)| < \epsilon) \lor \sim (|x - a| < \delta)];$$
$$(\exists a)(\exists \epsilon)(\forall \delta)(\exists x) [\sim (\ |f(x) - f(a)| < \epsilon) \land \sim\sim (|x - a| < \delta)];$$
$$(\exists a)(\exists \epsilon)(\forall \delta)(\exists x) \quad [(\ |f(x) - f(a)| \geq \epsilon) \land (|x - a| < \delta)].$$

To obtain the third line from the end, use Exercise D.4.4; the penultimate line is a consequence of Example D.4.1; and the last line makes use of the obvious fact that a sentence P is always logically equivalent to $\sim\sim P$.

D.4.6. Problem. Two students, Smith and Jones, are asked to prove a mathematical theorem of the form, "If P then if Q then R." Smith assumes that Q is a consequence of P and tries to prove R. Jones assumes both P and Q are true and tries to prove R. Is either of these students doing the right thing? Explain carefully.

D.4.7. Problem. The CONTRAPOSITIVE of the implication $P \Rightarrow Q$ is the implication $(\sim Q) \Rightarrow (\sim P)$. Without using truth tables or assigning truth values, show that an implication is logically equivalent to its contrapositive. (This is a very important fact. In many cases when you are asked to prove a theorem of the form $P \Rightarrow Q$, rather than assuming P and proving Q you will find it easier to assume that Q is false and conclude that P must also be false.) *Hint.* Use Exercise D.4.4. You may also use the obvious facts that disjunction is a commutative operation ($P \lor Q$ is logically equivalent to $Q \lor P$) and that P is logically equivalent to $\sim\sim P$.)

D.4.8. Problem. Use the formal definition of restricted quantifiers given in section D.3 together with the fact mentioned in Appendix A that the order of unrestricted universal quantifiers does not matter to show that the order of restricted universal quantifiers does not matter. That is, show that if S and T are sets and $P(x, y)$ is an open sentence, then $(\forall x \in S)(\forall y \in T)P(x, y)$ holds if and only if $(\forall y \in T)(\forall x \in S)P(x, y)$ does.

D.4.9. Problem. Let S be a set, and let $P(x)$ be an open sentence. Show that

(a) $\sim (\forall x \in S)P(x)$ if and only if $(\exists x \in S) \sim P(x)$.

(b) $\sim (\exists x \in S)P(x)$ if and only if $(\forall x \in S) \sim P(x)$.

Hint. Use the corresponding facts (given in the two paragraphs following Exercise D.4.4) for unrestricted quantifiers.

Writing mathematics

E.1. Proving theorems

Mathematical results are called *theorems*—or *propositions* or *lemmas* or *corollaries* or *examples*. All of these are intended to be mathematical facts. The different words reflect only a difference of emphasis. Theorems are more important than propositions. A lemma is a result made use of in a (usually more important) subsequent result. The German word for lemma is particularly suggestive: "*Hilfsatz*", meaning "helping statement". A corollary (to a theorem or proposition) is an additional result we get (almost) for free. All of these results are typically packaged in the form "If P, then Q". The assertion P is the *hypothesis* (or *premise* or *assumption* or *supposition*). The assertion Q is the *conclusion*. Notice that the result "Every object of type A is of type B" is in this form. It can be rephrased as "If x is an object of type A, then x is of type B."

The statements P and Q themselves may be complicated conjunctions, or disjunctions, or conditionals of simpler statements. One common type of theorem, for example, is "If P_1, P_2, ..., and P_m, then Q_1, Q_2, ..., and Q_n." (Personally, I think such a theorem is clumsily stated if m and n turn out to be large.)

A *proof* of a result is a sequence of statements, each with justification, that leads to the conclusion(s) of the desired result. The statements that constitute a proof may be definitions or hypotheses or statements that result from applying to previous steps of the proof a valid rule of inference. *Modus ponens* is the basic rule of inference. It says that if you know a proposition P and you also know that P implies Q, then you can conclude that Q is true. Another important rule of inference (sometimes called "universal instantiation") is that if you know a proposition $P(x)$ to be true for every x in a set S and you know that a is a member of S, then you can conclude that $P(a)$ is true.

Other rules of inference can be derived from *modus ponens*. Let's look at an example. Certainly, if we know that the proposition $P \wedge Q$ is true, we should be able to conclude that P is true. The reason is simple: we know (or can easily check) that

$$(84) \qquad\qquad (P \wedge Q) \Rightarrow P$$

is a tautology (true for all truth values of P and Q). Since $P \wedge Q$ is known to be true, P follows from (84) by *modus ponens*. No attempt is made here to list every rule of inference that it is appropriate to use. Most of them should be entirely obvious. For those that are not, truth tables may help. (As an example consider Problem G.1.13: If the product of two numbers x and y is zero, then either x or y must be zero. Some students feel obligated to prove two things: that if $xy = 0$ and $x \neq 0$, then $y = 0$ **AND** that if $xy = 0$ and $y \neq 0$, then $x = 0$. Examination of truth tables shows that this is not necessary.)

A proof in which you start with the hypotheses and reason until you reach the conclusion is a DIRECT PROOF. There are two other proof formats, which are known as INDIRECT PROOFS. The first comes about by observing that the proposition $\sim Q \Rightarrow \sim P$ is logically equivalent to $P \Rightarrow Q$. (We say that $\sim Q \Rightarrow \sim P$ is the CONTRAPOSITIVE of $P \Rightarrow Q$.) To prove that P implies Q, it suffices to assume that Q is false and prove, using this assumption that P is false. Some find it odd that to prove something is true, one starts by assuming it to be false. A slight variant of this is the PROOF BY CONTRADICTION. Here, to prove that P implies Q, assume two things: that P is true and that Q is false. Then attempt to show that these lead to a contradiction. We like to believe that the mathematical system we work in is consistent (although we know we can't prove it), so when an assumption leads us to a contradiction, we reject it. Thus in a proof by contradiction, when we find that P and $\sim Q$ can't both be true, we conclude that if P is true Q must also be true.

E.1.1. Problem. Prove that in an inconsistent system, everything is true. That is, prove that if P and Q are propositions and that if both P and $\sim P$ are true, then Q is true. *Hint.* Consider the proposition $(P \wedge \sim P) \Rightarrow Q$.

E.1.2. Problem. What is wrong with the following proof that 1 is the largest natural number.

> Let N be the largest natural number. Since N is a natural number, so is N^2. We see that $N^2 = N \cdot N \geq N \cdot 1 = N$. Clearly, the reverse inequality $N^2 \leq N$ holds because N is the largest natural number. Thus $N^2 = N$. This equation has only two solutions: $N = 0$ and $N = 1$. Since 0 is not a natural number, we have $N = 1$. That is, 1 is the largest natural number.

E.2. Checklist for writing mathematics

After you have solved a problem or discovered a counterexample or proved a theorem, there arises the problem of writing up your result. You want to do this in a way that will be as clear and as easy to digest as possible. Check your work against the following list of suggestions.

(a) *Have you clearly stated the problem you are asked to solve or the result you are trying to prove?*

> Have an audience in mind. Write to someone. And don't assume the person you are writing to remembers the problem. He or she may have gone on vacation or been fired; or maybe he or she just has a bad memory. You need not include every detail of the problem, but there should be enough explanation that a person not familiar with the situation can understand what you are talking (writing) about.

(b) *Have you included a paragraph at the beginning explaining the method you are going to use to address the problem?*

> No one is happy being thrown into a sea of mathematics with no clue as to what is going on or why. Be nice. Tell the reader what you are doing, what steps you intend to take, and what advantages you see to your particular approach to the problem.

(c) *Have you defined all the variables you use in your writeup?*

> *Never* be so rude as to permit a symbol to appear that has not been properly introduced. You may have a mental picture of a triangle with vertices labeled A, B, and C. When you use those letters, no one will know what they stand for unless you tell them. (Even if you have included an appropriately labeled graph, still tell the reader *in the text* what the letters denote.) Similarly, you may be consistent in always using the letter j to denote a natural number. But how would you expect the reader to know?
>
> It is standard practice to italicize variables so that they can be easily distinguished from regular text.

(d) *Is the logic of your report entirely clear and entirely correct?*

> It is an unfortunate fact of life that the slightest error in logic can make a "solution" to a problem totally worthless. It is also unfortunate that even a technically correct argument can be so badly expressed that no one will believe it.

(e) *In your writeup are your mathematical symbols and mathematical terms all correctly used in a standard fashion? And are all abbreviations standard?*

> Few things can make mathematics more confusing than misused or eccentrically used symbols. Symbols should *clarify* arguments not create yet another level of difficulty. By the way, symbols such as "$=$" and "$<$" are used *only* in formulas. They are not substitutes for the words "equals" and "less than" in ordinary text. Logical symbols such as $\Rightarrow$ are rarely appropriate in mathematical exposition: write "If A, then B," not "$A \Rightarrow B$." Occasionally they may be used in displays.

(f) *Are the spelling, punctuation, diction, and grammar of your report all correct?*

(g) *Is every word, every symbol, and every equation part of a sentence? And is every sentence part of a paragraph?*

> For some reason this seems hard for many students. Scratchwork, of course, tends to be full of free floating symbols and formulas. When you write up a result, get rid of all this clutter. Keep only what is necessary for a logically complete report of your work. And make sure any formula you keep becomes (an intelligible) part of a sentence. Study how the author of any good mathematics text deals with the problem of incorporating symbols and formulas into text.

(h) *Does every sentence start correctly and end correctly?*

> Sentences start with capital letters. *Never* start a sentence with a number or with a mathematical or logical symbol. Every declarative sentence ends with a period. Other sentences may end with a question mark or (rarely) an exclamation mark.

(i) *Is the* function *of every sentence of your report clear?*

> Every sentence has a function. It may be a definition. Or it may be an assertion you are about to prove. Or it may be a consequence of the preceding statement. Or it may be a standard result your argument depends on. Or it may be a summary of what you have just proved. **Whatever function a sentence serves, that function should be entirely clear to your reader.**

(j) *Have you avoided all unnecessary clutter?*

> Mindless clutter is one of the worst enemies of clear exposition. No one wants to see all the details of your arithmetic or algebra or trigonometry or calculus. Either your reader knows this stuff and could do it more easily than read it, or he or she doesn't know it and will find it meaningless. In either case, get rid of it. If you solve an equation, for example, state what the solutions are; don't show how you used the quadratic formula to find them. Write only things that inform. Logical argument informs, reams of routine calculations do not. Be ruthless in rooting out useless clutter.

(k) *Is the word "any" used unambiguously? And is the order in which quantifiers act entirely clear?*

> Be careful with the word "any", especially when taking negations. It can be surprisingly treacherous. "Any" may be used to indicate universal quantification. (In the assertion, "It is true for any x that $x^2 - 1 = (x + 1)(x - 1)$", the word "any" means "every".) It may also be used for existential quantification. (In the question, "Does $x^2 + 2 = 2x$ hold for any x?" the word "any" means "for some". Also notice that answering "yes" to this question does not mean that you believe it is true that "$x^2 + 2 = 2x$ holds for any x".) Negations

are worse. (What does "It is not true that $x^2 + 2 = 2x$ for any x"
mean?) One good way to avoid trouble: don't use "any".

And be careful of word order. The assertions "Not every element of
A is an element of B" and "Every element of A is not an element
of B" say quite different things. I recommend avoiding the second
construction entirely. Putting (some or all) quantifiers at the end of
a sentence can be a dangerous business. (The statement, "$P(x)$ or
$Q(x)$ fails to hold for all x," has at least two possible interpretations.
So does "$x \neq 2$ for all $x \in A$," depending on whether we read $\neq$ as
"is not equal to" or as "is different from".)

E.3. Fraktur and Greek alphabets

Fraktur is a typeface widely used until the middle of the twentieth century in
Germany and a few other countries. In this text it is used to denote families of sets
and families of linear maps. In general it is not a good idea to try to reproduce
these letters in handwritten material or when writing on the blackboard. In such
cases English script letters are both easier to read and easier to write. Here is a list
of uppercase Fraktur letters.

$\mathfrak{A}$ (A) $\quad$ $\mathfrak{B}$ (B) $\quad$ $\mathfrak{C}$ (C) $\quad$ $\mathfrak{D}$ (D) $\quad$ $\mathfrak{E}$ (E) $\quad$ $\mathfrak{F}$ (F) $\quad$ $\mathfrak{G}$ (G)

$\mathfrak{H}$ (H) $\quad$ $\mathfrak{I}$ (I) $\quad$ $\mathfrak{J}$ (J) $\quad$ $\mathfrak{K}$ (K) $\quad$ $\mathfrak{L}$ (L) $\quad$ $\mathfrak{M}$ (M) $\quad$ $\mathfrak{N}$ (N)

$\mathfrak{O}$ (O) $\quad$ $\mathfrak{P}$ (P) $\quad$ $\mathfrak{Q}$ (Q) $\quad$ $\mathfrak{R}$ (R) $\quad$ $\mathfrak{S}$ (S) $\quad$ $\mathfrak{T}$ (T) $\quad$ $\mathfrak{U}$ (U)

$\mathfrak{V}$ (V) $\quad$ $\mathfrak{W}$ (W) $\quad$ $\mathfrak{X}$ (X) $\quad$ $\mathfrak{Y}$ (Y) $\quad$ $\mathfrak{Z}$ (Z)

The following is a list of standard Greek letters used in mathematics. Notice
that in some cases both upper case and lower case are commonly used.

α (alpha)	β (beta)	Γ, γ (gamma)	Δ, δ (delta)	ϵ (epsilon)
ζ (zeta)	η (eta)	Θ, θ (theta)	ι (iota)	κ (kappa)
Λ, λ (lambda)	μ (mu)	ν (nu)	Ξ, ξ (xi)	Π, π (pi)
ρ (rho)	Σ, σ (sigma)	τ (tau)	Φ, ϕ (phi)	Ψ, ψ (psi)
χ (chi)	Ω, ω (omega)			

Set operations

F.1. Unions

Recall that if S and T are sets, then the UNION of S and T, denoted by $S \cup T$, is defined to be the set of all those elements x such that $x \in S$ or $x \in T$.

That is,

$$S \cup T := \{x : x \in S \text{ or } x \in T\} .$$

F.1.1. Example. If $S = [0, 3]$ and $T = [2, 5]$, then $S \cup T = [0, 5]$.

The operation of taking unions of sets has several essentially obvious properties. In the next proposition we list some of these.

F.1.2. Proposition. *Let S, T, U, and V be sets. Then*

(a) $S \cup (T \cup U) = (S \cup T) \cup U$ (*associativity*);

(b) $S \cup T = T \cup S$ (*commutativity*);

(c) $S \cup \emptyset = S$;

(d) $S \subseteq S \cup T'$;

(e) $S = S \cup T$ *if and only if $T \subseteq S$; and*

(f) *if $S \subseteq T$ and $U \subseteq V$, then $S \cup U \subseteq T \cup V$.*

We prove parts (a), (c), (d), and (e). Ordinarily, one would probably regard these results as too obvious to require proof. The arguments here are presented only to display some techniques used in writing formal proofs. Elsewhere in the text, references will not be given to this proposition when the facts (a)–(f) are used. When results are considered obvious, they may be mentioned but are seldom cited. The proofs of the remaining parts (b) and (f) are left as problems.

Proof of (a). A standard way to show that two sets are equal is to show that an element x belongs to one if and only if it belongs to the other. In the present case,

$$x \in S \cup (T \cup U) \text{ iff } x \in S \text{ or } x \in T \cup U$$
$$\text{iff } x \in S \text{ or } (x \in T \text{ or } x \in U)$$
$$\text{iff } (x \in S \text{ or } x \in T) \text{ or } x \in U$$
$$\text{iff } x \in S \cup T \text{ or } x \in U$$
$$\text{iff } x \in (S \cup T) \cup U.$$

Notice that the proof of the associativity of union $\cup$ depends on the associativity of "or" as a logical connective.

Since we are asked to show that two sets are equal, some persons feel it necessary to write a chain of equalities between sets:

$$S \cup (T \cup U) = \{x \colon x \in S \cup (T \cup U)\}$$
$$= \{x \colon x \in S \text{ or } x \in T \cup U\}$$
$$= \{x \colon x \in S \text{ or } (x \in T \text{ or } x \in U)\}$$
$$= \{x \colon (x \in S \text{ or } x \in T) \text{ or } x \in U\}$$
$$= \{x \colon x \in S \cup T \text{ or } x \in U\}$$
$$= \{x \colon x \in (S \cup T) \cup U\}$$
$$= (S \cup T) \cup U. \qquad \square$$

This second proof is virtually identical to the first; it is just a bit more cluttered. Try to avoid clutter; mathematics is hard enough without it.

Proof of (c). An element x belongs to $S \cup \emptyset$ if and only if $x \in S$ or $x \in \emptyset$. Since $x \in \emptyset$ is never true, $x \in S \cup \emptyset$ if and only if $x \in S$. That is, $S \cup \emptyset = S$. $\qquad \square$

Proof of (d). To prove that $S \subseteq S \cup T$, show that $x \in S$ implies $x \in S \cup T$. Suppose $x \in S$. Then it is certainly true that $x \in S$ or $x \in T$; that is, $x \in S \cup T$. $\qquad \square$

Proof of (e). First show that $S = S \cup T$ implies $T \subseteq S$. Then prove the converse—if $T \subseteq S$, then $S = S \cup T$. To prove that $S = S \cup T$ implies $T \subseteq S$, it suffices to prove the contrapositive. We suppose that $T \nsubseteq S$ and show that $S \neq S \cup T$. If $T \nsubseteq S$, then there is at least one element t in T that does not belong to S. Thus (by parts (d) and (b))

$$t \in T \subseteq T \cup S = S \cup T;$$

but $t \notin S$. Since t belongs to $S \cup T$ but not to S, these sets are not equal.

Now for the converse. Suppose $T \subseteq S$. Since we already know that $S \subseteq S \cup T$ (by part (d)), we need only show that $S \cup T \subseteq S$ in order to prove that the sets S and $S \cup T$ are identical. To this end suppose that $x \in S \cup T$. Then $x \in S$ or $x \in T \subseteq S$. In either case $x \in S$. Thus $S \cup T \subseteq S$. $\qquad \square$

F.1.3. Problem. Prove parts (b) and (f) of Proposition F.1.2.

On numerous occasions it is necessary for us to take the union of a large (perhaps infinite) family of sets. When we consider a family of sets (that is, a set whose members are themselves sets), it is important to keep one thing in mind. If x is a member of a set S and S is in turn a member of a family $\mathfrak{S}$ of sets, it does not follow that $x \in \mathfrak{S}$. For example, let $S = \{0, 1, 2\}$, $T = \{2, 3, 4\}$, $U = \{5, 6\}$, and $\mathfrak{S} = \{S, T, U\}$. Then 1 is a member of S and S belongs to $\mathfrak{S}$; but 1 is not a member of $\mathfrak{S}$ (because $\mathfrak{S}$ has only three members: S, T, and U).

F.1.4. Definition. Let $\mathfrak{S}$ be a family of sets. We define the UNION of the family $\mathfrak{S}$ to be the set of all x such that $x \in S$ for at least one set S in $\mathfrak{S}$. We denote the union of the family $\mathfrak{S}$ by $\bigcup \mathfrak{S}$ (or by $\bigcup_{S \in \mathfrak{S}} S$ or by $\bigcup \{S : S \in \mathfrak{S}\}$). Thus $x \in \bigcup \mathfrak{S}$, if and only if there exists $S \in \mathfrak{S}$ such that $x \in S$.

F.1.5. Notation. If $\mathfrak{S}$ is a finite family of sets $S_1, \ldots, S_n$, then we may write $\bigcup_{k=1}^n S_k$ or $S_1 \cup S_2 \cup \cdots \cup S_n$ for $\bigcup \mathfrak{S}$.

F.1.6. Example. Let $S = \{0, 1, 3\}$, $T = \{1, 2, 3\}$, $U = \{1, 3, 4, 5\}$, and $\mathfrak{S} = \{S, T, U\}$. Then

$$\bigcup \mathfrak{S} = S \cup T \cup U = \{0, 1, 2, 3, 4, 5\}.$$

The following very simple observations are worthy of note.

F.1.7. Proposition. *If $\mathfrak{S}$ is a family of sets and $T \in \mathfrak{S}$, then $T \subseteq \bigcup \mathfrak{S}$.*

Proof. If $x \in T$, then x belongs to at least one of the sets in $\mathfrak{S}$, namely T. $\square$

F.1.8. Proposition. *If $\mathfrak{S}$ is a family of sets and each member of $\mathfrak{S}$ is contained in a set U, then $\bigcup \mathfrak{S} \subseteq U$.*

Proof. Problem.

F.2. Intersections

F.2.1. Definition. Let S and T be sets. The INTERSECTION of S and T is the set of all x such that $x \in S$ and $x \in T$.

F.2.2. Example. If $S = [0, 3]$ and $T = [2, 5]$, then $S \cap T = [2, 3]$.

F.2.3. Proposition. *Let S, T, U, and V be sets. Then*

(a) $S \cap (T \cap U) = (S \cap T) \cap U$ *(associativity)*;

(b) $S \cap T = T \cap S$ *(commutativity)*;

(c) $S \cap \emptyset = \emptyset$;

(d) $S \cap T \subseteq S$;

(e) $S = S \cap T$ *if and only if* $S \subseteq T$;

(f) *if* $S \subseteq T$ *and* $U \subseteq V$, *then* $S \cap U \subseteq T \cap V$.

Proof. Problem.

There are two distributive laws for sets: union distributes over intersection (Proposition F.2.4 below) and intersection distributes over union (Proposition F.2.5).

F.2.4. Proposition. *Let S, T, and U be sets. Then*

$$S \cup (T \cap U) = (S \cup T) \cap (S \cup U).$$

Proof. Exercise. *Hint.* Use Problem D.1.4.

F.2.5. Proposition. *Let S, T, and U be sets. Then*
$$S \cap (T \cup U) = (S \cap T) \cup (S \cap U).$$

Proof. Problem.

Just as we may take the union of an arbitrary family of sets, we may also take its intersection.

F.2.6. Definition. Let $\mathfrak{S}$ be a family of sets. We define the INTERSECTION of the family $\mathfrak{S}$ to be the set of all x such that $x \in S$ for every S in $\mathfrak{S}$. We denote the intersection of $\mathfrak{S}$ by $\bigcap \mathfrak{S}$ (or by $\bigcap_{S \in \mathfrak{S}} S$ or by $\bigcap \{S : S \in \mathfrak{S}\}$).

F.2.7. Notation. If S is a finite family of sets $S_1, \ldots, S_n$, then we may write $\bigcap_{k=1}^{n} S_k$ or $S_1 \cap S_2 \cap \cdots \cap S_n$ for $\bigcap \mathfrak{S}$. Similarly, if $\mathfrak{S} = \{S_1, S_2, \ldots\}$, then we may write $\bigcap_{k=1}^{\infty} S_k$ or $S_1 \cap S_2 \cap \ldots$ for $\bigcap \mathfrak{S}$.

F.2.8. Example. Let $S = \{0, 1, 3\}$, $T = \{1, 2, 3\}$, $U = \{1, 3, 4, 5\}$, and $S = \{S, T, U\}$. Then
$$\bigcap S = S \cap T \cap U = \{1, 3\}.$$

Proposition F.2.4 may be generalized to say that union distributes over the intersection of an arbitrary family of sets. Similarly, there is a more general form of Proposition F.2.5 that says that intersection distributes over the union of an arbitrary family of sets. These two facts, which are stated precisely in the next two propositions, are known as GENERALIZED DISTRIBUTIVE LAWS.

F.2.9. Proposition. *Let T be a set, and let $\mathfrak{S}$ be a family of sets. Then*
$$T \cup \left(\bigcap \mathfrak{S} \right) = \bigcap \{T \cup S : S \in \mathfrak{S}\}.$$

Proof. Exercise.

F.2.10. Proposition. *Let T be a set, and let $\mathfrak{S}$ be a family of sets. Then*
$$T \cap \left(\bigcup \mathfrak{S} \right) = \bigcup \{T \cap S : S \in \mathfrak{S}\}.$$

Proof. Problem.

F.2.11. Definition. Sets S and T are said to be DISJOINT if $S \cap T = \emptyset$. More generally, a family $\mathfrak{S}$ of sets is a DISJOINT FAMILY (or a PAIRWISE DISJOINT FAMILY) if $S \cap T = \emptyset$ whenever S and T are distinct (that is, not equal) sets that belong to $\mathfrak{S}$.

CAUTION. Let $\mathfrak{S}$ be a family of sets. Do not confuse the following two statements.

(a) $\mathfrak{S}$ is a (pairwise) disjoint family.

(b) $\bigcap \mathfrak{S} = \emptyset$.

Certainly, if $\mathfrak{S}$ contains more than a single set, then (a) implies (b). But if $\mathfrak{S}$ contains three or more sets, the converse need not hold. For example, let $S = \{0, 1\}$, $T = \{3, 4\}$, $U = \{0, 2\}$, and $\mathfrak{S} = \{S, T, U\}$. Then $\mathfrak{S}$ is not a disjoint family (because $S \cap U$ is nonempty), but $\bigcap \mathfrak{S} = \emptyset$.

F.2.12. Example. Let S, T, U, and V be sets.

(a) Then $(S \cap T) \cup (U \cap V) \subseteq (S \cup U) \cap (T \cup V)$.

(b) Give an example to show that equality need not hold in (a).

Proof. Problem. *Hint.* Use Propositions F.1.2(d) and F.2.3(f) to show that $S \cap T$ and $U \cap V$ are contained in $(S \cup U) \cap (T \cup V)$. Then use Proposition F.1.2(f).

F.3. Complements

Recall that we regard all the sets with which we work in a particular situation as being subsets of some appropriate *universal* set. For each set S we define the COMPLEMENT of S, denoted by S^c, to be the set of all members of our universal set that do not belong to S. That is, we write $x \in S^c$ if and only if $x \notin S$.

F.3.1. Example. Let S be the closed interval $(-\infty, 3]$. If nothing else is specified, we think of this interval as being a subset of the real line $\mathbb{R}$ (our universal set). Thus S^c is the set of all x in $\mathbb{R}$ such that x is not less than or equal to 3. Thus S^c is the interval $(3, \infty)$.

F.3.2. Example. Let S be the set of all points (x, y) in the plane such that $x \geq 0$ and $y \geq 0$. Then S^c is the set of all points (x, y) in the plane such that either $x < 0$ or $y < 0$. That is,

$$S^c = \{(x, y) \colon x < 0\} \cup \{(x, y) \colon y < 0\}\,.$$

The two following propositions are DE MORGAN'S LAWS for sets. As you may expect, they are obtained by translating into the language of sets the facts of logic sharing that name. (See Example D.4.1 and Problem D.4.2.)

F.3.3. Proposition. *Let S and T be sets. Then*
$$(S \cup T)^c = S^c \cap T^c\,.$$

Proof. Exercise. *Hint.* Use Example D.4.1.

F.3.4. Proposition. *Let S and T be sets. Then*
$$(S \cap T)^c = S^c \cup T^c\,.$$

Proof. Problem.

Just as the distributive laws can be generalized to arbitrary families of sets, so too can De Morgan's laws. The complement of the union of a family is the intersection of the complements (Proposition F.3.5), and the complement of the intersection of a family is the union of the complements (Proposition F.3.6).

F.3.5. Proposition. *Let $\mathfrak{S}$ be a family of sets. Then*
$$\left(\bigcup \mathfrak{S}\right)^c = \bigcap \{S^c \colon S \in \mathfrak{S}\}\,.$$

Proof. Exercise.

F.3.6. Proposition. *Let $\mathfrak{S}$ be a family of sets. Then*
$$\left(\bigcap\mathfrak{S}\right)^c = \bigcup\{S^c : S \in \mathfrak{S}\}\,.$$

Proof. Problem.

F.3.7. Definition. If S and T are sets we define the COMPLEMENT OF T RELATIVE TO S, denoted by $S \setminus T$, to be the set of all x that belong to S but not to T. That is,
$$S \setminus T := S \cap T^c\,.$$
The operation $\setminus$ is usually called SET SUBTRACTION and $S \setminus T$ is read as "S minus T".

F.3.8. Example. Let $S = [0, 5]$ and $T = [3, 10]$. Then $S \setminus T = [0, 3)$.

It is a frequently useful fact that the union of two sets can be rewritten as a disjoint union (that is, the union of two disjoint sets).

F.3.9. Proposition. *Let S and T be sets. Then $S \setminus T$ and T are disjoint sets whose union is $S \cup T$.*

Proof. Exercise.

F.3.10. Exercise. Show that $(S \setminus T) \cup T = S$ if and only if $T \subseteq S$.

F.3.11. Problem. Let $S = (3, \infty)$, $T = (0, 10]$, $U = (-4, 5)$, $V = [-2, 8]$, and $\mathfrak{S} = \{S^c, T, U, V\}$.

(a) Find $\bigcup \mathfrak{S}$.

(b) Find $\bigcap \mathfrak{S}$.

F.3.12. Problem. If S, T, and U are sets, then
$$(S \cap T) \setminus U = (S \setminus U) \cap (T \setminus U)\,.$$

F.3.13. Problem. If S, T, and U are sets, then
$$S \cap (T \setminus U) = (S \cap T) \setminus (S \cap U)\,.$$

F.3.14. Problem. If S and T are sets, then $T \setminus S$ and $T \cap S$ are disjoint and
$$T = (T \setminus S) \cup (T \cap S)\,.$$

F.3.15. Problem. If S and T are sets, then $S \cap T = S \setminus (S \setminus T)$.

F.3.16. Definition. A family $\mathfrak{S}$ of sets COVERS (or is a COVER FOR or is a COVERING FOR) a set T if $T \subseteq \bigcup \mathfrak{S}$.

F.3.17. Problem. Find a family of open intervals that covers the set $\mathbb{N}$ of natural numbers and has the property that the sum of the lengths of the intervals is 1. *Hint.* $\sum_{k=1}^{\infty} 2^{-k} = 1$.

Arithmetic

G.1. The field axioms

The set $\mathbb{R}$ of real numbers is the cornerstone of calculus. It is remarkable that all of its properties can be derived from a very short list of axioms. We will not travel the rather lengthy road of deriving from these axioms all the properties (arithmetic of fractions, rules of exponents, etc.) of $\mathbb{R}$ that we use in this text. This journey, although interesting enough in itself, requires a substantial investment of time and effort. Instead we discuss briefly one standard set of axioms for $\mathbb{R}$ and, with the aid of these axioms, give sample derivations of some familiar properties of $\mathbb{R}$. In the present chapter we consider the first four axioms, which govern the operations on $\mathbb{R}$ of addition and multiplication. The name we give to the collective consequences of these axioms is ARITHMETIC.

G.1.1. Definition. A BINARY OPERATION $*$ on a set S is a rule that associates with each pair x and y of elements in S one and only one element $x * y$ in S. (More precisely, $*$ is a function from $S \times S$ into S; see Appendices K and N.)

The first four axioms say that the set $\mathbb{R}$ of real numbers under the binary operations of addition and multiplication (denoted, as usual, by $+$ and $\cdot$) form a FIELD. We will follow conventional practice by allowing xy as a substitute notation for $x \cdot y$.

G.1.2. Axiom (I). *The operations $+$ and $\cdot$ on $\mathbb{R}$ are associative (that is, $x + (y + z) = (x + y) + z$ and $x(yz) = (xy)z$ for all x, y, $z \in \mathbb{R}$) and commutative ($x + y = y + x$ and $xy = yx$ for all x, $y \in \mathbb{R}$).*

G.1.3. Axiom (II). *There exist distinct additive and multiplicative identities (that is, there are elements 0 and 1 in $\mathbb{R}$ with $1 \neq 0$ such that $x + 0 = x$ and $x \cdot 1 = x$ for all $x \in \mathbb{R}$).*

G.1.4. Axiom (III). *Every element x in $\mathbb{R}$ has an additive inverse (that is, a number $-x$ such that $x + (-x) = 0$), and every element $x \in \mathbb{R}$ different from 0 has a multiplicative inverse (that is, a number x^{-1} such that $x\,x^{-1} = 1$).*

G.1.5. Axiom (IV). *Multiplication distributes over addition (that is, $x(y + z) = xy + xz$ for all x, y, $z \in \mathbb{R}$).*

G.1.6. Example. Multiplication is not a binary operation on the set $\mathbb{R}' = \{x \in \mathbb{R}: x \neq -1\}$.

Proof. The numbers 2 and $-\frac{1}{2}$ belong to $\mathbb{R}'$, but their product does not. □

G.1.7. Example. Subtraction is a binary operation on the set $\mathbb{R}$ of real numbers, but it is neither associative nor commutative.

Proof. Problem.

G.1.8. Problem. Let $\mathbb{P}$ be the set of all real numbers x such that $x > 0$. On $\mathbb{P}$ define
$$x * y = \frac{xy}{x + y}.$$
Determine whether $*$ is a binary operation on $\mathbb{P}$. Determine whether $*$ is associative and whether it is commutative. Does $\mathbb{P}$ have an identity element with respect to $*$? (That is, is there a member e of $\mathbb{P}$ such that $x * e = x$ and $e * x = x$ for all x in $\mathbb{P}$?)

Subtraction and division are defined in terms of addition and multiplication by
$$x - y := x + (-y)$$
and, for $y \neq 0$,
$$\frac{x}{y} := xy^{-1}.$$

We use the familiar rule for avoiding an excess of parentheses: multiplication takes precedence over addition. Thus, for example, $wx + yz$ means $(wx) + (yz)$.

G.1.9. Problem. The rule given in Axiom IV is the LEFT DISTRIBUTIVE LAW. The RIGHT DISTRIBUTIVE LAW, $(x + y)z = xz + yz$, is also true. Use Axioms (I)–(IV) to prove it.

G.1.10. Exercise. Show that if x is a real number such that $x + x = x$, then $x = 0$. *Hint.* Simplify both sides of $(x + x) + (-x) = x + (x + (-x))$.

G.1.11. Problem. Show that the additive identity 0 annihilates everything in $\mathbb{R}$ under multiplication. That is, show that $0 \cdot x = 0$ for every real number x. *Hint.* Consider $(0 + 0)x$. Use Problem G.1.9 and Exercise G.1.10.

G.1.12. Exercise. Give a careful proof using only the axioms above that if w, x, y, and z are real numbers, then
$$(w + x) + (y + z) = z + (x + (y + w)).$$

Hint. Since we are to make explicit use of the associative law, be careful not to write expressions such as $w + x + (y + z)$. Another set of parentheses is needed to indicate the order of operations. Both $(w + x) + (y + z)$ and $w + (x + (y + z))$, for example, do make sense.

G.1.13. Problem. Show that if the product xy of two numbers is zero, then either $x = 0$ or $y = 0$. (Here the word "or" is used in its inclusive sense; both x and y may be 0. It is always used that way in mathematics.) *Hint.* Convince yourself that, as a matter of logic, it is enough to show that if y is not equal to 0, then x must be. Consider $(xy)y^{-1}$ and use Problem G.1.11.

G.2. Uniqueness of identities

Axiom (II) guarantees only the *existence* of additive and multiplicative identities 0 and 1. It is natural to inquire about their *uniqueness*. Could there be two real numbers that act as additive identities? That is, could we have numbers $0' \neq 0$ that satisfy

$$(85) \qquad\qquad x + 0 = x$$

and

$$(86) \qquad\qquad x + 0' = x$$

for all x in $\mathbb{R}$? The answer as you would guess is "no": there is only one additive identity in $\mathbb{R}$. The proof is very short.

G.2.1. Proposition. *The additive identity in $\mathbb{R}$ is unique.*

Proof. Suppose that the real numbers 0 and $0'$ satisfy (85) and (86) for all real numbers x. Then

$$
\begin{aligned}
0 &= 0 + 0' \\
  &= 0' + 0 \\
  &= 0'.
\end{aligned}
$$

The three equalities are justified, respectively, by (86), Axiom (I), and (85).   □

G.2.2. Proposition. *The multiplicative identity 1 on $\mathbb{R}$ is unique.*

Proof. Problem.

G.3. Uniqueness of inverses

The question of uniqueness also arises for inverses; only their existence is guaranteed by Axiom (III). Is it possible for a number to have more than one additive inverse? That is, if x is a real number, is it possible that there are two different numbers, say $-x$ and $\overline{x}$, such that the equations

$$(87) \qquad\qquad x + (-x) = 0 \qquad \text{and} \qquad x + \overline{x} = 0$$

both hold? The answer is "no".

G.3.1. Proposition. *Additive inverses in $\mathbb{R}$ are unique.*

Proof. Assume that the equations (87) are true. We show that $-x$ and $\bar{x}$ are the same number.

$$\begin{aligned}
\bar{x} &= \bar{x} + 0 \\
&= \bar{x} + (x + (-x)) \\
&= (\bar{x} + x) + (-x) \\
&= (x + \bar{x}) + (-x) \\
&= 0 + (-x) \\
&= (-x) + 0 \\
&= -x.
\end{aligned}$$
□

G.3.2. Problem. The proof of Proposition G.3.1 contains seven equal signs. Justify each one.

G.3.3. Problem. Prove that in $\mathbb{R}$, multiplicative inverses are unique.

G.3.4. Example. Knowing that identities and inverses are unique is helpful in deriving additional properties of the real numbers. For example, the familiar fact that

$$-(-x) = x$$

follows immediately from the equation

(88) $(-x) + x = 0.$

What Proposition G.3.1 tells us is that if $a + b = 0$, then b must be the additive inverse of a. So from (88) we conclude that x must be the additive inverse of $-x$; in symbols, $x = -(-x)$.

G.3.5. Problem. Show that if x is a nonzero real number, then

$$\left(x^{-1}\right)^{-1} = x.$$

G.4. Another consequence of uniqueness

We can use Proposition G.3.1 to show that in $\mathbb{R}$

(89) $-(x + y) = -x - y.$

Before looking at the proof of this assertion, it is well to note the two uses of the "$-$" sign on the right-hand side of (89). The first, attached to x, indicates the additive inverse of x; the second indicates subtraction. Thus $-x - y$ means $(-x) + (-y)$. The idea behind the proof is to add the right-hand side of (89) to $x + y$. If the result is 0, then the uniqueness of additive inverses (Proposition G.3.1), tells us that $-x - y$ is the additive inverse of $x + y$. And that is exactly what we get:

$$\begin{aligned}
(x + y) + (-x - y) &= (x + y) + ((-x) + (-y)) \\
&= (y + x) + ((-x) + (-y)) \\
&= y + (x + ((-x) + (-y))) \\
&= y + ((x + (-x)) + (-y)) \\
&= y + (0 + (-y)) \\
&= (y + 0) + (-y) \\
&= y + (-y) \\
&= 0.
\end{aligned}$$

G.4.1. Problem. Justify each step in the proof of equation (89).

G.4.2. Problem. Prove that if x and y are nonzero real numbers, then
$$(xy)^{-1} = y^{-1}x^{-1} \,.$$

G.4.3. Problem. Show that
$$(-1)x = -x$$
for every real number x. *Hint.* Use the uniqueness of additive inverses.

G.4.4. Problem. Show that
$$-(xy) = (-x)y = x(-y)$$
and that
$$(-x)(-y) = xy$$
for all x and y in $\mathbb{R}$. *Hint.* For the first equality add $(-x)y$ to xy.

G.4.5. Problem. Use the first four axioms for $\mathbb{R}$ to develop the rules for adding, multiplying, subtracting, and dividing fractions. Show for example that
$$\frac{a}{b} + \frac{c}{d} = \frac{ad + bc}{bd}$$
if b and d are not zero. (Remember that, by definition, $\frac{a}{b} + \frac{c}{d}$ is $ab^{-1} + cd^{-1}$ and $\frac{ad+bc}{bd}$ is $(ad + bc)(bd)^{-1}$.)

Order properties of $\mathbb{R}$

The second group of axioms are the order axioms. They concern a subset $\mathbb{P}$ of $\mathbb{R}$ (call this the set of STRICTLY POSITIVE numbers).

H.1.1. Axiom (V). *The set $\mathbb{P}$ is closed under addition and multiplication. (That is, if x and y belong to $\mathbb{P}$, so do $x + y$ and xy.)*

H.1.2. Axiom (VI). *For each real number x exactly one of the following is true: $x = 0$, $x \in \mathbb{P}$, or $-x \in \mathbb{P}$. This is the* axiom of trichotomy.

Define the relation $<$ on $\mathbb{R}$ by

$$x < y \text{ if and only if } y - x \in \mathbb{P}.$$

Also define $>$ on $\mathbb{R}$ by

$$x > y \text{ if and only if } y < x.$$

We write $x \leq y$ if $x < y$ or $x = y$, and $x \geq y$ if $y \leq x$.

H.1.3. Proposition. *On $\mathbb{R}$ the relation $<$ is transitive (that is, if $x < y$ and $y < z$, then $x < z$).*

Proof. If $x < y$ and $y < z$, then $y - x$ and $z - y$ belong to $\mathbb{P}$. Thus

$$
\begin{aligned}
z - x &= z + (-x) \\
&= (z + 0) + (-x) \\
&= (z + (y + (-y))) + (-x) \\
&= (z + ((-y) + y)) + (-x) \\
&= ((z + (-y)) + y) + (-x) \\
&= (z + (-y)) + (y + (-x)) \\
&= (z - y) + (y - x) \in \mathbb{P}.
\end{aligned}
$$

This shows that $x < z$. $\qquad\square$

H.1.4. Problem. Justify each of the seven equal signs in the proof of Proposition H.1.3.

H.1.5. Exercise. Show that a real number x belongs to the set $\mathbb{P}$ if and only if $x > 0$.

H.1.6. Proposition. *If $x > 0$ and $y < z$ in* $\mathbb{R}$*, then $xy < xz$.*

Proof. Exercise. *Hint.* Use Problem G.4.4.

H.1.7. Proposition. *If x, y, $z \in \mathbb{R}$ and $y < z$, then $x + y < x + z$.*

Proof. Problem. *Hint.* Use equation (89).

H.1.8. Proposition. *If $w < x$ and $y < z$, then $w + y < x + z$.*

Proof. Problem.

H.1.9. Problem. Show that $1 > 0$. *Hint.* Keep in mind that 1 and 0 are assumed to be distinct. (Look at the axiom concerning additive and multiplicative identities.) If 1 does not belong to $\mathbb{P}$, what can you say about the number -1? What about $(-1)(-1)$? Use Problem G.4.4.

H.1.10. Proposition. *If $x > 0$, then $x^{-1} > 0$.*

Proof. Problem.

H.1.11. Proposition. *If $0 < x < y$, then $1/y < 1/x$.*

Proof. Problem.

H.1.12. Proposition. *If $0 < w < x$ and $0 < y < z$, then $wy < xz$.*

Proof. Exercise.

H.1.13. Problem. Show that $x < 0$ if and only if $-x > 0$.

H.1.14. Problem. Show that if $y < z$ and $x < 0$, then $xz < xy$.

H.1.15. Problem. Show that $x < y$ if and only if $-y < -x$.

H.1.16. Problem. Suppose that x, $y \geq 0$ and $x^2 = y^2$. Show that $x = y$.

H.1.17. Problem. Show in considerable detail how the preceding results can be used to solve the inequality

$$\frac{5}{x + 3} < 2 - \frac{1}{x - 1}.$$

H.1.18. Problem. Let $\mathbb{C} = \{(a, b) \colon a, b \in \mathbb{R}\}$. On $\mathbb{C}$ define two binary operations $+$ and $\cdot$ by

$$(a, b) + (c, d) = (a + c, b + d)$$

and

$$(a, b) \cdot (c, d) = (ac - bd, ad + bc).$$

Show that $\mathbb{C}$ under these operations is a field. (That is, $\mathbb{C}$ satisfies Axioms (I)–(IV).) This is the field of COMPLEX NUMBERS.

Determine whether it is possible to make $\mathbb{C}$ into an ordered field. (That is, determine whether it is possible to choose a subset $\mathbb{P}$ of $\mathbb{C}$ that satisfies Axioms (V) and (VI).)

The axioms presented thus far define an ordered field. To obtain the particular ordered field $\mathbb{R}$ of real numbers, we require one more axiom. We assume that $\mathbb{R}$ is order complete; that is, $\mathbb{R}$ satisfies the *least upper bound axiom*. This axiom will be stated (and discussed in some detail) in Appendix J (in particular, see Axiom J.3.1).

There is a bit more to the axiomatization of $\mathbb{R}$ than we have indicated in the preceding discussion. For one thing, how do we know that the axioms are consistent? That is, how do we know that they will not yield a contradiction? For this purpose one constructs a MODEL for $\mathbb{R}$, that is, a concrete mathematical object that satisfies all the axioms for $\mathbb{R}$. One standard procedure is to define the positive integers in terms of sets: 0 is the empty set $\emptyset$, the number 1 is the set whose only element is 0, the number 2 is the set whose only element is 1, and so on. Using the positive integers, we construct the set $\mathbb{Z}$ of all integers $\ldots, -2, -1, 0, 1, 2, \ldots$. From these we construct the set $\mathbb{Q}$ of rational numbers (that is, numbers of the form p/q, where p and q are integers and $q \neq 0$). Finally, the reals are constructed from the rationals.

Another matter that requires attention is the use of the definite article in the expression "*the* real numbers". This makes sense only if the axioms are shown to be categorical; that is, if there is *essentially* only one model for the axioms. This turns out to be correct about the axioms for $\mathbb{R}$ given an appropriate technical meaning of "essentially"—but we will not pursue this matter here. More about both the construction of the reals and their uniqueness can be found in [**Spi67**].

Natural numbers
and mathematical induction

The *principle of mathematical induction* is predicated on a knowledge of the natural numbers, which we introduce in this section. In what follows it is helpful to keep in mind that the axiomatic development of the real numbers sketched in the preceding chapter says nothing about the natural numbers; in fact, it provides explicit names for only three real numbers: 0, 1, and -1.

I.1.1. Definition. A collection J of real numbers is INDUCTIVE if

(a) $1 \in J$, and

(b) $x + 1 \in J$ whenever $x \in J$.

I.1.2. Example. The set $\mathbb{R}$ is itself inductive; so are the intervals $(0, \infty)$, $[-1, \infty)$, and $[1, \infty)$.

I.1.3. Proposition. *Let $\mathfrak{A}$ be a family of inductive subsets of $\mathbb{R}$. Then $\bigcap \mathfrak{A}$ is inductive.*

Proof. Exercise.

I.1.4. Definition. Let J be the family of all inductive subsets of $\mathbb{R}$. Define

$$\mathbb{N} := \bigcap J.$$

We call $\mathbb{N}$ the set of NATURAL NUMBERS.

Notice that according to Proposition I.1.3, the set $\mathbb{N}$ is inductive. It is the *smallest* inductive set, in the sense that it is contained in *every* inductive set. The elements of $\mathbb{N}$ have standard names. Define $2 := 1 + 1$. Since 1 belongs to $\mathbb{N}$ and $\mathbb{N}$ is inductive, 2 belongs to $\mathbb{N}$. Define $3 := 2 + 1$. Since 2 belongs to $\mathbb{N}$, so does 3. Define $4 := 3 + 1$; etc.

I.1.5. Definition. The set of INTEGERS, denoted by $\mathbb{Z}$, is defined to be

$$-\mathbb{N} \cup \{0\} \cup \mathbb{N},$$

where $-\mathbb{N} := \{-n : n \in \mathbb{N}\}$.

The next proposition is practically obvious, but as it is an essential ingredient of several subsequent arguments (e.g., Problem I.1.16), we state it formally.

I.1.6. Proposition. *If $n \in \mathbb{N}$, then $n \geq 1$.*

Proof. Since the set $[1, \infty)$ is inductive, it must contain $\mathbb{N}$.   □

The observation made previously that $\mathbb{N}$ is the smallest inductive set, clearly implies that no proper subset of $\mathbb{N}$ can be inductive. This elementary fact has a rather fancy name: it is the PRINCIPLE OF MATHEMATICAL INDUCTION.

I.1.7. Theorem (Principle of mathematical induction)**.** *Every inductive subset of $\mathbb{N}$ equals $\mathbb{N}$.*

By spelling out the definition of "inductive set" in the preceding theorem, we obtain a longer, perhaps more familiar, statement of the principle of mathematical induction.

I.1.8. Corollary. *If S is a subset of $\mathbb{N}$ that satisfies*

(a) $1 \in S$ *and*

(b) $n + 1 \in S$ *whenever $n \in S$,*

then $S = \mathbb{N}$.

Perhaps even more familiar is the version of the preceding that refers to "a proposition (or assertion, or statement) concerning the natural number n".

I.1.9. Corollary. *Let $P(n)$ be a proposition concerning the natural number n. If $P(1)$ is true and if $P(n + 1)$ is true whenever $P(n)$ is true, then $P(n)$ is true for all $n \in \mathbb{N}$.*

Proof. In Corollary I.1.8 let $S = \{n \in \mathbb{N} : P(n) \text{ is true}\}$. Then $1 \in S$ and $n + 1$ belongs to S whenever n does. Thus $S = \mathbb{N}$. That is, $P(n)$ is true for all $n \in \mathbb{N}$.   □

I.1.10. Exercise. Use mathematical induction to prove the following assertion: The sum of the first n natural numbers is $\frac{1}{2}n(n + 1)$. *Hint.* Recall that if p and q are integers with $p \leq q$ and if $c_p, c_{p+1}, \ldots, c_q$ are real numbers, then the sum $c_p + c_{p+1} + \cdots + c_q$ may be denoted by $\sum_{k=p}^{q} c_k$. Using this summation notation, we may write the desired conclusion as $\sum_{k=1}^{n} k = \frac{1}{2}n(n + 1)$.

It is essentially obvious that there is nothing crucial about starting inductions with $n = 1$. Let m be any integer, and let $P(n)$ be a proposition concerning integers $n \geq m$. If we prove that $P(m)$ is true and that $P(n + 1)$ is true whenever $P(n)$ is true and $n \geq m$, then we may conclude that $P(n)$ is true for all $n \geq m$. (*Proof.* Apply Corollary I.1.9 to the proposition Q where $Q(n) = P(n + m - 1)$.)

I.1.11. Problem. Let $a, b \in \mathbb{R}$ and $m \in \mathbb{N}$. Then

$$a^m - b^m = (a - b) \sum_{k=0}^{m-1} a^k b^{m-k-1} \,.$$

Hint. Multiply out the right-hand side. This is not an induction problem.

I.1.12. Problem. If $r \in \mathbb{R}$, $r \neq 1$, and $n \in \mathbb{N}$, then

$$\sum_{k=0}^{n} r^k = \frac{1 - r^{n+1}}{1 - r} \,.$$

Hint. Use Problem I.1.11

I.1.13. Definition. Let $m, n \in \mathbb{N}$. We say that m is a FACTOR of n if $n/m \in \mathbb{N}$. Notice that 1 and n are always factors of n; these are the TRIVIAL FACTORS of n. The number n is COMPOSITE if $n > 1$ and if it has at least one nontrivial factor. (For example, 20 has several nontrivial factors: 2, 4, 5, and 10. Therefore, it is composite.) If $n > 1$ and it is not composite, it is PRIME. (For example, 7 is prime; its only factors are 1 and 7.)

I.1.14. Problem. Prove that if $n \in \mathbb{N}$ and $2^n - 1$ is prime, then so is n. (*Hint.* Prove the contrapositive. Use Problem I.1.11.) Illustrate your technique by finding a nontrivial factor of $2^{403} - 1$.

I.1.15. Problem. Show that $\sum_{k=1}^{n} k^2 = \frac{1}{6}n(n+1)(2n+1)$ for every $n \in \mathbb{N}$.

I.1.16. Problem. Use only the definition of $\mathbb{N}$ and the results given in this section to prove (a) and (b).

(a) If $m, n \in \mathbb{N}$ and $m < n$, then $n - m \in \mathbb{N}$. *Hint.* One proof of this involves an induction within an induction. Restate the assertion to be proved as follows. For every $m \in \mathbb{N}$ it is true that

(90)   if $n \in \mathbb{N}$ and $n > m$, then $n - m \in \mathbb{N}$.

Prove this assertion by induction on m. That is, show that (90) holds for $m = 1$, and then show that it holds for $m = k + 1$, provided that it holds for $m = k$. To show that (90) holds for $m = 1$, prove that the set

$$J := \{1\} \cup \{n \in \mathbb{N} \colon n - 1 \in \mathbb{N}\}$$

is an inductive set.

(b) Let $n \in \mathbb{N}$. There does not exist a natural number k such that $n < k < n + 1$. *Hint.* Argue by contradiction. Use part (a).

I.1.17. Problem (The binomial theorem). If $x, y \in \mathbb{R}$ and $n \in \mathbb{N}$, then

$$(x + y)^n = \sum_{k=0}^{n} \binom{n}{k} x^{n-k} y^k \,.$$

Hint. Use induction. Recall that $0! = 1$, that $n! = n(n-1)(n-2) \cdots 1$ for $n \in \mathbb{N}$, and that

$$\binom{n}{k} = \frac{n!}{k!(n-k)!}$$

for $0 \leq k \leq n$.

The final result of this section is the PRINCIPLE OF WELL-ORDERING. It asserts that every nonempty subset of the natural numbers has a smallest element.

I.1.18. Proposition. *If $\emptyset \neq K \subseteq \mathbb{N}$, then there exists $a \in K$ such that $a \leq k$ for every $k \in K$.*

Proof. Exercise. *Hint.* Assume that K has no smallest member. Show that this implies $K = \emptyset$ by proving that the set

$$J \equiv \{n \in \mathbb{N} \colon n < k \text{ for every } k \in K\}$$

is inductive. In the inductive step problem, Problem I.1.16(b) may prove useful.

I.1.19. Problem. (This slight modification of the principle of mathematical induction is occasionally useful.) Let $P(n)$ be a proposition concerning the natural number n. If $P(n)$ is true whenever $P(k)$ is true for all $k \in \mathbb{N}$ such that $k < n$, then $P(n)$ is true for all n. *Hint.* Use the well-ordering principle.

Least upper bounds
and greatest lower bounds

The last axiom for the set $\mathbb{R}$ of real numbers is the *least upper bound axiom*. Before stating it we make some definitions.

J.1. Upper and lower bounds

J.1.1. Definition. A number u is an UPPER BOUND for a set A of real numbers if $u \geq a$ for every $a \in A$. If the set A has at least one upper bound, it is said to be BOUNDED ABOVE. Similarly, v is a LOWER BOUND for A if $v \leq a$ for every $a \in A$, and a set with at least one lower bound is BOUNDED BELOW. The set A is BOUNDED if it is bounded both above and below. (Perhaps it should be emphasized that when we say, for example, that A *has* an upper bound, we mean only that there is a real number u that is greater than or equal to each member of A; we do *not* mean that u necessarily belongs to A—although of course it may.)

J.1.2. Example. The set $A = \{x \in \mathbb{R} \colon |x - 2| < 5\}$ is bounded.

Proof. Problem.

J.1.3. Example. The open interval $(-1, 1)$ has infinitely many upper bounds. In fact, any set that is bounded above has infinitely many upper bounds.

Proof. Problem.

J.1.4. Example. The set $A = \{x \in \mathbb{R} \colon x^3 - x \leq 0\}$ is not bounded.

Proof. Problem.

J.2. Least upper and greatest lower bounds

J.2.1. Definition. A number ℓ is the SUPREMUM (or LEAST UPPER BOUND) of a set A if

(a) ℓ is an upper bound for A, and

(b) $\ell \leq u$ whenever u is an upper bound for A.

If ℓ is the least upper bound of A, we write $\ell = \sup A$. Similarly, a lower bound g of a set is the INFIMUM (or GREATEST LOWER BOUND) of a set A if it is greater than or equal to every lower bound of the set. If g is the greatest lower bound of A, we write $g = \inf A$.

If A is not bounded above (and consequently, $\sup A$ does not exist), then it is common practice to write $\sup A = \infty$. Similarly, if A is not bounded below, we write $\inf A = -\infty$.

CAUTION. The expression "$\sup A = \infty$" does *not* mean that $\sup A$ exists and equals some object called ∞; it *does* mean that A is not bounded above.

It is clear that least upper bounds and greatest lower bounds, when they exist, are unique. If, for example, ℓ and m are both least upper bounds for a set A, then $\ell \leq m$ and $m \leq \ell$; so $\ell = m$.

J.2.2. Definition. Let $A \subseteq \mathbb{R}$. If there exists a number M *belonging to* A such that $M \geq a$ for every $a \in A$, then this element is the LARGEST ELEMENT (or GREATEST ELEMENT, or MAXIMUM) of A. We denote this element (when it exists) by $\max A$.

Similarly, if there exists a number m *belonging to* A such that $m \leq a$ for every $a \in A$, then this element is the SMALLEST ELEMENT (or LEAST ELEMENT, or MINIMUM) of A. We denote this element (when it exists) by $\min A$.

J.2.3. Example. Although the largest element of a set (when it exists) is always a least upper bound, the converse is not true. It is possible for a set to have a least upper bound but no maximum. The interval $(-2, 3)$ has a least upper bound (namely, 3), but it has no largest element.

J.2.4. Example. If $A = \{x \in \mathbb{R} : |x| < 4\}$, then $\inf A = -4$ and $\sup A = 4$. But A has no maximum or minimum. If $B = \{|x| : x < 4\}$, then $\inf B = 0$ but $\sup B$ does not exist. (It is correct to write $\sup B = \infty$.) Furthermore, B has a smallest element, $\min B = 0$, but no largest element.

Incidentally, the words "maximum", "supremum", "minimum", and "infimum" are all singular. The preferred plurals are, respectively, "maxima", "suprema", "minima", and "infima".

J.2.5. Problem. For each of the following sets find the least upper bound and the greatest lower bound (if they exist).

(a) $A = \{x \in \mathbb{R} : |x - 3| < 5\}$.

(b) $B = \{|x - 3| : x < 5\}$.

(c) $C = \{|x - 3| : x > 5\}$.

J.2.6. Problem. Show that the set $\mathbb{P}$ of positive real numbers has an infimum but no smallest element.

J.2.7. Exercise. Let $f(x) = x^2 - 4x + 3$ for every $x \in \mathbb{R}$, let $A = \{x \colon f(x) < 3\}$, and let $B = \{f(x) \colon x < 3\}$.

(a) Find $\sup A$ and $\inf A$ (if they exist).

(b) Find $\sup B$ and $\inf B$ (if they exist).

J.2.8. Example. Let $A = \left\{ x \in \mathbb{R} \colon \frac{5}{x-3} - 3 \geq 0 \right\}$. Then $\sup A = \max A = 14/3$, $\inf A = 3$, and $\min A$ does not exist.

Proof. Problem.

J.2.9. Example. Let $f(x) = -\frac{1}{2} + \sin x$ for $x \in \mathbb{R}$.

(a) If $A = \{f(x) \colon x \in \mathbb{R}\}$, then $\inf A = -\frac{3}{2}$ and $\sup A = \frac{1}{2}$.

(b) If $B = \{|f(x)| \colon x \in \mathbb{R}\}$, then $\inf B = 0$ and $\sup B = \frac{3}{2}$.

Proof. Problem.

J.2.10. Example. Let $f(x) = x^{20} - 2$ for $0 < x < 1$.

(a) If $A = \{f(x) \colon 0 < x < 1\}$, then $\inf A = -2$ and $\sup A = -1$.

(b) If $B = \{|f(x)| \colon 0 < x < 1\}$, then $\inf B = 1$ and $\sup B = 2$.

Proof. Problem.

J.2.11. Example. Let $f(x) = x^{20} - \frac{1}{4}$ for $0 \leq x \leq 1$.

(a) If $A = \{f(x) \colon 0 \leq x \leq 1\}$, then $\inf A = -\frac{1}{4}$ and $\sup A = \frac{3}{4}$.

(b) If $B = \{|f(x)| \colon 0 \leq x \leq 1\}$, then $\inf B = 0$ and $\sup B = \frac{3}{4}$.

Proof. Problem.

J.2.12. Problem. Let $f(x) = -4x^2 - 4x + 3$ for every $x \in \mathbb{R}$, let $A = \{x \in \mathbb{R} \colon f(x) > 0\}$, and let $B = \{f(x) \colon -2 < x < 2\}$.

(a) Find $\sup A$ and $\inf A$ (if they exist).

(b) Find $\sup B$ and $\inf B$ (if they exist).

J.2.13. Problem. For $c > 0$ define a function f on $[0, \infty)$ by $f(x) = x\,e^{-cx}$. Find $\sup\{|f(x)| \colon x \geq 0\}$.

J.2.14. Problem. For each $n = 1, 2, 3, \ldots$, define a function f_n on $\mathbb{R}$ by $f_n(x) = \dfrac{x}{1 + nx^2}$. For each $n \in \mathbb{N}$, let $A_n = \{f_n(x) \colon x \in \mathbb{R}\}$. For each n find $\inf A_n$ and $\sup A_n$.

J.3. The least upper bound axiom for $\mathbb{R}$

We now state our last assumption concerning the set $\mathbb{R}$ of real numbers. This is the *least upper bound* (or *order completeness*) *axiom*.

J.3.1. Axiom (VII). *Every nonempty set of real numbers that is bounded above has a least upper bound.*

J.3.2. Notation. If A and B are subsets of $\mathbb{R}$ and $\alpha \in \mathbb{R}$, then

$$A + B := \{a + b \colon a \in A \text{ and } b \in B\},$$
$$AB := \{ab \colon a \in A \text{ and } b \in B\},$$
$$\alpha B := \{\alpha\}B = \{\alpha b \colon b \in B\}, \text{ and}$$
$$-A := (-1)A = \{-a \colon a \in A\}.$$

J.3.3. Proposition. *If A is a nonempty subset of $\mathbb{R}$ that is bounded below, then A has a greatest lower bound. In fact,*

$$\inf A = -\sup(-A).$$

Proof. Let b be a lower bound for A. Then since $b \leq a$ for every $a \in A$, we see that $-b \geq -a$ for every $a \in A$. This says that $-b$ is an upper bound for the set $-A$. By the *least upper bound axiom* (Axiom J.3.1) the set $-A$ has a least upper bound, say ℓ. We show that $-\ell$ is the greatest lower bound for A. Certainly, it is a lower bound [$\ell \geq -a$ for all $a \in A$ implies $-\ell \leq a$ for all $a \in A$].

Again letting b be an arbitrary lower bound for A, we see, as above, that $-b$ is an upper bound for $-A$. Now $\ell \leq -b$, since ℓ is the least upper bound for $-A$. Thus $-\ell \geq b$. We have shown

$$\inf A = -\ell = -\sup(-A). \qquad \square$$

J.3.4. Corollary. *If A is a nonempty set of real numbers that is bounded above, then*

$$\sup A = -\inf(-A).$$

Proof. If A is bounded above, then $-A$ is bounded below. By the preceding proposition $\inf(-A) = -\sup A$. $\qquad \square$

J.3.5. Proposition. *Suppose $\emptyset \neq A \subseteq B \subseteq \mathbb{R}$.*

(a) *If B is bounded above, so is A and $\sup A \leq \sup B$.*

(b) *If B is bounded below, so is A and $\inf A \geq \inf B$.*

Proof. Problem.

J.3.6. Proposition. *If A and B are nonempty subsets of $\mathbb{R}$ that are bounded above, then $A + B$ is bounded above and*

$$\sup(A + B) = \sup A + \sup B.$$

Proof. Problem. *Hint.* It is easy to show that if ℓ is the least upper bound for A and m is the least upper bound for B, then $\ell + m$ is *an* upper bound for $A + B$.

One way to show that $\ell + m$ is the *least* upper bound for $A + B$, is to argue by contradiction. Suppose there exists an upper bound u for $A + B$ that is strictly less than $\ell + m$. Find numbers a in A and b in B that are close enough to ℓ and m, respectively, so that their sum exceeds u.

An even nicer proof results from taking u to be an arbitrary upper bound for $A + B$ and proving directly that $\ell + m \leq u$. Start by observing that $u - b$ is an upper bound for A for every $b \in B$, and consequently $l \leq u - b$ for every $b \in B$.

J.3.7. Proposition. *If A and B are nonempty subsets of $[0, \infty)$ that are bounded above, then the set AB is bounded above and*

$$\sup(AB) = (\sup A)(\sup B).$$

Proof. Exercise. *Hint.* The result is trivial if $A = \{0\}$ or if $B = \{0\}$. So suppose that both A and B contain elements strictly greater than 0, in which case $\ell := \sup A > 0$ and $m := \sup B > 0$. Show that the set AB is bounded above. (If $x \in AB$, there exist $a \in A$ and $b \in B$ such that $x = ab$.) Then AB has a least upper bound, say c. To show that $\ell m \leq c$, assume to the contrary that $c < \ell m$. Let $\epsilon = \ell m - c$. Since ℓ is the least upper bound for A, we may choose $a \in A$ so that $a > \ell - \epsilon(2m)^{-1}$. Having chosen this a, explain how to choose $b \in B$ so that $ab > \ell m - \epsilon$.

J.3.8. Proposition. *If B is a nonempty subset of $[0, \infty)$ that is bounded above and if $\alpha \geq 0$, then αB is bounded above and*

$$\sup(\alpha B) = \alpha \sup B.$$

Proof. Problem. *Hint.* This is a *very* easy consequence of one of the previous propositions.

J.4. The Archimedean property

One interesting property that distinguishes the set of real numbers from many other ordered fields is that for any real number a (no matter how large) and any positive number ϵ (no matter how small) it is possible by adding together enough copies of ϵ to obtain a sum greater than a. This is the ARCHIMEDEAN PROPERTY of the real number system. It is an easy consequence of the order completeness of the reals; that is, it follows from the least upper bound axiom (Axiom J.3.1).

J.4.1. Proposition (The Archimedean property of $\mathbb{R}$). *If $a \in \mathbb{R}$ and $\epsilon > 0$, then there exists $n \in \mathbb{N}$ such that $n\epsilon > a$.*

Proof. Problem. *Hint.* Argue by contradiction. Assume that the set $A := \{n\epsilon : n \text{ belongs to } \mathbb{N}\}$ is bounded above.

It is worth noting that the preceding proposition shows that the set $\mathbb{N}$ of natural numbers is not bounded above. (Take $\epsilon = 1$.)

Another useful consequence of the *least upper bound axiom* is the existence of nth roots of numbers $a \geq 0$. Below we establish the existence of square roots;

but the proof we give can be modified without great difficulty to show that every number $a \geq 0$ has an nth root (see Problem J.4.5).

J.4.2. Proposition. *Let $a \geq 0$. There exists a unique number $x \geq 0$ such that $x^2 = a$.*

Proof. Exercise. *Hint.* Let $A = \{t > 0 : t^2 < a\}$. Show that A is not empty and that it is bounded above. Let $x = \sup A$. Show that assuming $x^2 < a$ leads to a contradiction. (Choose ϵ in $(0,1)$ so that $\epsilon < 3^{-1}x^{-2}(a - x^2)$ and prove that $x(1+\epsilon)$ belongs to A.) Also show that assuming $x^2 > a$ produces a contradiction. (Choose ϵ in $(0,1)$ so that $\epsilon < (3a)^{-1}(x^2 - a)$, and prove that the set $A \cap \left(x(1+\epsilon)^{-1}, x\right)$ is not empty. What can be said about $x(1+\epsilon)^{-1}$?)

J.4.3. Notation. The unique number x guaranteed by the preceding proposition is denoted by $\sqrt{a}$ or by $a^{\frac{1}{2}}$. Similarly, nth roots are denoted by either $\sqrt[n]{a}$ or $a^{\frac{1}{n}}$.

J.4.4. Problem. Prove the following properties of the square root function.

(a) If $x, y \geq 0$, then $\sqrt{xy} = \sqrt{x}\sqrt{y}$.

(b) If $0 < x < y$, then $\sqrt{x} < \sqrt{y}$. *Hint.* Consider $(\sqrt{y})^2 - (\sqrt{x})^2$.

(c) If $0 < x < 1$, then $x^2 < x$ and $x < \sqrt{x}$.

(d) If $x > 1$, then $x < x^2$ and $\sqrt{x} < x$.

J.4.5. Problem. Restate the assertions of the preceding problem for nth roots (and nth powers). Explain what alterations in the proofs must be made to accommodate this change.

J.4.6. Definition. Let $x \in \mathbb{R}$. The ABSOLUTE VALUE of x, denoted by $|x|$, is defined to be $\sqrt{x^2}$. In light of the preceding proposition it is clear that if $x \geq 0$, then $|x| = x$; and if $x < 0$, then $|x| = -x$. From this observation it is easy to deduce two standard procedures for establishing an inequality of the form $|x| < c$ (where $c > 0$). One is to show that $x^2 < c^2$. The other is to show that $-c < x < c$ (or, what is the same thing, that both $x < c$ and $-x < c$ hold). Both methods are used extensively throughout the text, especially in Chapter 3 when we discuss continuity of real valued functions.

J.4.7. Problem. Prove that if $a, b \in \mathbb{R}$, then

(a) $|ab| = |a|\,|b|$;

(b) $|a + b| \leq |a| + |b|$; and

(c) $||a| - |b|| \leq |a - b|$.

J.4.8. Problem. Prove that if $a, b \in \mathbb{R}$, then $|ab| \leq \frac{1}{2}(a^2 + b^2)$. *Hint.* Consider the square of $a - b$ and of $a + b$.

Products, relations, and functions

K.1. Cartesian products

Ordered pairs are familiar objects. They are used, among other things, for coordinates of points in the plane. In the first sentence of Appendix B it was promised that all subsequent mathematical objects would be defined in terms of sets. So here just for the record is a formal definition of "ordered pair".

K.1.1. Definition. Let x and y be elements of arbitrary sets. Then the ORDERED PAIR (x, y) is defined to be $\{\{x, y\}, \{x\}\}$. This definition reflects our intuitive attitude: an ordered pair is a set $\{x, y\}$ with one of the elements, here x, designated as being "first". Thus we specify two things: $\{x, y\}$ and $\{x\}$.

Ordered pairs have only one interesting property: two of them are equal if and only if both their first coordinates and their second coordinates are equal. As you will discover by proving the next proposition, this fact follows easily from the definition.

K.1.2. Proposition. *Let x, y, u, and v be elements of arbitrary sets. Then $(x, y) = (u, v)$ if and only if $x = u$ and $y = v$.*

Proof. Exercise. *Hint.* Do not assume that the set $\{x, y\}$ has two elements. If $x = y$, then $\{x, y\}$ has only one element.

K.1.3. Problem. Asked to define an "ordered triple", one might be tempted to try a definition analogous to the definition of ordered pairs: let (a, b, c) be $\{\{a, b, c\}, \{a, b\}, \{a\}\}$. This appears to specify the entries, to pick out a "first" element, and to identify the "first two" elements. Explain why this won't work. (See Definition K.1.8 for a definition that does work.)

K.1.4. Definition. Let S and T be sets. The CARTESIAN PRODUCT of S and T, denoted by $S \times T$, is defined to be $\{(x, y) : x \in S \text{ and } y \in T\}$. The set $S \times S$ is often denoted by S^2.

K.1.5. Example. Let $S = \{1, x\}$ and $T = \{x, y, z\}$. Then
$$S \times T = \{(1, x), (1, y), (1, z), (x, x), (x, y), (x, z)\}.$$

K.1.6. Problem. Let $S = \{0, 1, 2\}$ and $T = \{1, 2, 3\}$. List all members of $(T \times S) \setminus (S \times T)$.

K.1.7. Problem. Let S, T, U, and V be sets. Then

(a) $(S \times T) \cap (U \times V) = (S \cap U) \times (T \cap V)$;

(b) $(S \times T) \cup (U \times V) \subseteq (S \cup U) \times (T \cup V)$; and

(c) equality need not hold in (b).

The proofs of (a) and (b) in the preceding problem are not particularly difficult. Nonetheless, before one can write down a proof, one must have a conjecture as to what is true. How could we have *guessed* initially that equality holds in (a) but not in (b)? The answer is, as it frequently is in mathematics, *by looking at pictures.* Try the following: Make a sketch where S and U are overlapping intervals on the x-axis and T and V are overlapping intervals on the y-axis. Then $S \times T$ and $U \times V$ are overlapping rectangles in the plane. Are not (a) and (b) almost obvious from your sketch?

We will also have occasion to use ordered n-tuples and n-fold Cartesian products for n greater than 2.

K.1.8. Definition. Let $n \geq 3$. We define ORDERED n-TUPLES inductively. Suppose ordered $(n-1)$-tuples $(x_1, \ldots, x_{n-1})$ have been defined. Let $(x_1, \ldots, x_n) := ((x_1, \ldots, x_{n-1}), x_n)$. An easy inductive proof shows that $(x_1, \ldots, x_n) = (y_1, \ldots, y_n)$ if and only if $x_k = y_k$ for $k = 1, \ldots, n$.

K.1.9. Definition. If $S_1, \ldots, S_n$ are sets, we define the CARTESIAN PRODUCT $S_1 \times \cdots \times S_n$ to be the set of all ordered n-tuples $(x_1, \ldots, x_n)$, where $x_k \in S_k$ for $1 \leq k \leq n$. We write S^n for $S \times \cdots \times S$ (n factors).

K.1.10. Example. The n-fold Cartesian product of the set $\mathbb{R}$ of real numbers is the set $\mathbb{R}^n$ of all n-tuples of real numbers and is often called (Euclidean) n-space.

K.2. Relations

Calculus is primarily about functions. We differentiate functions, we integrate them, we represent them as infinite series. A function is a special kind of relation. So it is convenient before introducing functions to make a few observations concerning the more general concept—*relations.*

K.2.1. Definition. A RELATION from a set S to a set T is a subset of the Cartesian product $S \times T$. A relation from the set S to itself is often called a relation *on S* or a relation *among members of S.*

There is a notational oddity concerning relations. To indicate that an ordered pair (a, b) belongs to a relation $R \subseteq S \times T$, we almost always write something like aRb rather than $(a, b) \in R$, which we would expect from the definition. For example, the relation "less than" is a relation on the real numbers. (We discussed this relation in Appendix H.) Technically then, since $<$ is a subset of $\mathbb{R} \times \mathbb{R}$, we could (correctly) write expressions such as $(3, 7) \in <$. Of course we don't. We write $3 < 7$ instead. And we say, "3 is less than 7", not "the pair $(3, 7)$ belongs to the relation *less than*". This is simply a matter of convention; it has no mathematical or logical content.

K.3. Functions

Functions are familiar from beginning calculus. Informally, a function consists of a pair of sets and a "rule" that associates with each member of the first set (the *domain*) one and only one member of the second (the *codomain*). While this informal "definition" is certainly adequate for most purposes and seldom leads to any misunderstanding, it is nevertheless sometimes useful to have a more precise formulation. This is accomplished by defining a *function* to be a special type of relation between two sets.

K.3.1. Definition. A FUNCTION f is an ordered triple (S, T, G) where S and T are sets and G is a subset of $S \times T$ satisfying the following:

(a) for each $s \in S$ there is a $t \in T$ such that $(s, t) \in G$; and

(b) if (s, t_1) and (s, t_2) belong to G, then $t_1 = t_2$.

In this situation we say that f is a *function from S into T* (or that *f maps S into T*) and write $f : S \to T$. The set S is the DOMAIN (or the INPUT SPACE) of f. The set T is the CODOMAIN (or TARGET SPACE or the OUTPUT SPACE) of f. And the relation G is the GRAPH of f. In order to avoid explicit reference to the graph G, it is usual to replace the expression "$(x, y) \in G$" by "$y = f(x)$"; the element $f(x)$ is the IMAGE of x under f. In this text (but not everywhere!) the words "transformation", "map", and "mapping" are synonymous with "function". The domain of f is denoted by $\operatorname{dom} f$.

K.3.2. Example. There are many ways of specifying a function. Statements (a)–(d) below define exactly the same function. We will use these (and other similar) notations interchangeably.

(a) For each real number x we let $f(x) = x^2$.

(b) Let $f = (S, T, G)$ where $S = T = \mathbb{R}$ and $G = \{(x, x^2) : x \in \mathbb{R}\}$.

(c) Let $f : \mathbb{R} \to \mathbb{R}$ be defined by $f(x) = x^2$.

(d) Consider the function $f : \mathbb{R} \to \mathbb{R} : x \mapsto x^2$.

K.3.3. Notation. If S and T are sets, we denote by $\mathcal{F}(S, T)$ the family of all functions from S into T.

K.3.4. Convention. A REAL VALUED function is a function whose codomain lies in $\mathbb{R}$. A function OF A REAL VARIABLE is a function whose domain is contained in $\mathbb{R}$. Some real valued functions of a real variable may be specified simply by writing down a formula. When the domain and codomain are not specified, the understanding is that the domain of the function is the largest set of real numbers for which the formula makes sense and the codomain is taken to be $\mathbb{R}$.

In the case of real valued functions on a set S, we frequently write $\mathcal{F}(S)$ instead of $\mathcal{F}(S, \mathbb{R})$.

K.3.5. Example. Let $f(x) = (x^2 + x)^{-1}$. Since this formula is meaningful for all real numbers except -1 and 0, we conclude that the domain of f is $\mathbb{R} \setminus \{-1, 0\}$.

K.3.6. Example. Let $f(x) = (x^2 + x)^{-1}$ for $x > 0$. Here the domain of f is specified: it is the interval $(0, \infty)$.

K.3.7. Exercise. Let $f(x) = (1 - 2(1 + (1 - x)^{-1})^{-1})^{-1}$.

(a) Find $f(\frac{1}{2})$.

(b) Find the domain of f.

K.3.8. Exercise. Let $f(x) = (-x^2 - 4x - 1)^{-1/2}$. Find the domain of f.

K.3.9. Problem. Let $f(x) = (1 - (2 + (3 - (1 + x)^{-1})^{-1})^{-1})^{-1}$.

(a) Find $f(\frac{1}{2})$.

(b) Find the domain of f.

K.3.10. Problem. Let $f(x) = (-x^2 - 7x - 10)^{-1/2}$.

(a) Find $f(-3)$.

(b) Find the domain of f.

K.3.11. Problem. Let $f(x) = \dfrac{\sqrt{x^2 - 4}}{5 - \sqrt{36 - x^2}}$. Find the domain of f. Express your answer as a union of intervals.

K.3.12. Problem. Explain carefully why two functions f and g are equal if and only if their domains and codomains are equal and $f(x) = g(x)$ for every x in their common domain.

Properties of functions

L.1. Images and inverse images

L.1.1. Definition. If $f\colon S \to T$ and $A \subseteq S$, then $f^{\to}(A)$ (the IMAGE of A under f) is $\{f(x)\colon x \in A\}$. It is common practice to write $f(A)$ for $f^{\to}(A)$. The set $f^{\to}(S)$ is the RANGE (or IMAGE) of f; usually we write ran f for $f^{\to}(S)$.

L.1.2. Exercise. Let

$$f(x) = \begin{cases} -1 & \text{for } x < -2, \\ 7 - x^2 & \text{for } -2 \le x < 1, \\ \dfrac{1}{x} & \text{for } x \ge 1, \end{cases}$$

and $A = (-4, 4)$. Find $f^{\to}(A)$.

L.1.3. Exercise. Let $f(x) = 3x^4 + 4x^3 - 36x^2 + 1$. Find ran f.

L.1.4. Definition. Let $f\colon S \to T$ and $B \subseteq T$. Then $f^{\leftarrow}(B)$ (the INVERSE IMAGE of B under f) is $\{x \in S\colon f(x) \in B\}$. In many texts $f^{\leftarrow}(B)$ is denoted by $f^{-1}(B)$. This may cause confusion by suggesting that functions always have inverses (see section M.2).

L.1.5. Exercise. Let $f(x) = \arctan x$ and $B = (\frac{\pi}{4}, 2)$. Find $f^{\leftarrow}(B)$.

L.1.6. Exercise. Let $f(x) = -\sqrt{9 - x^2}$ and $B = (1, 3)$. Find $f^{\leftarrow}(B)$.

L.1.7. Problem. Let

$$f(x) = \begin{cases} -x - 4 & \text{for } x \le 0, \\ x^2 + 3 & \text{for } 0 < x \le 2, \\ (x - 1)^{-1} & \text{for } x > 2, \end{cases}$$

and $A = (-3, 4)$. Find $f^{\to}(A)$.

L.1.8. Problem. Let $f(x) = 4 - x^2$ and $B = (1, 3]$. Find $f^{\leftarrow}(B)$.

L.1.9. Problem. Let $f(x) = \dfrac{x}{1-x}$.

(a) Find $f^{\leftarrow}([0,a])$ for $a > 0$.

(b) Find $f^{\leftarrow}([-\frac{3}{2}, -\frac{1}{2}])$.

L.1.10. Problem. Let $f(x) = -x^2 + 4\arctan x$. Find $\operatorname{ran} f$.

L.1.11. Problem. Let

$$f(x) = \begin{cases} x+1 & \text{for } x < 1, \\ 8 + 2x - x^2 & \text{for } x \geq 1. \end{cases}$$

Let $A = (-2, 3)$ and $B = [0, 1]$. Find $f^{\rightarrow}(A)$ and $f^{\leftarrow}(B)$.

L.2. Composition of functions

Let $f\colon S \to T$ and $g\colon T \to U$. The COMPOSITE of g and f, denoted by $g \circ f$, is the function taking S to U defined by

$$(g \circ f)(x) = g(f(x))$$

for all x in S. The operation $\circ$ is COMPOSITION. We again make a special convention for real valued functions of a real variable: the domain of $g \circ f$ is the set of all x in $\mathbb{R}$ for which the expression $g(f(x))$ makes sense.

L.2.1. Example. Let $f(x) = (x-1)^{-1}$ and $g(x) = \sqrt{x}$. Then the domain of $g \circ f$ is the interval $(1, \infty)$, and for all x in that interval

$$(g \circ f)(x) = g(f(x)) = \frac{1}{\sqrt{x-1}}.$$

Proof. The square root of $x - 1$ exists only when $x \geq 1$; and since we take its reciprocal, we exclude $x = 1$. Thus $\operatorname{dom}(g \circ f) = (1, \infty)$. $\qquad\square$

L.2.2. Exercise. Let

$$f(x) = \begin{cases} 0 & \text{for } x < 0, \\ 3x & \text{for } 0 \leq x \leq 2, \\ 2 & \text{for } x > 2, \end{cases}$$

and

$$g(x) = \begin{cases} x^2 & \text{for } 1 < x < 3, \\ -1 & \text{otherwise.} \end{cases}$$

Sketch the graph of $g \circ f$.

L.2.3. Proposition. *Composition of functions is associative but not necessarily commutative.*

Proof. Exercise. *Hint.* Let $f\colon S \to T$, $g\colon T \to U$, and $h\colon U \to V$. Show that $h \circ (g \circ f) = (h \circ g) \circ f$. Give an example to show that $f \circ g$ and $g \circ f$ may fail to be equal.

L.2.4. Problem. Let $f(x) = x^2 + 2x^{-1}$, $g(x) = 2(2x+3)^{-1}$, and $h(x) = \sqrt{2x}$. Find $(h \circ g \circ f)(4)$.

L.2.5. Problem. If $f\colon S \to T$ and $g\colon T \to U$, then

(a) $(g \circ f)^{\rightarrow}(A) = g^{\rightarrow}(f^{\rightarrow}(A))$ for every $A \subseteq S$.

(b) $(g \circ f)^{\leftarrow}(B) = f^{\leftarrow}(g^{\leftarrow}(B))$ for every $B \subseteq U$.

L.3. The identity function

The family of all functions mapping a set S into a set T is denoted by $\mathcal{F}(S,T)$. One member of $\mathcal{F}(S,S)$ is particularly noteworthy, the IDENTITY FUNCTION on S. It is defined by

$$I_S \colon S \to S \colon x \mapsto x.$$

When the set S is understood from context, we write I for I_S.

The identity function is characterized algebraically by the following conditions:

$$\text{if } f \colon R \to S, \text{ then } I_S \circ f = f$$

and

$$\text{if } g \colon S \to T, \text{ then } g \circ I_S = g.$$

L.3.1. Definition. More general than the identity function are the inclusion maps. If $A \subseteq S$, then the INCLUSION MAP taking A into S is defined by

$$\iota_{A,S} \colon A \to S \colon x \mapsto x.$$

When no confusion is likely to result, we abbreviate $\iota_{A,S}$ to ι. Notice that $\iota_{S,S}$ is just the identity map I_S.

L.4. Diagrams

It is frequently useful to think of functions as arrows in diagrams. For example, the situation $f \colon R \to S$, $h \colon R \to T$, $j \colon T \to U$, $g \colon S \to U$ may be represented by the following diagram.

$$
\begin{array}{ccc}
R & \xrightarrow{\;\;f\;\;} & S \\
\downarrow{\scriptstyle h} & & \downarrow{\scriptstyle g} \\
T & \xrightarrow[\;\;j\;\;]{} & U
\end{array}
$$

The diagram is said to COMMUTE if $j \circ h = g \circ f$.

Diagrams need not be rectangular. For instance,

$$
\begin{array}{ccc}
 & R & \\
{\scriptstyle f}\downarrow & & \searrow{\scriptstyle k} \\
S & \xrightarrow[\;\;g\;\;]{} & T
\end{array}
$$

is a commutative diagram if $k = g \circ f$.

L.4.1. Example. Here is a diagrammatic way of stating the associative law for composition of functions. If the triangles in the diagram

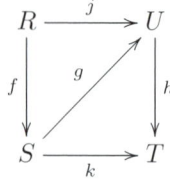

$$
\begin{array}{ccc}
R & \xrightarrow{\;j\;} & U \\
\downarrow{\scriptstyle f} & \nearrow{\scriptstyle g} & \uparrow{\scriptstyle h} \\
S & \xrightarrow[\;k\;]{} & T
\end{array}
$$

commute, then so does the rectangle.

L.5. Restrictions and extensions

If $f\colon S \to T$ and $A \subseteq S$, then the RESTRICTION of f to A, denoted by $f\big|_A$, is the function $f \circ \iota_{A,S}$. That is, it is the mapping from A into T whose value at each x in A is $f(x)$.

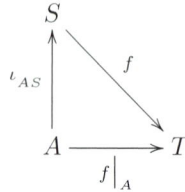

$$
\begin{array}{ccc}
 & S & \\
\iota_{AS}\big\uparrow & & \searrow{\scriptstyle f} \\
A & \xrightarrow[\;f|_A\;]{} & T
\end{array}
$$

Suppose that $g\colon A \to T$ and $A \subseteq S$. A function $f\colon S \to T$ is an EXTENSION of g to S if $f\big|_A = g$, that is, if the diagram

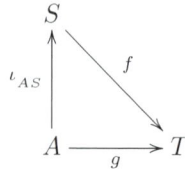

$$
\begin{array}{ccc}
 & S & \\
\iota_{AS}\big\uparrow & & \searrow{\scriptstyle f} \\
A & \xrightarrow[\;g\;]{} & T
\end{array}
$$

commutes.

Functions that have inverses

M.1. Injections, surjections, and bijections

A function f is INJECTIVE (or ONE-TO-ONE) if $x = y$ whenever $f(x) = f(y)$. That is, f is injective if no two distinct elements in its domain have the same image. For a real valued function of a real variable, this condition may be interpreted graphically: A function is one-to-one if and only if each horizontal line intersects the graph of the function at most once. An injective map is called an INJECTION.

M.1.1. Example. The sine function is not injective (since, for example, $\sin 0 = \sin \pi$).

M.1.2. Example. Let $f(x) = \dfrac{x+2}{3x-5}$. The function f is injective.

Proof. Exercise.

M.1.3. Exercise. Find an injective mapping from $\{x \in \mathbb{Q} : x > 0\}$ into $\mathbb{N}$.

M.1.4. Problem. Let $f(x) = \dfrac{2x-5}{3x+4}$. Show that f is injective.

M.1.5. Problem. Let $f(x) = 2x^2 - x - 15$. Show that f is *not* injective.

M.1.6. Problem. Let $f(x) = x^3 - 2x^2 + x - 3$. Is f injective?

M.1.7. Problem. Let $f(x) = x^3 - x^2 + x - 3$. Is f injective?

M.1.8. Definition. A function is SURJECTIVE (or ONTO) if its range is equal to its codomain.

M.1.9. Example. Define $f \colon \mathbb{R} \to \mathbb{R}$ by $f(x) = x^2$ and $g \colon \mathbb{R} \to [0, \infty)$ by $g(x) = x^2$. Then the function g is surjective while f is not (even though the two functions have the same graph!).

M.1.10. Exercise. Find a surjection (that is, a surjective map) from $[0, 1]$ onto $[0, \infty)$.

M.1.11. Definition. A function is BIJECTIVE if it is both injective and surjective. A bijective map is called a BIJECTION or a ONE-TO-ONE CORRESPONDENCE.

M.1.12. Exercise. Give an explicit formula for a bijection between $\mathbb{Z}$ and $\mathbb{N}$.

M.1.13. Exercise. Give an explicit formula for a bijection between $\mathbb{R}$ and the open interval $(0, 1)$.

M.1.14. Exercise. Give a formula for a bijection between the interval $[0, 1)$ and the unit circle $x^2 + y^2 = 1$ in the plane.

M.1.15. Exercise. Let $f \colon [1, 2] \to [0, 3]$ be defined by $f(x) = 1/x$. Find an extension g of f to the interval $[0, 3]$ such that $g \colon [0, 3] \to [0, 3]$ is a bijection.

M.1.16. Exercise. Let $f \colon R \to [-1, 1]$ be defined by $f(x) = \sin x$. Find a set $A \subseteq R$ such that the restriction of f to A is a bijection from A onto $[-1, 1]$.

M.1.17. Problem. Let $f \colon S \to T$ and $g \colon T \to U$. Prove the following.

(a) If f and g are injective, so is $g \circ f$.

(b) If f and g are surjective, so is $g \circ f$.

(c) If f and g are bijective, so is $g \circ f$.

M.1.18. Problem. Find a bijection between the open intervals $(0, 1)$ and $(-8, 5)$. Prove that the function you use really *is* a bijection between the two intervals.

M.1.19. Problem. Find a bijection between the open intervals $(0, 1)$ and $(3, \infty)$. (Proof is not required.)

M.1.20. Problem. Find a bijection between the interval $(0, 1)$ and the parabola $y = x^2$ in the plane. (Proof is not required.)

M.1.21. Problem. Let $f \colon [1, 2] \to [0, 11]$ be defined by $f(x) = 3x^2 - 1$. Find an extension g of f to $[0, 3]$ that is a bijection from $[0, 3]$ onto $[0, 11]$. (Proof is not required.)

It is important for us to know how the image $f^{\to}$ and the inverse image $f^{\leftarrow}$ of a function f behave with respect to unions, intersections, and complements of sets. The basic facts are given in the next ten propositions. Although these results are quite elementary, we make extensive use of them in studying continuity.

M.1.22. Proposition. *Let $f \colon S \to T$ and $B \subseteq T$.*

(a) $f^{\to}(f^{\leftarrow}(B)) \subseteq B$.

(b) *Equality need not hold in* (a).

(c) *Equality does hold in* (a) *if f is surjective.*

Proof. Exercise.

M.1.23. Proposition. *Let $f \colon S \to T$ and $A \subseteq S$.*

(a) $A \subseteq f^{\leftarrow}(f^{\to}(A))$.

(b) *Equality need not hold in* (a).

(c) *Equality does hold in* (a) *if f is injective.*

Proof. Problem.

M.1.24. Problem. Prove the converse of Proposition M.1.23. That is, show that if $f: S \to T$ and $f^{\leftarrow}(f^{\to}(A)) = A$ for all $A \subseteq S$, then f is injective. *Hint.* Suppose $f(x) = f(y)$. Let $A = \{x\}$. Show that $y \in f^{\leftarrow}(f^{\to}(A))$.

M.1.25. Proposition. *Let $f: S \to T$ and $A, B \subseteq S$. Then*
$$f^{\to}(A \cup B) = f^{\to}(A) \cup f^{\to}(B).$$

Proof. Exercise.

M.1.26. Proposition. *Let $f: S \to T$ and $C, D \subseteq T$. Then*
$$f^{\leftarrow}(C \cup D) = f^{\leftarrow}(C) \cup f^{\leftarrow}(D).$$

Proof. Problem.

M.1.27. Proposition. *Let $f: S \to T$ and $C, D \subseteq T$. Then*
$$f^{\leftarrow}(C \cap D) = f^{\leftarrow}(C) \cap f^{\leftarrow}(D).$$

Proof. Exercise.

M.1.28. Proposition. *Let $f: S \to T$ and $A, B \subseteq S$.*

(a) $f^{\to}(A \cap B) \subseteq f^{\to}(A) \cap f^{\to}(B)$.

(b) *Equality need not hold in* (a).

(c) *Equality does hold in* (a) *if f is injective.*

Proof. Problem.

M.1.29. Proposition. *Let $f: S \to T$ and $D \subseteq T$. Then*
$$f^{\leftarrow}(D^c) = \left(f^{\leftarrow}(D)\right)^c.$$

Proof. Problem.

M.1.30. Proposition. *Let $f: S \to T$ and $A \subseteq S$.*

(a) *If f is injective, then $f^{\to}(A^c) \subseteq \left(f^{\to}(A)\right)^c$.*

(b) *If f is surjective, then $f^{\to}(A^c) \supseteq \left(f^{\to}(A)\right)^c$.*

(c) *If f is bijective, then $f^{\to}(A^c) = \left(f^{\to}(A)\right)^c$.*

Proof. Problem. *Hints.* For part (a), let $y \in f^{\to}(A^c)$. To prove that $y \in \left(f^{\to}(A)\right)^c$, assume to the contrary that $y \in f^{\to}(A)$ and derive a contradiction. For part (b), let $y \in \left(f^{\to}(A)\right)^c$. Since f is surjective, there exists $x \in S$ such that $y = f(x)$. Can x belong to A?.

M.1.31. Proposition. *Let $f: S \to T$ and $\mathfrak{A} \subseteq \mathfrak{P}(S)$.*

(a) $f^{\to}(\bigcap \mathfrak{A}) \subseteq \bigcap \{f^{\to}(A) \colon A \in \mathfrak{A}\}$.

(b) *If f is injective, equality holds in* (a).

(c) $f^{\to}(\bigcup \mathfrak{A}) = \bigcup \{f^{\to}(A) \colon A \in \mathfrak{A}\}$.

Proof. Exercise.

M.1.32. Proposition. *Let* $f\colon S \to T$ *and* $\mathfrak{B} \subseteq \mathfrak{P}(T)$.

(a) $f^{\leftarrow}(\bigcap \mathfrak{B}) = \bigcap \{f^{\leftarrow}(B)\colon B \in \mathfrak{B}\}$.

(b) $f^{\leftarrow}(\bigcup \mathfrak{B}) = \bigcup \{f^{\leftarrow}(B)\colon B \in \mathfrak{B}\}$.

Proof. Problem.

M.2. Inverse functions

Let $f\colon S \to T$ and $g\colon T \to S$. If $g \circ f = I_S$, then g is a LEFT INVERSE of f and, equivalently, f is a RIGHT INVERSE of g. We say that f is INVERTIBLE if there exists a function from T into S that is both a left and a right inverse for f. Such a function is denoted by f^{-1} and is called the INVERSE of f. (Notice that the last "the" in the preceding sentence requires justification; see Proposition M.2.1 below.) A function is INVERTIBLE if it has an inverse. According to the definition just given, the inverse f^{-1} of a function f must satisfy

$$f \circ f^{-1} = I_T \qquad \text{and} \qquad f^{-1} \circ f = I_S\,.$$

A simple, but important, consequence of this is that for an invertible function, $y = f(x)$ if and only if $x = f^{-1}(y)$. (*Proof.* If $y = f(x)$, then $f^{-1}(y) = f^{-1}(f(x)) = I_S(x) = x$. Conversely, if $x = f^{-1}(y)$, then $f(x) = f(f^{-1}(y)) = I_T(y) = y$.)

M.2.1. Proposition. *A function can have at most one inverse.*

Proof. Exercise.

M.2.2. Proposition. *If a function has both a left inverse and a right inverse, then the left and right inverses are equal* (*and therefore the function is invertible*).

Proof. Problem.

M.2.3. Exercise. The *arcsine* function is defined to be the inverse of what function? (*Hint.* The answer is not *sine*.) What about *arccosine*? *arctangent*?

The next two propositions tell us that a necessary and sufficient condition for a function to have right inverse is that it be surjective and that a necessary and sufficient condition for a function to have a left inverse is that it be injective. Thus, in particular, a function is invertible if and only if it is bijective. In other words, the invertible members of $\mathcal{F}(S, T)$ are the bijections.

M.2.4. Proposition. *Let* $S \neq \emptyset$. *A function* $f\colon S \to T$ *has a right inverse if and only if it is surjective.*

Proof. Exercise.

M.2.5. Proposition. *Let* $S \neq \emptyset$. *A function* $f\colon S \to T$ *has a left inverse if and only if it is injective.*

Proof. Problem.

M.2.6. Problem. Prove that if a function f is bijective, then $\left(f^{-1}\right)^{\leftarrow} = f^{\rightarrow}$.

M.2.7. Problem. Let $f(x) = \dfrac{ax + b}{cx + d}$ where a, b, c, $d \in \mathbb{R}$ and not both c and d are zero.

(a) Under what conditions on the constants $a, b, c,$ and d is f injective?

(b) Under what conditions on the constants $a, b, c,$ and d is f its own inverse?

Products

The Cartesian product of two sets, which was defined in Appendix K, is best thought of not just as a collection of ordered pairs but as this collection together with two distinguished *projection* mappings.

N.1.1. Definition. Let S_1 and S_2 be nonempty sets. For $k = 1$, 2, define the COORDINATE PROJECTIONS $\pi_k \colon S_1 \times S_2 \to S_k$ by $\pi_k(s_1, s_2) = s_k$. We notice two simple facts:

(a) π_1 and π_2 are surjections, and

(b) $z = (\pi_1(z), \pi_2(z))$ for all $z \in S_1 \times S_2$.

If T is a nonempty set and if $g \colon T \to S_1$ and $h \colon T \to S_2$, then we define the function $(g, h) \colon T \to S_1 \times S_2$ by

$$(g, h)(t) = (g(t), h(t)).$$

N.1.2. Example. If $g(t) = \cos t$ and $h(t) = \sin t$, then (g, h) is a map from $\mathbb{R}$ to the unit circle in the plane. (This is a PARAMETRIZATION of the unit circle.)

N.1.3. Definition. Let S_1, S_2, and T be nonempty sets, and let $f \colon T \to S_1 \times S_2$. For $k = 1$, 2, we define functions $f^k \colon T \to S_k$ by $f^k = \pi_k \circ f$; these are the COMPONENTS of f. (The superscripts have nothing to do with powers. We use them because we wish later to attach subscripts to functions to indicate partial differentiation.) Notice that $f(t) = (\pi_1(f(t)), \pi_2(f(t)))$ for all $t \in T$, so that

$$f = (\pi_1 \circ f, \pi_2 \circ f) = (f^1, f^2).$$

If we are given the function f, the components f^1 and f^2 have been defined so as to make the following diagram commute.

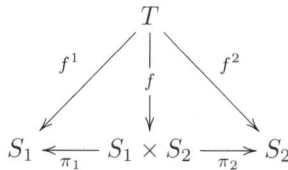

$$
\begin{array}{ccc}
 & T & \\
f^1 \swarrow & \downarrow f & \searrow f^2 \\
S_1 \xleftarrow{\pi_1} & S_1 \times S_2 & \xrightarrow{\pi_2} S_2
\end{array}
$$

On the other hand if the functions $f^1\colon T \to S_1$ and $f^2\colon T \to S_2$ are given, then there exists a function f, namely (f^1, f^2), that makes the diagram commute. Actually, (f^1, f^2) is the *only* function with this property, a fact that we prove in the next exercise.

N.1.4. Exercise. Suppose that $f^1 \in \mathcal{F}(T, S_1)$ and $f^2 \in \mathcal{F}(T, S_2)$. Then there exists a unique function $g \in \mathcal{F}(T, S_1 \times S_2)$ such that $\pi_1 \circ g = f^1$ and $\pi_2 \circ g = f^2$.

The following problem, although interesting, is not needed elsewhere in this text, so it is listed as optional. It says, roughly, that any set that "behaves like a Cartesian product" must be in one-to-one correspondence with the Cartesian product.

N.1.5. Problem (Optional). Let S_1, S_2, and P be nonempty sets, and let $\rho_k\colon P \to S_k$ be surjections. Suppose that for every set T and every pair of functions $f^k \in \mathcal{F}(T, S_k)$ ($k = 1, 2$), there exists a unique function $g \in \mathcal{F}(T, P)$ such that $f^k = \rho_k \circ g$ ($k = 1, 2$). Then there exists a bijection from P onto $S_1 \times S_2$. *Hint.* Consider the following diagrams.

(91)

$$
\begin{array}{c}
P \\
\rho_1 \swarrow \quad \downarrow \rho \quad \searrow \rho_2 \\
S_1 \xleftarrow{\ \pi_1\ } S_1 \times S_2 \xrightarrow{\ \pi_2\ } S_2
\end{array}
$$

(92)

$$
\begin{array}{c}
S_1 \times S_2 \\
\pi_1 \swarrow \quad \downarrow \pi \quad \searrow \pi_2 \\
S_1 \xleftarrow{\ \rho_1\ } P \xrightarrow{\ \rho_2\ } S_2
\end{array}
$$

(93)

$$
\begin{array}{c}
P \\
\rho_1 \swarrow \quad \downarrow \rho \quad \searrow \rho_2 \\
S_1 \xleftarrow{\ \rho_1\ } P \xrightarrow{\ \rho_2\ } S_2
\end{array}
$$

Exercise N.1.4 tells us that there exists a unique map ρ that makes diagram (91) commute, and by hypothesis there exists a unique map π that makes diagram (92) commute. Conclude from (91) and (92) that (93) commutes when $g = \pi \circ \rho$. It is obvious that (93) commutes when $g = I_P$. Then use the uniqueness part of the hypothesis to conclude that π is a left inverse for ρ. Now construct a new diagram replacing P by $S_1 \times S_2$ and ρ_k by π_k in (93).

Finite and infinite sets

There are a number of ways of comparing the *sizes* of sets. In this chapter and the next we examine perhaps the simplest of these, cardinality. Roughly speaking, we say that two sets have the "same number of elements" if there is a one-to-one correspondence between the elements of the sets. In this sense the open intervals $(0, 1)$ and $(0, 2)$ have the same number of elements. (The map $x \mapsto 2x$ is a bijection.) Clearly, this is only one sense of the idea of "size". It is certainly also reasonable to regard $(0, 2)$ as being bigger than $(0, 1)$ because it is twice as long.

We derive only the most basic facts concerning cardinality. In this appendix we discuss some elementary properties of finite and infinite sets, and in the next we distinguish between countable and uncountable sets. This is all we will need.

O.1.1. Definition. Two sets S and T are CARDINALLY EQUIVALENT if there exists a bijection from S onto T, in which case we write $S \sim T$. It is easy to see that cardinal equivalence is indeed an equivalence relation; that is, it is reflexive, symmetric, and transitive.

O.1.2. Proposition. *Cardinal equivalence is an equivalence relation. Let S, T, and U be sets. Then*

(a) $S \sim S$;

(b) *if $S \sim T$, then $T \sim S$; and*

(c) *if $S \sim T$ and $T \sim U$, then $S \sim U$.*

Proof. Problem.

O.1.3. Definition. A set S is FINITE if it is empty or if there exists $n \in \mathbb{N}$ such that $S \sim \{1, \ldots, n\}$. A set is INFINITE if it is not finite. The next few facts concerning finite and infinite sets probably appear obvious, but writing down the proofs may in several instances require a bit of thought. Here is a question that merits some reflection: if one is unable to explain exactly why a result is obvious, is it really obvious?

One *obvious* result is that if two initial segments $\{1, \ldots, m\}$ and $\{1, \ldots, n\}$ of the set of natural numbers are cardinally equivalent, then $m = n$. We prove this next.

O.1.4. Proposition. *Let* $n \in \mathbb{N}$. *If* $m \in \mathbb{N}$ *and* $\{1, \ldots, m\} \sim \{1, \ldots, n\}$, *then* $m = n$.

Proof. Exercise. *Hint.* Use induction on n.

O.1.5. Definition. We define $\operatorname{card} \emptyset$, the cardinal number of the empty set, to be 0. If S is a nonempty finite set, then by Proposition O.1.4 there exists only one positive integer n such that $S \sim \{1, \ldots, n\}$. This integer, $\operatorname{card} S$, is the CARDINAL NUMBER of the set S, or the NUMBER OF ELEMENTS in S. Notice that if S is finite with cardinal number n and $T \sim S$, then T is also finite and has cardinal number n. Thus for finite sets the expressions "cardinally equivalent" and "have the same cardinal number" are interchangeable.

O.1.6. Example. Let $S = \{a, b, c, d\}$ (see Convention B.1.5). Then $\operatorname{card} S = 4$, since $\{a, b, c, d\} \sim \{1, 2, 3, 4\}$.

One simple fact about cardinal numbers is that the number of elements in the union of two disjoint finite sets is the sum of the numbers of elements in each.

O.1.7. Proposition. *If* S *and* T *are disjoint finite sets, then* $S \cup T$ *is finite and*

$$\operatorname{card}(S \cup T) = \operatorname{card} S + \operatorname{card} T.$$

Proof. Exercise.

A variant of the preceding result is given in Problem O.1.10. We now take a preliminary step toward proving that subsets of finite sets are themselves finite (Proposition O.1.9).

O.1.8. Lemma. *If* $C \subseteq \{1, \ldots, n\}$, *then* C *is finite and* $\operatorname{card} C \leq n$.

Proof. Exercise. *Hint.* Use mathematical induction. If $C \subseteq \{1, \ldots, k+1\}$, then $C \setminus \{k+1\} \subseteq \{1, \ldots, k\}$. Examine the cases $k + 1 \notin C$ and $k + 1 \in C$ separately.

O.1.9. Proposition. *Let* $S \subseteq T$. *If* T *is finite, then* S *is finite and* $\operatorname{card} S \leq \operatorname{card} T$.

Proof. Problem. *Hint.* The case $T = \emptyset$ is trivial. Suppose $T \neq \emptyset$. Let $\iota : S \to T$ be the inclusion map of S into T (see Appendix L). There exist $n \in \mathbb{N}$ and a bijection $f : T \to \{1, \ldots, n\}$. Let $C = \operatorname{ran}(f \circ \iota)$. The map from S to C defined by $x \mapsto f(x)$ is a bijection. Use Lemma O.1.8.)

The preceding proposition "subsets of finite sets are finite" has a useful contrapositive: "sets that contain infinite sets are themselves infinite."

O.1.10. Problem. Let S be a set, and let T be a finite set. Prove that

$$\operatorname{card}(T \setminus S) = \operatorname{card} T - \operatorname{card}(T \cap S).$$

Hint. Use Problem F.3.14 and Proposition O.1.7.

Notice that it is a consequence of the preceding result of Problem O.1.10 that if $S \subseteq T$ (where T is finite), then

$$\operatorname{card}(T \setminus S) = \operatorname{card} T - \operatorname{card} S.$$

How do we show that a set S is infinite? If our only tool were the definition, we would face the prospect of proving that there does *not* exist a bijection from S onto an initial segment of the natural numbers. It would be pleasant to have a more direct approach than establishing the nonexistence of maps. This is the point of our next proposition.

O.1.11. Proposition. *A set is infinite if and only if it is cardinally equivalent to a proper subset of itself.*

Proof. Exercise. *Hint.* Suppose a set S is infinite. Show that it is possible to choose inductively a sequence of distinct elements $a_1, a_2, a_3, \ldots$ in S. (Suppose $a_1, \ldots, a_n$ have already been chosen. Can $S \setminus \{a_1, \ldots, a_n\}$ be empty?) Map each a_k to a_{k+1}, and map each member of S that is not an a_k to itself.

For the converse argue by contradiction. Suppose that $T \sim S$ where T is a proper subset of S, and assume that $S \sim \{1, \ldots, n\}$. Prove that $S \setminus T \sim \{1, \ldots, p\}$ for some $p \in \mathbb{N}$. Write T as $S \setminus (S \setminus T)$, and obtain $n = n - p$ by computing the cardinality of T in two ways. (Make use of Problem O.1.10.) What does $n = n - p$ contradict?

O.1.12. Example. The set $\mathbb{N}$ of natural numbers is infinite.

Proof. The map $n \mapsto n + 1$ is a bijection from $\mathbb{N}$ onto $\mathbb{N} \setminus \{1\}$, which is a proper subset of $\mathbb{N}$. $\qquad\qquad\square$

O.1.13. Exercise. The interval $(0, 1)$ is infinite.

O.1.14. Example. The set $\mathbb{R}$ of real numbers is infinite.

The next two results tell us that functions take finite sets to finite sets and that for injective functions finite sets come from finite sets.

O.1.15. Proposition. *If T is a set, S is a finite set, and $f : S \to T$ is surjective, then T is finite.*

Proof. Exercise. *Hint.* Use Propositions M.2.4 and M.2.5.

O.1.16. Proposition. *If S is a set, T is a finite set, and $f : S \to T$ is injective, then S is finite.*

Proof. Exercise.

O.1.17. Problem. Let S and T be finite sets. Prove that $S \cup T$ is finite and that

$$\operatorname{card}(S \cup T) = \operatorname{card} S + \operatorname{card} T - \operatorname{card}(S \cap T).$$

Hint. Use Problem O.1.10.

O.1.18. Problem. If S is a finite set with cardinal number n, what is the cardinal number of $\mathfrak{P}(S)$?

Countable
and uncountable sets

There are many sizes of infinite sets—infinitely many in fact. In our subsequent work we need only distinguish between countably infinite and uncountable sets. A set is countably infinite if it is in one-to-one correspondence with the set of positive integers; if it is neither finite nor countably infinite, it is uncountable. In this section we present some basic facts about and examples of both countable and uncountable sets. This is all we will need. Except for Problem P.1.21, which is presented for general interest, we ignore the many intriguing questions that arise concerning various sizes of uncountable sets. For a very readable introduction to such matters, see [**Kap77**, chapter 2].

P.1.1. Definition. A set is COUNTABLY INFINITE (or DENUMERABLE) if it is cardinally equivalent to the set $\mathbb{N}$ of natural numbers. A bijection from $\mathbb{N}$ onto a countably infinite set S is an ENUMERATION of the elements of S. A set is COUNTABLE if it is either finite or countably infinite. If a set is not countable it is UNCOUNTABLE.

P.1.2. Example. The set $\mathbb{E}$ of even integers in $\mathbb{N}$ is countable.

Proof. The map $n \mapsto 2n$ is a bijection from $\mathbb{N}$ onto $\mathbb{E}$. $\qquad\square$

The first proposition of this section establishes the fact that the *smallest* infinite sets are the countable ones.

P.1.3. Proposition. *Every infinite set contains a countably infinite subset.*

Proof. Problem. *Hint.* Review the proof of Proposition O.1.11.

If we are given a set S that we believe to be countable, it may be extremely difficult to prove this by exhibiting an explicit bijection between $\mathbb{N}$ and S. Thus it is of great value to know that certain constructions performed with countable sets result in countable sets. The next five propositions provide us with ways of generating new countable sets from old ones. In particular, we show that each of the following is countable:

(a) any subset of a countable set.

(b) the range of a surjection with countable domain.

(c) the domain of an injection with countable codomain.

(d) the product of any finite collection of countable sets.

(e) the union of a countable family of countable sets.

P.1.4. Proposition. *If $S \subseteq T$ where T is countable, then S is countable.*

Proof. Exercise. *Hint.* Show first that every subset of $\mathbb{N}$ is countable.

The preceding has an obvious corollary: if $S \subseteq T$ and S is uncountable, then so is T.

P.1.5. Proposition. *If $f\colon S \to T$ is injective and T is countable, then S is countable.*

Proof. Problem. *Hint.* Adapt the proof of Proposition O.1.16.

P.1.6. Proposition. *If $f\colon S \to T$ is surjective and S is countable, then T is countable.*

Proof. Problem. *Hint.* Adapt the proof of Proposition O.1.15.

P.1.7. Lemma. *The set $\mathbb{N} \times \mathbb{N}$ is countable.*

Proof. Exercise. *Hint.* Consider the map $(m, n) \mapsto 2^{m-1}(2n - 1)$.

P.1.8. Example. The set $\mathbb{Q}^+ \setminus \{0\} = \{x \in \mathbb{Q} : x > 0\}$ is countable.

Proof. Suppose that the rational number m/n is written in lowest terms. (That is, m and n have no common factors greater than 1.) Define $f(m/n) = (m, n)$. It is easy to see that the map $f\colon \mathbb{Q}^+ \setminus \{0\} \to \mathbb{N} \times \mathbb{N}$ is injective. By Proposition P.1.5 and Lemma P.1.7, Q^+ is countable. $\square$

P.1.9. Proposition. *If S and T are countable sets, then so is $S \times T$.*

Proof. Problem. *Hint.* Either $S \sim \{1, \ldots, n\}$ or else $S \sim \mathbb{N}$. In either case there exists an injective map $f\colon S \to \mathbb{N}$. Similarly, there exists an injection $g\colon T \to \mathbb{N}$. Define the function $f \times g\colon S \times T \to \mathbb{N} \times \mathbb{N}$ by $(f \times g)(x, y) = (f(x), g(y))$.

P.1.10. Corollary. *If $S_1, \ldots, S_n$ are countable sets, then $S_1 \times \cdots \times S_n$ is countable.*

Proof. Proposition P.1.9 and induction. $\square$

Finally we show that a countable union of countable sets is countable.

P.1.11. Proposition. *Suppose that $\mathfrak{A}$ is a countable family of sets and that each member of $\mathfrak{A}$ is itself countable. Then $\bigcup \mathfrak{A}$ is countable.*

Proof. Exercise. *Hint.* Use Lemma P.1.7 and Proposition P.1.6.

P.1.12. Example. The set $\mathbb{Q}$ of rational numbers is countable.

Proof. Let $A = \mathbb{Q}^+ \setminus \{0\}$ and $B = -A = \{x \in \mathbb{Q} \colon x < 0\}$. Then $\mathbb{Q} = A \cup B \cup \{0\}$. The set A is countable by Example P.1.8. Clearly, $A \sim B$ (the map $x \mapsto -x$ is a bijection) so B is countable. Since Q is the union of three countable sets, it is itself countable by Proposition P.1.11. $\square$

By virtue of items P.1.4–P.1.11, we have a plentiful supply of countable sets. We now look at an important example of a set that is not countable.

P.1.13. Example. The set $\mathbb{R}$ of real numbers is uncountable.

Proof. We take it to be known that if we exclude decimal expansions that end in an infinite string of 9's, then every real number has a unique decimal expansion. (For an excellent and thorough discussion of this matter, see Stromberg's beautiful text *Introduction to classical real analysis* [**Str81**], especially Theorem 2.57.) By (the corollary to) Proposition P.1.4 it will suffice to show that the open unit interval $(0, 1)$ is uncountable. Argue by contradiction: assume that $(0, 1)$ is countably infinite. (We know, of course, from Example O.1.13 that it is not finite.) Let $r_1, r_2, r_3, \ldots$ be an enumeration of $(0, 1)$. For each $j \in \mathbb{N}$ the number r_j has a unique decimal expansion

$$0.r_{j1}\, r_{j2}\, r_{j3} \cdots .$$

Construct another number $x = 0.x_1\, x_2\, x_3 \cdots$ as follows. For each k choose $x_k = 1$ if $r_{kk} \neq 1$ and $x_k = 2$ if $r_{kk} = 1$. Then x is a real number between 0 and 1, and it cannot be any of the numbers r_k in our enumeration (since it differs from r_k at the kth decimal place). But this contradicts the assertion that $r_1, r_2, r_3, \ldots$ is an enumeration of $(0, 1)$. $\square$

P.1.14. Problem. Prove that the set of irrational numbers is uncountable.

P.1.15. Problem. Show that if S is countable and T is uncountable, then $T \setminus S \sim T$.

P.1.16. Problem. Let ϵ be an arbitrary number greater than zero. Show that the rationals in $[0, 1]$ can be covered by a countable family of open intervals the sum of whose lengths is no greater than ϵ. (Recall that a family $\mathfrak{U}$ of sets is said to *cover* a set A if $A \subseteq \bigcup \mathfrak{U}$.) Is it possible to cover the set $\mathbb{Q}$ of all rationals in $\mathbb{R}$ by such a family? *Hint.* $\sum_{k=1}^{\infty} 2^{-k} = 1$.

P.1.17. Problem. (*Definition.* The OPEN DISK in $\mathbb{R}^2$ with radius $r > 0$ and center (p, q) is defined to be the set of all points (x, y) in $\mathbb{R}^2$ such that $(x-p)^2 + (y-q)^2 < r^2$.) Prove that the family of all open disks in the plane whose centers have rational coordinates and whose radii are rational is countable.

P.1.18. Problem. (*Definition.* A real number is ALGEBRAIC if it is a root of some polynomial of degree greater than 0 with integer coefficients. A real number that is not algebraic is TRANSCENDENTAL. It can be shown that the numbers π and e, for example, are transcendental.) Show that the set of all transcendental numbers in $\mathbb{R}$ is uncountable. *Hint.* Start by showing that the set of polynomials with integer coefficients is countable.

P.1.19. Problem. Prove that the set of all sequences whose terms consist of only 0's and 1's is uncountable. *Hint.* Something like the argument in Example P.1.13 works.

P.1.20. Problem. Let $\mathfrak{J}$ be a disjoint family of intervals in $\mathbb{R}$ each with length greater than 0. Show that $\mathfrak{J}$ is countable.

P.1.21. Problem. Find an uncountable set that is not cardinally equivalent to $\mathbb{R}$. *Hint.* Let $\mathcal{F} = \mathcal{F}(\mathbb{R}, \mathbb{R})$. Assume there exists a bijection $\phi\colon \mathbb{R} \to \mathcal{F}$. What about the function f defined by

$$f(x) = 1 + \big(\phi(x)\big)(x)$$

for all $x \in \mathbb{R}$?

Bibliography

[Apo74]   Tom M. Apostol, *Mathematical analysis*, 2nd ed., Addison-Wesley Publishing Co., Reading, Mass.–London–Don Mills, Ont., 1974. MR0344384

[Die62]   J. Dieudonné, *Foundations of modern analysis*, Pure and Applied Mathematics, Vol. X, Academic Press, New York–London, 1960. MR0120319

[Hal58]   Paul R. Halmos, *Finite-dimensional vector spaces*, The University Series in Undergraduate Mathematics, D. Van Nostrand Co., Inc., Princeton–Toronto–New York–London, 1958. 2nd ed. MR0089819

[Hal67]   Paul R. Halmos, *A Hilbert space problem book*, D. Van Nostrand Co., Inc., Princeton, N.J.–Toronto, Ont.–London, 1967. MR0208368

[HK71]   Kenneth Hoffman and Ray Kunze, *Linear algebra*, Second edition, Prentice-Hall, Inc., Englewood Cliffs, N.J., 1971. MR0276251

[Kap77]   Irving Kaplansky, *Set theory and metric spaces*, 2nd ed., Chelsea Publishing Co., New York, 1977. MR0446980

[LS90]   Lynn H. Loomis and Shlomo Sternberg, *Advanced calculus*, Jones and Bartlett Publishers, Boston, MA, 1990. MR1140004

[Moo32]   C. N. Moore, *Summability of Series*, Amer. Math. Monthly **39** (1932), no. 2, 62–71. MR1522454

[Som56]   Arnold Sommerfeld, *Thermodynamics and statistical mechanics*, Academic Press, New York, 1956.

[Spi67]   Michael Spivak, *Calculus*, W. A. Benjamin, New York, 1967.

[Str81]   Karl R. Stromberg, *Introduction to classical real analysis*, Wadsworth International Mathematics Series, Wadsworth International, Belmont, Calif., 1981. MR604364

Index

Published Titles in This Series